Electrodissolution Processes
Fundamentals and Applications

Madhav Datta

CRC Press
Taylor & Francis Group
Boca Raton London New York

CRC Press is an imprint of the
Taylor & Francis Group, an **informa** business

First edition published 2021
by CRC Press
6000 Broken Sound Parkway NW, Suite 300, Boca Raton, FL 33487-2742

and by CRC Press
2 Park Square, Milton Park, Abingdon, Oxon, OX14 4RN

Library of Congress Cataloging-in-Publication Data
Names: Datta, Madhav, author.
Title: Electrodissolution processes : fundamentals and applications / authored by Madhav Datta.
Description: First edition. | Boca Raton : CRC Press, 2021. | Includes bibliographical references and index.
Identifiers: LCCN 2020033128 (print) | LCCN 2020033129 (ebook) | ISBN 9780367407032 (hardback) | ISBN 9780367808594 (ebook)
Subjects: LCSH: Electrochemical cutting. | Dissolution (Chemistry) | Electrocatalysis.
Classification: LCC TJ1191 .D37 2021 (print) | LCC TJ1191 (ebook) | DDC 660/.297—dc23
LC record available at https://lccn.loc.gov/2020033128
LC ebook record available at https://lccn.loc.gov/2020033129

ISBN: 9780367407032 (hbk)
ISBN: 9780367808594 (ebk)

Typeset in Times
by codeMantra

Contents

Preface

Electrodissolution, in this book, refers to the material removal from a workpiece, which is made an anode in an electrolytic cell. Controlled material removal by electrodissolution forms the basis of Electrochemical Machining (ECM) and Electropolishing (EP). The application of ECM/EP in micromachining, microfabrication, and thin-film processing is referred to as Electrochemical Micromachining (EMM). These nonconventional machining and finishing processes are employed in the manufacturing and processing industries for a variety of applications ranging from machining, shaping, and finishing of complex-shaped large metallic parts such as turbine blades to the fabrication of microstructures in thin films and foils by through-mask EMM for sophisticated electronic devices. These high-rate anodic dissolution processes are expected to continue to play an ever-increasing role in the shaping and finishing of hard to machine advanced materials, and fabrication of microcomponents such as micromachines, microelectronic packages, and devices. Design and optimization of these processes for the production of reliable, reproducible, and high-yielding components require an understanding of the process parameters that influence the high-rate anodic dissolution reactions.

Electrochemical processing continues to play a decisive role in the advancement of many technology sectors including aerospace, energy, biomedical, automotive, and high-end processing. While there are numerous books available on electrodeposition, a book on Electrodissolution Processes does not exist to the best of the author's knowledge. Among the books written on related topics, two classic books, namely *Electrochemical Machining* by J. A. McGeough (Chapman Hall, 1974) and *Electropolishing* by W. J. McTegart (Pergamon Press, 1956), are now outdated and out of print. Since then, few other books and chapters or review papers on these topics have appeared in books and journals, but a book dealing with both fundamentals and applications of electrodissolution is missing. This is an attempt to write a book that is entirely dedicated to high-rate anodic dissolution processes addressing electrochemical, materials, and mechanical approaches to understanding ECM/EMM/EP processes.

In principle, ECM, EMM, and EP can be understood and theoretically modeled based on the same principles provided one considers the scaling effects related to each process. The book highlights the importance of the understanding of basic principles required for designing and optimization of these processes. This book reviews and discusses some of the basic principles involved in these processes and presents their application in advanced machining and finishing operations. The contents of the book have been designed in a manner to give equal emphasis on the fundamentals and applications of the electrodissolution processes. An effort has been made to provide up-to-date information on each topic. Due to page restrictions, an in-depth discussion of some of the topics was not possible. However, a sufficient number of references contained in each chapter should help the reader in her/his quest for an in-depth study of the topic. To avoid turning to previous chapters and facilitate easy reading, few equations and discussions have been repeated wherever required.

Whereas the book focuses on the understanding and applications of electrodissolution processes that are driven by an externally applied voltage/current, in the introductory chapter, few selected open-circuit-based metal dissolution processes are briefly described because of their relevance to the high-rate electrodissolution processes. The open-circuit processes described in Chapter 1 include corrosion, selective metal removal by chemical etching, and chemical mechanical polishing (CMP). The introductory chapter also provides an opportunity to introduce some of the basic principles that are of significant importance to the processes driven by the externally applied voltage/current.

Depending on the metal-electrolyte combination and the operating conditions, different electrochemical reactions take place at increasing anodic potentials. Active Dissolution, Passivity, and the phenomena of passive film breakdown that lead to the Transpassive Dissolution of metals are examined in detail in Chapters 2 and 3. Mass transport and current distribution aspects are discussed in Chapter 4. The important role of mass transport and surface films on metal removal rate and surface finish during high rate anodic dissolution of iron, nickel, chromium, and their alloys is discussed in Chapter 5. Recognizing the unique properties of difficult to machine titanium, tungsten, and their carbides and their increasing importance in many industries including aerospace, biomedical, electronics, and photovoltaics, Chapter 6 is devoted to reviewing the anodic dissolution behavior of these materials under ECM conditions. Chapter 7 deals with the electropolishing of metals in concentrated acid solutions, focusing on mass transport processes and identification of the rate-controlling species that lead to surface finishing.

The application section of the book provides essential elements of electrochemical techniques for machining, micromachining, and polishing. Chapters on these topics also include a few case studies that demonstrate their capabilities. Chapter 8 on Electrochemical Machining provides a detailed description of the process including the factors influencing metal removal rate, surface finish, and machining accuracy. Different ECM techniques and some of the assisted techniques that are incorporated for the enhancement of the ECM process performance are described. In Chapter 9, the different maskless methods such as Jet EMM, microdrilling, and wire EMM, used for localizing the anodic dissolution process to create precision microstructures are presented. Different assisted techniques including laser, abrasive, vibration, and ultrasonic are presented that are incorporated with EMM to enhance its performance. Chapter 10 deals with the Through-Mask Electrochemical Micromachining (TMEMM) process. The effectiveness of simulation studies in understanding and overcoming some of the challenges encountered in TMEMM is explored. The last two chapters are devoted to the application of the electropolishing process. Chapter 11 presents a process description of the industrial practice of electropolishing including some assisted techniques for micro/nanosmoothing. Chapter 12 presents a critical review of the use of electropolishing and Electrochemical Mechanical Polishing (ECMP) for the planarization of copper interconnects.

It is hoped that this book will be a useful reference for researchers and graduate students working in the field. It should serve as a textbook for specialized graduate and senior undergraduate courses in science and engineering. In particular, the book could be a part of electrochemistry, electrochemical engineering, chemical engineering, mechanical engineering, materials engineering, and other related engineering curricula. Also, the book should be a good addition to the engineering and science library catalogs.

I would like to express my sincere thanks to Dr. Gagandeep Singh, Publisher (Engineering) of Taylor and Francis, who contacted and encouraged me to write this book. Dr. Singh's interest in the making of the book was evident from his frequent inquiry about the progress of the book writing. I gratefully acknowledge his valuable help in getting permission for the reproduction of figures from published literature. I would also like to thank the editorial team consisting of Ms. Jyotsna Jangra, Mr. Lakshay Gaba, Ms. Mouli Sharma, and Mr. Aswini K for providing me the needed information and help through the various stages of the production of the book.

Madhav Datta
Coimbatore, Tamil Nadu, India
Tracy, California
madhavdatta@amrita.edu

Author

Madhav Datta is an R&D professional with extensive experience in academic and industrial research. He received a BS in chemical engineering from H.B. Technological Institute, Kanpur, an MS in chemical engineering from the University of California at Los Angeles, and a PhD from the Materials Department of Ecole Polytechnique Fédérale de Lausanne (EPFL), Switzerland in 1975. Dr. Datta's affiliations include the Materials Department of EPFL (1973–1984), IBM's T.J. Watson Research Center, Yorktown Heights, New York (1985–1999), Intel's Logic Technology Development, Portland, Oregon (1999–2002); and Emerson Network Power's Cooligy Precision Cooling, Mountain View, California (2003–2013). In 2015, he joined the Coimbatore campus of Amrita Vishwa Vidyapeetham (Amrita University) as the Chairman of Amrita Center for Industrial Research and Innovation and a Distinguished Professor in the Chemical Engineering and Materials Science Department. His research interests include electrochemical dissolution and deposition, and their application in microelectronic packaging, including flip-chip technology, microcooling devices, and electronic materials. For his contribution to electrochemical processing and especially to electrodissolution processes as applied to microfabrication in the electronics industry, he was awarded the Electrodeposition Research Award of the Electrochemical Society in 1998. Dr. Datta is an innovator with 49 issued US Patents. He has over 100 scientific publications to his credit and is the author or editor of several books on electrochemical processing and micro/nanoelectronics. He coedited a series on New Trends in Electrochemical Technology. He has held several administrative positions, including divisional chairs in the Electrochemical Society (ECS) and the International Society of Electrochemistry (ISE). Dr. Datta chaired the Technology Working Group and Technology Implementation Group of National Electronics Manufacturing Initiative (NEMI) and developed a first-of-its-kind technology road map of energy storage systems for the electronics industry.

1 Open-Circuit Metal Dissolution Processes

1.1 INTRODUCTION

Several nonconventional machining and finishing processes such as electrochemical machining and electropolishing involve metal dissolution at high rates from a workpiece, which is made an anode in an electrolytic cell. Chemical etching and chemical polishing are other widely used manufacturing processes which operate under open-circuit conditions, the driving force for metal dissolution being derived from the reactivity of the solution. In the latter processes, although referred to as chemical processes, the partial oxidation and reduction reactions involved are electrochemical such that at the open-circuit potential (OCP), the values of the anodic and cathodic partial currents are equal. In this regard, the well-known corrosion process also involves material dissolution (oxidation) reaction from anodic sites, while a cathodic reduction reaction takes place on another site on the same sample. Whereas corrosion prevention is of crucial importance for the technology-based society, the corrosion process is also advantageously exploited in electrochemical machining and electropolishing by separating the anodic reaction from the cathodic reaction in a properly designed electrochemical cell. Indeed, controlled high-rate anodic dissolution-based electrochemical metal shaping and finishing processes are widely employed in the manufacturing of complicated shaped large parts and they offer new opportunities for micro/nanofabrication. This book focuses on the understanding and applications of electrodissolution processes that are driven by an externally applied voltage/current. However, in this introductory chapter, open-circuit-based processes are briefly described because of their relevance to electrodissolution processes. These processes include corrosion; selective metal removal by chemical etching, chemical milling, and chemical micromachining; chemical polishing; and chemical mechanical polishing (CMP). In this chapter, some of these open-circuit processes are discussed which provide an opportunity to introduce some of the basic principles that are also of significant importance to the processes driven by the externally applied voltage/current.

1.2 CORROSION

Corrosion is the degradation of a material mainly caused by an irreversible electrochemical reaction on the surface with its environment. It is a complex process so that its prediction and control require a thorough understanding of the thermodynamic and kinetic fundamentals which govern the process. An enormous amount of information is available on the subject, and several books are available that deal with the fundamentals and practical aspects of corrosion [1–5]. This section on corrosion briefly describes the essentials of the process, including its various types, some fundamental aspects related to thermodynamics and kinetics, and some of the methods of its prevention.

Both anodic and cathodic reactions occur on a corroding part. Corrosion occurs at the anode, where the material dissolves by an oxidation reaction. The anodic site is separated from the cathodic site, where a reduction reaction takes place. An electrical potential difference exists between these sites and current flows through the solution from the anode to the cathode. This is accompanied by the flow of electrons from the anode to the cathode through the metal. A typical example is that of corroding steel for which the anodic oxidation reaction is as follows:

$$Fe \rightarrow Fe^{2+} + 2e^-. \tag{1.1}$$

The primary cathodic reaction is oxygen reduction:

$$O_2 + H_2O + 2e^- \rightarrow 2OH^-. \tag{1.2}$$

Oxygen dissolved in water reaches the surface by diffusion. The oxygen reduction reaction is generally a diffusion-controlled process, which in turn controls the rate of corrosion. The production of hydroxide ions according to Equation 1.2 creates a localized high pH at the cathode.

Another possible cathodic reaction is the proton reduction:

$$2H^+ + 2e^- \rightarrow H_2. \tag{1.3}$$

At neutral or higher pH, the concentration of H^+ ions is too low for this reaction to contribute significantly to the overall corrosion rate. However, as pH decreases, this reaction becomes more important until, at a pH of about 4 or above, it becomes the predominant cathodic reaction.

The dissolved Fe^{2+} ions react with OH^- ions according to the following reaction:

$$Fe^{2+} + 2OH^- \rightarrow Fe(OH)_2. \tag{1.4}$$

The ferrous hydroxide thus formed combines with oxygen and water to produce ferric hydroxide, $Fe(OH)_3$, which when dehydrated forms Fe_2O_3, the well-known common rust.

1.2.1 TYPES OF CORROSION

The corrosion resistance of the metal is not an intrinsic property of the metal, but a property of the metal-environment combined system. A metal may rapidly corrode in a certain environment, while under different conditions it may remain unaffected. Several reasons are responsible for the formation of anodic and cathodic sites that are necessary to produce corrosion. These include metallurgical properties such as metal grain size, impurities in the metal, surface defects, and composition differences; chemical composition of the environment such as oxygen, salt concentration, temperature, and convection; and the presence of thermomechanical cycles. When these local differences are not large and the anodic and cathodic sites can shift from place to place on the metal surface, the corrosion is uniform. Uniform corrosion is the most basic form of corrosion during which a metallic object is more or less uniformly consumed and converted to ionic species. Much of our fundamental understanding of corrosion, such as thermodynamics and kinetics, is based on uniform corrosion.

Localized corrosion is the most serious industrial problem, which occurs when the anodic sites remain stationary. Different types of localized corrosion include pitting, galvanic corrosion, selective leaching, crevice corrosion, intergranular corrosion, stress corrosion cracking (SCC), and microbiologically influenced corrosion. Other forms of corrosion include erosion-corrosion and formicary corrosion. In the following, a brief description of different types of corrosion is presented, and photographs of some selected corrosion types are shown in Figure 1.1.

Uniform corrosion is the corrosive attack leading to uniform loss of metal and general thinning of the corroding surface. It is not regarded as the most serious form of corrosion since it is relatively easily measured and predicted, making disastrous failures relatively rare. If surface corrosion is permitted to continue, the surface may become rough and surface corrosion can lead to more serious types of corrosion. Uniform corrosion can be practically controlled by cathodic protection or the use of coatings or paints. In some cases, equipment designers take into consideration uniform corrosion by simply specifying a corrosion allowance. With uniform corrosion, fouling is usually a more serious problem than equipment failure.

Galvanic corrosion is the corrosion damage induced when two dissimilar materials are coupled in a corrosive environment. It occurs when two dissimilar metals are brought into electrical contact in an electrolyte. The driving force for corrosion is a potential difference between the different materials. When a galvanic couple forms, one of the metals in the couple becomes the anode and corrodes faster than it would all by itself, while the other becomes the cathode and corrodes much slower than it would alone. Galvanic corrosion cells can be set up on the macroscopic level or the microscopic level. On the microstructural level, different phases or other microstructural features can be subject to galvanic currents.

Pitting corrosion is the localized corrosive attack of a passive metal that leads to the development of small cavities known as pits. Their growth eventually leads to perforation of a metal plate or pipe wall. Pitting corrosion damage, therefore, can be quite important, even if the absolute amount of corroded material is small. Pitting corrosion requires the presence of aggressive anions, most often chloride ions, and

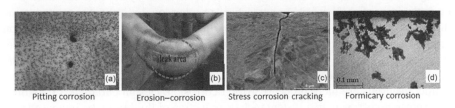

| Pitting corrosion | Erosion–corrosion | Stress corrosion cracking | Formicary corrosion |

FIGURE 1.1 Examples of some selected corrosion types: (a) pitting corrosion, (b) erosion-corrosion [6], (c) stress corrosion cracking [7], and (d) formicary (ant nest) corrosion [9].

of an oxidizing agent such as oxygen or ferric ions. A corrosion cell forms between the growing pit which is the anode and the passive surface surrounding the pit which serves as the cathode. Because the anode/cathode surface ratio is small, dissolution inside the pit can be very fast. Corrosion products often cover the pits. A small, narrow pit with minimal overall metal loss can lead to the failure of an entire engineering system. Pitting corrosion is a dangerous form of corrosion damage since it is difficult to detect and predict. Figure 1.1a shows an example of pitting corrosion in cast iron water mains.

Crevice corrosion occurs within localized volumes of stagnant solution trapped in pockets, corners, or beneath a shield such as a seal, gasket, fastener, etc., and where oxygen cannot freely circulate. Deposits of sand, dust, scale, and corrosion products can all create zones where the liquid remains stationary and cannot be renewed. These conditions can create a situation where two different environments are in contact on the same metal surface such as the creation of a differential aeration cell. Crevice corrosion is encountered particularly in metals and alloys which owe their resistance to the stability of a passive film. Since these films are unstable in the presence of high concentrations of Cl^- and H^+ ions, gradual acidification of the electrolyte caused by insufficient oxygen penetration under a shield causes localized metal dissolution leading to the formation of a crevice. Crevice corrosion is highly accelerated if chloride, sulfate, or bromide ions are present in the electrolyte solution.

Intergranular corrosion is usually related to the segregation of specific elements or the formation of a compound in the grain boundary. Corrosion then occurs by a preferential attack on the grain-boundary phase, or in a zone adjacent to it that has lost an element necessary for adequate corrosion resistance – thus making the grain-boundary zone anodic relative to the remainder of the surface. The attack usually progresses along a narrow path along the grain boundary and, in an extreme case of grain-boundary corrosion, an entire grain may be dislodged due to the complete deterioration of its boundaries. In such a case, the mechanical properties of the structure will be seriously affected. If materials with incorrect heat treatment enter service, they are liable to crack or fail by intergranular corrosion much more rapidly than properly treated materials.

Erosion-corrosion is generally a non-Faradic acceleration in the rate of corrosion attack of metal due to the movement of corrosive fluid. For a given material, a critical fluid velocity must be exceeded for erosion-corrosion to take place. The presence of abrasives in the form of suspended solid particles in the moving fluid can further enhance erosion-corrosion. It is often localized at areas where water changes direction. The morphology of surfaces affected by erosion-corrosion may be in the form of shallow pits or horseshoes or other local phenomena related to the flow direction. In a system susceptible to pitting corrosion, a combination of erosion and corrosion can lead to extremely high pitting. Erosion-corrosion is most prevalent in soft alloys such as copper, aluminum, and lead alloys. Materials selection and operating conditions play an important role in the management of erosion-corrosion. Harder materials are more prone to erosion-corrosion. Design parameters that minimize erosion-corrosion in a system include reduction of fluid velocity, increased pipe diameter, and prevention of burrs. Figure 1.1b is an example of erosion-corrosion case study where a leakage was encountered in the 90° elbow in a pipe system of a geothermal production facility after 2 weeks in service. According to design specifications, the elbow was made of carbons steel [6].

Cavitation is a special case of erosion-corrosion and is caused by the formation and collapse of tiny vapor bubbles that result in pits on the metal surface. Cavitation removes protective surface scales by the implosion of gas bubbles in a fluid. The implosions produce shock waves with extremely high pressures. The subsequent corrosion attack is the result of hydro-mechanical effects from liquids in the regions of low pressure where flow velocity changes, disruptions, or alterations in flow direction have occurred. Cavitation damage often appears as a collection of closely spaced, sharp-edged pits or craters on the surface.

SCC is the damage induced by the combined influence of tensile stress and a corrosive environment. The stresses may be in the form of directly applied stresses or in the form of residual stresses. Residual stresses are introduced in the material during forming, heat treatment, welding, machining, and grinding. In SCC, most of the surface usually remains unattacked, but with fine cracks penetrating the material. These cracks can have an intergranular or a transgranular morphology. SCC is classified as a catastrophic form of corrosion, as the detection of such fine cracks can be very difficult and the damage not easily predicted. SCC can lead to a disastrous failure with minimal overall material loss. Figure 1.1c shows the effect of grinding on SCC initiation on Type 304L stainless steel during slow strain rate tensile tests [7]. In the picture the cracks initiated from the machined troughs and penetrated beyond the ultrafine-grained layer intergranularly.

Selective leaching is the preferential corrosion of one or more constituents of an alloy in a corrosive environment. The preferential dissolution of the more active element in an alloy is caused by the potential difference between the alloying elements. Dezincification of brass is a typical example of selective leaching in which zinc is preferentially leached out of a copper-zinc alloy leaving behind a copper-rich surface layer that is porous and brittle. Graphitic corrosion is also a preferential dissolution process which leads to deterioration of gray cast iron by selective leaching of metallic constituents in the alloy leaving behind graphite. During cast iron graphitic corrosion, the porous graphite network, which makes up 4%–5% of the total mass of the alloy, is impregnated with insoluble corrosion products. As a result, the cast iron retains its appearance and shape but is weaker structurally. Similarly, decarburization is the selective loss of carbon from the surface of a carbon-containing alloy. Decobaltification, denickelification, etc. are other examples of selective leaching.

Microbiologically Influenced Corrosion (MIC) is the formation of "biofilms" on cooling system surfaces. Biofilms consist of organisms and their hydrated polymeric secretions. Numerous types of organisms may exist in any particular biofilm deposits. The biofilms may consist of aerobic bacteria at the water interface, or they may be anaerobic bacteria such as sulfate-reducing bacteria at the oxygen-depleted metal surface. These deposits can cause accelerated localized corrosion by creating differential aeration cells. The nonuniform formation of biofilm creates an inherent differential, which is enhanced by the oxygen consumption of organisms in the biofilm. Many of the by-products of microbial metabolism, including organic acids and hydrogen sulfide, are corrosive. These materials can concentrate in the biofilm causing an accelerated metal attack.

Formicary corrosion is a unique form of copper corrosion which is due to the formation of a network of tiny pinhole-type pits [8]. Multiplication and interconnection of such tiny pits form tunnels through the copper wall. Corrosion initiates from the exterior of the copper tube/pipe and progresses inward until penetration occurs and leak results. Formicary corrosion is specific to cooper and occurs in the presence of three essential constituents: oxygen, moisture, and organic acids. The absence of any one of these three items will not allow formicary corrosion to occur [8]. Commonly known organic acids that attack copper include formic and acetic acids. These acids are formed from the conversion of a variety of volatile organic compounds (VOCs) including formaldehyde. VOC-producing products include packing materials (foam insulation, plywood, paper); solvents, oil and oil-based paints, and building materials. Formicary corrosion has been reported mainly in copper heat exchangers used in cooling systems. Figure 1.1d is an example of formicary corrosion showing a series of ant nest types of attacks on the exterior of a copper tube in a tube-fin heat exchanger installed in a cooling system [9].

1.2.2 FUNDAMENTALS OF CORROSION

This section intends to briefly provide the essentials of thermodynamic and kinetic aspects of corrosion. Let us first define a few basic electrochemical terminologies:

- The OCP is the potential set up spontaneously by an electrode when immersed in an electrolyte in the absence of an external current. For a single electrode, the open-circuit potential is equal to the equilibrium potential, also known as reversible potential, E_{rev}.
- When a potential is imposed on an electrode such that its potential differs from the open-circuit potential, an electric current passes through the electrode-electrolyte interface. This overpotential, η, is defined as the difference between the electrode potential, E, and the equilibrium or reversible potential of an electrode reaction: $\eta = E - E_{rev}$.
- A polarization curve establishes the functional dependence between current density and potential. A polarization curve can be experimentally determined by controlling either the potential or the current. One thus obtains a potentiostatic polarization curve, $i = f(E)$, or a galvanostatic polarization curve, $E = f(i)$, respectively.
- The OCP of a mixed electrode undergoing corrosion is called the corrosion potential.
- Electrochemical reactions typically involve the transfer of charge across the interface. There are two types of charge transfer reactions. Ion transfer reactions involve the transfer of ions from the electrode to the electrolyte or vice versa. Electron transfer reactions involve the transfer of charge between ions in the electrolyte and typically occur heterogeneously at an electrode surface.
- Mass transport determines the concentration of the reactants and products at the electrode surface. The electrolyte layer close to the electrode

surface, in which the concentration of the reactants or products differs from that in the bulk electrolyte, is called the diffusion layer. The thickness of the diffusion layer depends on the prevailing hydrodynamic conditions. This topic will be discussed in detail elsewhere in the book.

1.2.2.1 Corrosion Thermodynamics

Thermodynamics provides a basis for the understanding of the energy changes associated with the corrosion reaction. In general, it can predict whether a corrosion reaction is possible. However, thermodynamic principles cannot predict corrosion rates. The rate at which the reaction proceeds is governed by kinetics. Gibb's free energy expression provides a tool with which to predict whether a corrosion reaction is thermodynamically possible. Gibb's free energy, ΔG, is given by:

$$\Delta G = -nFE, \tag{1.5}$$

where F is the Faraday constant and E is the standard emf potential expressed by the standard potential for anode half-cell, E_{oxd}, and standard potential for cathode half-cell, E_{red},

$$E = E_{oxd} + E_{red} \tag{1.6}$$

A reaction will take place only if the value of ΔG is negative. If the ΔG value is positive, the reaction will not proceed.

The extent of oxidation and reduction reactions depends on the concentration of anodic and cathodic reactants. The Nernst equation expresses the exact potential (emf) of a cell in terms of activities of reactants and products of the cell. Considering that all reactants and products are in their standard states, i.e., activities equal to unity, the Nernst expression for the cell potential, E, is written as follows:

$$E = E^o + 2.3 \frac{RT}{nF} \log_{10} \frac{a_{oxd}}{a_{red}} \tag{1.7}$$

where E^o is the potential when all reactants and products are in their standard states (unit activity), a_{oxd} is the activity of the oxidized species, and a_{red} is the activity of the reduced species.

A summary of thermodynamic data in the form of potential-pH diagrams, devised by Pourbaix, is popularly known as the Pourbaix diagram [10]. The diagrams are generally constructed using equilibrium constants, solubility data, and a form of the Nernst equation which includes a pH term. These diagrams provide useful information about the ability to predict the occurrence of corrosion, the nature of possible corrosion products, and predicting the influence of environmental changes on the corrosion reaction. However, since these diagrams do not provide any information about the kinetics of reactions, these diagrams cannot be used to assess the extent of corrosion damage.

1.2.2.2 Corrosion Kinetics

In the following, the kinetics of corrosion is presented by considering the reaction to be controlled (i) by the kinetics of a charge-transfer reaction, either anodic or cathodic, at the metal-electrolyte interface, and (ii) by the rate of mass transport of the cathodic reaction or of the anodic reaction products.

- **Charge transfer-controlled corrosion rate**

  When an electrode is immersed in an electrolyte, it attains a potential known as the corrosion potential. The corrosion potential is also called the open-circuit potential. The corrosion potential is a mixed potential, the value of which depends on the rate of the anodic as well as the cathodic reactions. The total current density, i, is the sum of the partial anodic current density, i_a, and the partial cathodic current density, i_c:

$$i = i_a + i_c \tag{1.8}$$

  At equilibrium, the anodic current density is equal to the cathodic current density such that the net current flowing is zero:

$$i = i_a + i_c = 0, \text{ and} \tag{1.9}$$

$$i = i_a = -i_c = i_o \tag{1.10}$$

  In Equation 1.10, i_o is the exchange current density, which characterizes the activation-controlled kinetics, the rate of which is controlled by the charge transfer reaction.

  For a reaction in which the rate is limited by the activation overvoltage, the relationship between the rate of reaction, expressed by current density, i, and the potential, E, is given by the Butler-Volmer equation, which in its simple form for a sufficiently large value of anodic polarization (overpotential >50 mV), can be expressed as:

$$i = i_o \exp\left[\frac{\alpha n F \eta_a}{RT}\right] \tag{1.11}$$

  where α is the charge transfer coefficient (value ranges between 0 and 1, it is usually taken as 0.5), n is the ionic charge equivalent, F is the Faraday constant, η_a is the anodic overpotential, R is the gas constant, and T is the temperature.

  Rearrangement of Equation 1.11 gives the well-known Tafel equation:

$$\eta_a = b_a \log\left(\frac{i}{i_o}\right) \tag{1.12}$$

where $b_a = 2.3\left(RT/\alpha nF\right)$ is the anodic Tafel slope. For cathodic activation polarization, a similar equation can be written as:

$$\eta_c = -b_c \log\left(\frac{|i|}{i_o}\right) \tag{1.13}$$

Tafel expressions generally provide an accurate description of corrosion kinetics for charge transfer-controlled reactions.

As stated above, in a corroding system, the corrosion potential is a mixed potential such that its value depends on the rate of the anodic as well as the cathodic reactions. Figure 1.2, known as the Evans diagram, shows the anodic and cathodic polarization curves where the electrode potential is plotted as a function of the logarithm of the current density. Such curves can be generated by potentiodynamic or potentiostatic polarization of a metal in an appropriate electrolyte. As the potential is increased from the corrosion potential in either anodic or cathodic directions, the rate of these reactions increases. The log of current density vs. potential relationship follows a straight line for both the reactions corresponding to the Tafel kinetics. The intersection point of the Tafel lines gives the values of the corrosion potential (E_{corr}) and the corrosion current (i_{corr}). The value of the corrosion current thus obtained by the extrapolation of Tafel lines can be converted to corrosion rate by using the Faraday law. However, such calculations are based on several assumptions which include: (i) the corrosion is uniform, (ii) corrosion products are in the form of dissolved ions such that there is no accumulation of surface products, and (iii) the mass transport effects are absent.

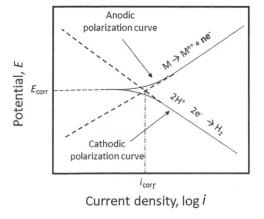

FIGURE 1.2 Determination of corrosion rate from polarization curves by extrapolation of anodic and cathodic Tafel lines.

- **Mass transport-controlled corrosion rate**

  Mass transport effects become very important in influencing the corrosion rate under two different conditions involving concentration gradient in the environment (i) limited by cathodic reactant or (ii) limited by metal ion concentration gradient at the metal-electrolyte interface of the corroding surface as shown in Figure 1.3. Reactions that are controlled by mass transport generally exhibit a limiting current, i_l, defined by:

$$i_l = nFD \frac{C_b}{\delta} \tag{1.14}$$

where D is the diffusion coefficient, C_b is the bulk concentration of the reacting species, and δ is the diffusion layer thickness, the value of which depends on the prevailing hydrodynamic conditions in the system.

Corrosion limited by transport of a cathodic reactant: When a corrosive environment contains only small concentrations of a cathodic reactant such as typically found for corrosion by oxygen in neutral media or corrosion by protons in weakly acidic environments according to the following reactions, the oxygen reduction reaction is:

$$O_2 + 2H_2O + 4e^- \rightarrow 4OH^-, \tag{1.15}$$

and the proton reduction reaction is:

$$H^+ + e^- \rightarrow \frac{1}{2}H_2 \tag{1.16}$$

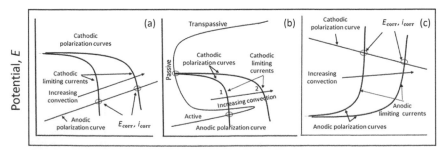

FIGURE 1.3 Determination of corrosion rates from polarization curves for corrosion reaction controlled by mass transport of a cathodic reactant (a) for an actively dissolving metal and (b) for a metal exhibiting active-passive-transpassive behavior. Part (c) shows the case of a corrosion reaction controlled by mass transport of the anodic reaction products. Curves 1 and 2 in case (b) refer to two different convection (flow rate) conditions; corrosion rate increases with increasing flow (curve 1) but at very high flow rates the cathodic curve 2 intersects at the passive state of the metal indicating insignificant corrosion.

The very low solubility of oxygen in the water and even lower in salt solutions limits the oxygen flux at the metal surface that governs the rate of corrosion. The cathodic oxygen reduction reaction shows a limiting current. Figure 1.3a shows the partial anodic and cathodic current densities during the corrosion of a metal by oxygen. The corrosion potential is located within the region of the limiting current plateau for oxygen reduction, which is defined by the following expression:

$$i_l = \frac{4FD_{O_2} C_{b,O_2}}{\delta}$$

(1.17)

The corrosion current density, i_{corr}, in this case is equal to the limiting current density for oxygen reduction, i_{l,O_2}, and therefore it varies with the thickness of the diffusion layer which is controlled by hydrodynamic conditions.

In weakly acidic environments, the reaction rate may be limited by the rate of proton transport. Under these conditions, the pH and hydrodynamic conditions determine the value of the limiting current:

$$i_l = \frac{FD_{H^+} C_{b,H^+}}{\delta}$$

(1.18)

Thus, when the corrosion potential lies in the plateau region, the corrosion rate is entirely limited by the rate of mass transport of protons.

According to Equations 1.17 and 1.18, the cathodic limiting current density is determined by the rate of convection in the corroding system. The influence of an increase in convection on the corrosion parameters of an actively dissolving metal is depicted by the Evans diagram shown in Figure 1.3a. The intersection of the cathodic limiting current density and the anodic curve for active dissolution provides information about E_{corr} and i_{corr}; the higher the convection rate the higher is the susceptibility to corrosion. For a metal that exhibits an active-passive-transpassive behavior, the situation is somewhat different as shown in Figure 1.3b. An increase in convection rate increases the corrosion rate (curve 1) up to a certain level beyond which the intersection of the cathodic polarization curve resides in the passive state of the metal (curve 2). Under these conditions, the corrosion rate is insignificant.

Corrosion limited by the transport of reaction products: During the metal dissolution process in a corroding system, a metal ion concentration gradient is established at the surface. With an increasing rate of dissolution, the local metal ion concentration at the surface increases until the saturation concentration (C_{sat}) is reached, which leads to precipitation of a salt film that is formed at the surface. The mass transport of metal ions from the surface of the film into the bulk electrolyte then limits the metal dissolution rate. The limiting current density in the presence of a salt film is given by the following equation:

$$i_l = \frac{nFD_{M^{n+}} C_{\text{sat}}}{\delta} \tag{1.19}$$

where $D_{M^{n+}}$ is the diffusion coefficient of the dissolving metal ions. Figure 1.3c shows an Evans diagram for corrosion controlled by mass transport of the anodic reaction products. When the corrosion potential of metal is located within the plateau of the anodic limiting current, the corrosion rate equals the anodic limiting current density. It is therefore limited by the transport of reaction products from the surface of the metal to the bulk electrolyte. Its value can be estimated from the above equation. Similar to the conditions discussed above for cathodic limiting currents, an increase in the convection rate leads to an increase in the anodic limiting current density. A higher convection rate effectively removes the reaction products, thereby exposing the bare metal surface to corrosion. These conditions, therefore, lead to increased corrosion rate.

- **Determination of corrosion rate by immersion**

    In addition to the electrochemical methods described above to determine the corrosion rates, a more direct and realistic approach to finding the corrosion rate is to use the method of measuring weight loss by immersion tests. Weight loss measurement is the simplest and well-established method of determining corrosion losses in equipment and plants. A sample in the form of a coupon of the metal or alloy to be tested is weighed and is immersed in a well-defined corrosive environment for a desired interval of time. The coupon is then removed and cleaned to remove all corrosion products and is reweighed. The weight loss is then converted to a corrosion rate. This technique in its simplest form does not require any complex equipment or procedures. However, this method can be complemented by a solution analysis, or in case of thin foil samples, by measuring the changes in ohmic resistance. The weight loss method provides a direct measurement of the corrosion rate without any assumptions or approximations. Furthermore, it applies to all corrosive environments and all forms of corrosion. Because of these attractive features, weight loss measurement continues to be the most widely used means of determining corrosion loss. Several standard immersion testing procedures established by ASTM, NACE, and MTI are frequently used wherever applicable. One of the key drawbacks of the immersion tests is that they generally require a long time to get a reliably measurable weight loss. On the other hand, electrochemical testing provides a faster and mechanistic determination of corrosion loss.

1.2.3   PITTING CORROSION

As stated above, pitting is a form of localized corrosion that produces a deep penetrating attack of the material. Pitting generally occurs on active-passive metals in oxidizing chloride-containing environments. A typical anodic polarization curve for an active-passive transition metal is shown in Figure 1.4a. The figure also shows that at a certain potential, E_{pit}, a sharp increase in current density takes place due to the localized breakdown of the passive film leading to pitting. The pitting potential, E_{pit}, depends on both material composition and environmental conditions, such as chloride

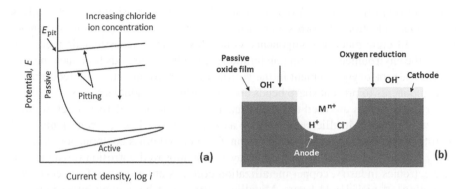

FIGURE 1.4 (a) Anodic polarization curve for a metal showing active-passive-transpassive transitions. In chloride-containing solutions, the localized breakdown of passive film takes place at a potential, E_{pit}, the value of which depends on the chloride ion concentration. (b) Schematic presentation of reactions taking place within and outside the corroding pit.

content, pH, and temperature. An increase in chloride concentration facilitates the pitting phenomenon, which is qualitatively shown in Figure 1.4a.

Once started, the penetration rate of pitting is so high that the whole metal thickness is attacked in a short time. Figure 1.4b shows schematically the reactions taking place during pitting. Pitting is initiated by the breakdown of the passive film. Metal dissolution occurs within the pit, which acts as the anode and the surrounding passive films act as the cathode, where oxygen reduction takes place. Inside the pit, the solution gradually becomes more aggressive as the hydrolysis reaction of metal ions proceeds. A series of reactions within the pit results in a decrease in pH and an increase in chloride concentration which stimulates the anodic attack. Outside the pit, pH increases, thus strengthening the passive film. As corrosion proceeds, dissolved metal ions migrate and diffuse towards the pit mouth, then reacting with hydroxyl ions and precipitating as hydroxide. High cathodic to anodic area ratio and high electrolyte conductivity within the pit result in a high penetration rate.

It is generally difficult to stop the corrosion reaction in deeply penetrated pits. If detected at the initial stage of pitting, the corroding materials with shallow pits can be recovered by washing with a sodium carbonate-type alkaline solution. However, it is generally not easy to detect pitting at its early stage. Application of cathodic protection and the selection of resistant materials are the possible ways of preventing pitting on susceptible metals. Material selection in the design phase must consider the expected operating condition, surface condition, and exposed environmental conditions such as chloride content, pH, oxidizing power, and bacterial activity. Some of the preventive measures are discussed below.

1.2.4 Corrosion Prevention

Metallic corrosion can be managed, slowed, or even stopped by using preventive measures such as appropriate design considerations or by protective measures such as coatings, corrosion inhibitors, or cathodic protection.

Corrosion prevention by design aims at reducing the corrosion risk and involves a judicious selection of corrosion-resistant materials, and prevention of contact between different metallic components. Other design approaches include modification of the environment by acting on humidity, fluid flow, methods to reduce the sulfur, chloride, or oxygen content in the surrounding environment, or by redesigning the components to prevent sharp bends and to minimize residual stresses.

Surface coatings are used to protect metals from the degradative effect of environmental gasses. Metallic coatings such as zinc and its alloys are commonly used to protect steel from atmospheric corrosion. Copper and brass components are commonly coated with nickel or chromium layers for improved corrosion resistance. In the electronics industry, copper metallization contacts are protected from corrosion by coatings of nickel/gold layers. Metallic coatings can be applied by various wet processes (electroplating or electroless plating), dry processes (PVD, CVD), or thermal spray. Inorganic nonmetallic coatings such as conversion coatings are obtained by anodizing, phosphatizing, or chromatizing. Other nonmetallic coatings include contact coatings such as ceramic or enamel coatings. Organic coatings include polymer coatings, paints, and varnishes. Aging of the resin in the paint, the formation of blisters, or delamination other defects in organic coatings can locally expose the surface thus causing localized corrosion.

Corrosion inhibitors are chemical products that are added to water or to any other process fluid to slow down the rate of corrosion. They are normally classified as anodic, cathodic, and film-forming agents depending on their mode of action. Anodic inhibitors cause a large anodic shift by forming a protective oxide film on the surface of the metal. This forces the metallic surface into the passivation region, which significantly reduces the corrosion current of the material. Some examples of anodic inhibitors include chromates, nitrates, phosphates, molybdates, and tungstates. Cathodic inhibitors slow down the cathodic reaction to limit the diffusion of reducing species to the metal surface. Cathodic poison and oxygen scavengers are examples of this type of inhibitor. Examples of cathodic inhibitors include calcium polyphosphate and calcium polyphosphonate. **Mixed inhibitors** are film-forming compounds that reduce both the cathodic and anodic reactions. The most commonly used mixed inhibitors are silicates, a combination of phosphates and polyphosphates, and combination of phosphates and phosphonates. The use of combined anodic and cathodic inhibitors reduces the overall amount required compared with the use of one single inhibitor. Organic inhibitors form a monomolecular of an inhibiting film between the metal and the water. These products are generally surfactants with hydrophobic and hydrophilic groups. Cooling systems use copper or copper alloys for which the most commonly used inhibitors are azole derivatives such as benzotriazole (BTA) and its derivatives and benzimidazole and its derivatives.

Cathodic protection is an electrochemical method that converts the anodic (active) sites on a metallic surface to cathodic (passive) sites through the application of opposing current. This can be done by two different means. The first is the introduction of a galvanic anode which in the corrosive electrolytic environment sacrifices itself by corroding to protect the cathode. Sacrificial anodes are generally made of zinc, aluminum, or magnesium, metals that have the most negative electropotential. The sacrificial anodes are required to be regularly replaced. Impressed current protection is another method of cathodic protection in which the negative terminal of a current source is connected to the metal, while the positive terminal is attached to an auxiliary anode, thus forming an electrical circuit through which a precalculated current is applied. Unlike in the sacrificial anode system, the auxiliary anode is not sacrificed in the cathodic protection system. This method is often used to protect buried pipelines and ship hulls.

1.3   CHEMICAL ETCHING

Wet chemical etching involves the removal of unwanted material by the exposure of a workpiece to an etchant whereby the exposed material is oxidized by the reactivity of the etchant to produce reaction products that are carried away from the surface by the medium. Wet chemical etching involves the conversion of a solid insoluble material to a soluble form. The extended lattice of metal atoms in the solid-state is broken down so that these atoms can enter the solution as soluble compounds. This is accompanied by the removal of electrons from the metal. These electrons are accepted by the etchant, which acts as an oxidizing agent. In general, oxidation-reduction (redox) and/or complexation-type reactions are involved in chemical etching. Although no external current is supplied, anodic and cathodic sites are present on the reactive surface such that the rate of material removal (oxidation) is balanced by the rate of reduction of the etchant species. The metal removal reaction typically involves several sequential steps. For example, in the chemical etching of copper by ferric chloride solution, the etchant is reduced to produce different complexes in solution. The complexation reaction is dependent on the amount of chloride ions and water molecules. The reactions are either activation controlled or mass transport controlled. In an activation-controlled process, the dissolution kinetics is dependent on the chemical reactivity of the species involved. In a mass transport-controlled process, the metal removal rate is determined by the speed at which the reaction product is removed from the surface and the fresh reactant is supplied to the surface. Temperature variations also profoundly influence the kinetics of metal removal reaction. An effective chemical etching process, therefore, requires a close control of the solution composition, temperature, and mass transport. Adequate stirring (e.g., by spraying or by part motion) is necessary to ensure the supply of fresh solution at the surface and to maintain uniformity of temperature of the solution and the workpiece.

1.3.1   PHOTOCHEMICAL ETCHING

Industrial application of chemical etching is mainly used in conjunction with photoresist masks to create microstructures on metallic films, sheets, and foils. The process known as photochemical etching (PCM) involves the following steps: art generation and photo tool generation, metal part cleaning to increase adhesion to photoresist, photoresist coating using dip coating of liquid resist or dry resist lamination by hot roller, photoresist exposure and development, through-mask chemical etching, and finally photoresist stripping and inspection. In PCM, the shape evolution of microstructure is controlled by several phenomena such as dissolution kinetics, surface films, and mass transport [11,12]. PCM is generally isotropic in nature, i.e., the material is etched both vertically and laterally at the same rate. The etch boundary, therefore, recedes at a 45° angle relative to the surface [13].

Chemical etching baths contain chemicals that are generally corrosive and toxic, thus posing safety and disposal issues. In many manufacturing processes, the waste treatment and disposal costs often surpass actual etching process costs [12]. Ever-increasing cost of incineration and imposition of landfilling restrictions are the main reasons behind the need for developing processes with particular emphasis on waste disposal issues. All these factors demand the development of improved chemical etching processes through the selection of nontoxic chemicals, and advanced regeneration technology.

A variety of metallic components employed in the electronics industry are fabricated by chemical etching. They range from removal of extremely thin seed layers in the fabrication of flip-chip bumps to two-sided through-mask etching of thick sheets of metal and alloys. Such items include metal screens, metal masks, recording heads, springs, lead frames, EMI shields, connectors, print bands, encoder disk, etc. Some of the products produced by chemical etching are illustrated in Figure 1.5, which demonstrates the capabilities of the process. For mass production, several repetitive designs are frequently joined by small tabs. Tabbing has major advantages of handling ease and one-step etching in high-speed, conveyorized systems. A wide range of usable formulations are available for etching any class of alloy. The specific choice of a given formulation takes into consideration the photoresist material, tolerance, metal removal rate, handling, and surface finish. The complicated selection process practiced in the industry has resulted in a long list of applicable formulations that are compiled in the literature [13,14]. The list generally includes ferric chloride, acids, acid mixtures, and alkaline solutions in combination with proprietary additives. The temperature in these applications generally varies between 300°C and 600°C. The material removal rate for most of the metals and alloys has been reported to be between 6 and 75 μm/min.

1.3.2   PRINTED CIRCUIT BOARDS

Chemical etching plays an important role in the fabrication of printed circuit boards (PCB). The PCB has a double function: first, it is the mounting panel for the components, and second, it contains the conductors which connect all components as desired. Typically, many conductors are laid out close to each other. One of the simplest processes used for PCB fabrication is the "print and etch" process where the

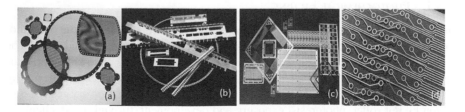

FIGURE 1.5   Examples of electronic and electromechanical parts microfabricated by photochemical etching: (a) Ti, stainless steel screens, (b) EMI shields, (c) lead frames, and (d) connectors. (Reproduced with the permission from Paul Richards, President, PCM Products.)

pattern is printed on the board and the unwanted copper is etched away. This method is primarily used in the fabrication of PCB for consumer item applications where high conductor density is not required (e.g., in inexpensive toys, low-cost transistor radio). In the more demanding PCBs, two or more levels of metal layers are required. The levels are insulated from one another, and they have to conduct connections only at certain points. These multilayered boards are produced by through-hole plating process where frequently a metallic resist, such as a tin-lead alloy, is plated on the circuit pattern and inside the holes which connect the top and bottom of the board. The etchant used in the chemical etching of these PCBs should be sufficiently selective to remove the base copper layer in the presence of a lead-tin mask. Some of the formulations of etchants that are used in PCB industry include ferric chloride, chromic-sulfuric acids, ammonium persulfate, hydrogen peroxide-sulfuric acid, cupric chloride, and ammoniacal or alkaline etchant [15]. Out of these, the most commonly used etchants are the cupric chloride and the ammoniacal etchants. Both etchants can be automatically regenerated or replenished to maintain steady-state etching conditions. The limitation of cupric chloride etchant is that it attacks lead-tin resist: hence, the etchant is only applicable in the fabrication of inexpensive "print and etch" PCBs in which only copper is present. On the other hand, ammoniacal etchant is widely used in multilayered PCBs because of its selectivity for copper in the presence of tin and lead-tin resists. The progress towards integration and miniaturization in silicon technology is also impacting the PCB technology. A modern PCB contains an ever-increasing number of contacts and conductors that are to be accommodated in the same area. A combination of circuit speeds and high density of packaging requires much tighter geometrical tolerances. This, in-turn, requires finer line geometry and more closely controlled feature tolerances. Conductor widths and spacings as narrow as 76 μm are already practiced in the PCB industry today, and the processes for finer line PCBs are being developed [15].

1.3.3 ANISOTROPY CONSIDERATIONS

As stated above, wet PCM of polycrystalline engineering materials is isotropic which means that the material is etched vertically and also sideways under the mask. The resulting cavity has rounded corners and edges, and is larger than the mask opening. In anisotropic etching, the etch rate in the vertical direction is faster than in the lateral directions. When properly designed, the anisotropic etching can produce cavities with straight sides and less undercutting of the mask. High-density patterning by wet etching requires fabrication of high aspect ratio features with fine dimensions. An understanding of the conditions leading to anisotropy is therefore critical in micro/nanofabrication by wet etching.

The etch rate in a wet etching process is determined by the ability of the process to transport the fresh etchant to the surface being etched and to remove the etched products away from the surface. To increase the etch rate, various methods, such as spray and impinging jet, are used so that the etchant is forced to flow along the surface, thus adding the effect of convective transport. Kuiken and Tijburg [16,17] investigated the anisotropic behavior in deep etching. They contended that in deep cavities, the etchant moves along the hole, the flow in the cavity being characterized by one or more trapped eddies. The emergence of every new vortex leads to a drastic reduction of the etch rate. This results in strong undercutting effects. Kuiken and Tijburg [17] proposed the use of centrifugal etching in which the formation of vortices within the etching cavity at larger etch depths is avoided. They designed a cell consisting of a rotatable double-walled cylinder and performed etching of masked patterned copper plates using an aqueous $FeCl_3$ solution under conditions of enhanced gravity. Their results show that centrifugal etching involving artificial gravity effect gives a markedly increased etch rate and a reduced undercutting of the mask edge.

Georgiadou and Alkire [18,19] investigated the role of reactive species and transport processes on the shape evolution during anisotropic wet etching of photoresist patterned Cu foil. They studied the etching of patterns with two different aspect ratios (width/depth): a shallow (5:1) and a deeper one (1:1) in $CuCl_2$ solutions containing HCl and KCl as the additive oxidants. Rotating disk electrode studies indicated that the anisotropic chemical etching in these solutions was related to the presence of a sparingly soluble surface film of CuCl. In this etchant, both KCl and HCl additions serve to increase the etch rate owing to the increased solubility of CuCl. The etch rate was found to be dependent on oxidant concentration, fluid velocity, and pattern geometry. In solutions containing high amounts of oxidizing agent, it was concluded that precipitation of a film occurred on the entire surface owing to mass transport limitations of the anodic dissolution product [Cu(I) chloro-complex]. The presence of the film caused the etch rate to be controlled by mass transport. Furthermore, due to restricted local mass transport in the vicinity of the corner, the presence of such films suppressed photoresist undercutting.

1.4   CHEMICAL MECHANICAL POLISHING

CMP involves intimate contact between a wafer surface and a pad that is charged with carefully formulated colloidal slurry. The relative motion between the wafer and the pad combined with applied pressure and chemical activity of the slurry results in the polishing of the wafer surface. CMP has found application in various aspects of semiconductor manufacturing from transistors to chip interconnects and involves polishing/planarization of a variety of materials that include silicon, silicon dioxide, silicon nitride, tungsten, tungsten nitride, tantalum, tantalum nitride, titanium, titanium nitride, platinum, copper, and polymers.

CMP differs from mechanical polishing processes such as grinding, lapping, and abrasion and chemical removal processes such as chemical etching, electroetching, and electropolishing, even though some overlap with one or more of these processes may always exist to some extent. The material removal rate is dependent on variables such as the type of pad, slurry composition, pad conditioning, and applied pressure. However, the effect of chemical action vs. mechanical action is not easily separable. Similar to electroplating, an understanding of the physics of CMP has been generated by modeling the process on wafer-scale and feature-scale [20–23]. The most common wafer-scale model for material removal is the classical Preston's equation [24], which relates material removal to the applied pressure and the linear velocity at the interface. Wafer-scale modeling also considered the tribological aspects of the fluid film between the wafer and the pad [21]. Feature-scale modeling approach assumed that the material removal is related to sharp edges that are defined as slurry particles embedded in the fibers of the polishing pads. Elastic modulus of the slurry and the material to be polished are the critical factors in determining the CMP removal rate [22]. These models, however, do not directly include the contribution of chemical action from the slurry. Based on the development of the CMP process for tungsten and copper, Kaufman et al. [25] suggested a model that includes the action of a metal etchant, a metal passivation agent, and an abrasive agent. According to the model, actions of these agents result in the removal of passivating films from the high spots while such films continue to protect the low spots. Continuous cycles of formation, removal, and reformation of the passivating layer continue during the CMP process. However, this model also fails to explain the exact role of the etchant. Indeed, the mechanisms involved in CMP are far from being understood. The material removal and polishing at a given point in CMP is determined by the topography and the relative heights of the surrounding features. Coplanarity and dishing are the critical issues that form the basis of a CMP development process. A detailed description of these and other aspects in CMP are given elsewhere (Watts Microelectronic packaging). Several reviews have been published on many aspects of CMP pads including materials development, thermomechanical properties, pad/slurry interaction, and hydrodynamics [26–30].

1.4.1 CMP IN COPPER INTERCONNECT TECHNOLOGY

Chip interconnections are dielectric embedded metal wiring that connects the various circuit elements and distribute power in a chip. They also function as the interface between the device and the package, and require excellent control of mechanical properties. The exceptional planarity provided by the CMP process enhanced the electrical and mechanical properties of the metal and dielectric films, and in turn enhanced their reliability. Also, enhanced planarity facilitated the continuous increase in circuit density, number of interconnect levels, and innovation of new design elements such as stacked vias. The electrical performance of the interconnect is determined by the resistive and capacitive elements that contribute to the gate delay and hence affect the transistor performance.

The development of the CMP process for Cu was one of the milestones that made the implementation of Cu interconnect possible. An important aspect of the Cu CMP process development involved consideration of the interactions between Cu plating and Cu CMP. These interactions are at both local and global levels and create challenges in process integration. Overfill effect of dense lines during electroplating is an example of such interactions. The overfill effects are known to leave residual Cu in dense lines during CMP resulting in electrical failure due to shorting. On the other hand, overpolishing of wafers to clear the Cu residues leads to excessive dishing in wide Cu lines. Another example is the need for significant excess plating to fill up the wide lines for geometric leveling during plating. However, such "overburden copper" places a heavy burden on the CMP process. Optimization of these processes, therefore, involves a closer interaction between CMP and electroplating teams.

In the multilevel copper interconnection, the hierarchical interconnect scheme consists of lower interconnects at minimum pitch and thickness to minimize capacitance and maximize interconnect density, and the upper-level interconnects are scaled uniformly both vertically and horizontally to maintain a constant capacitance while reducing resistance. The transistors and other devices in the chip are connected by an efficient wiring layout consisting of a stack of 6–13 metallization layers embedded in dielectric layers. The number of metallization layers depends on the product and its performance. Some advanced products have a higher number of metallization layers. For example, Intel's 14-nm finFET process consists of 9–12 metal layers, while Intel's 10-nm process features 13 metal layers consisting of a combination of cobalt and copper metallization layers [31].

Multilevel copper interconnection layers are fabricated using barrier materials, such as Ti/TiN, Ta/TaN, and Co, that are deposited between the wiring metal and dielectric. Ti/TiN is usually used to improve the adhesion of W to the dielectric such as SiO_2 owing to the poor adhesive properties of W on the SiO_2. The barrier metal is required to prevent the diffusion of Cu into surrounding materials. Initially, SiO_2 was the dielectric material followed by carbon-doped oxide (CDO or SiCOH), with a lower dielectric constant ($k = 2.7$–3) starting with a 90-nm technology node. Ultralow k porous SiCOH materials ($k \approx 2.4$–2.0) with porous SiCOH have been used in applications beyond 22-nm technology nodes. The copper CMP process involves two steps. The first step accomplishes the removal of Cu overburden and stops on TaN/Ta liner or slightly before the liner is exposed. The second step involves the removal of TaN/Ta liner, the hard or soft mask materials, and then about 30–40 nm of the dielectric material as well. Complete removal of the liner and hard mask materials should be accomplished without significant loss of Cu line thickness, and the dielectric removal target should be achieved without excessive thinning. These goals need to be achieved across all line widths, pattern densities, and feature sizes. Additional steps such as buffing and chemical rinsing may also be used to remove particulates and provide a protective layer on the Cu surface. The most important task in metal CMP is to minimize the polishing selectivity between the interconnection material, barrier material, and dielectric layers. The differences in the properties of the wiring metal and the barrier metal lead to variations in removal rates, and this results in selectivity problems during CMP. Chemical selectivity is the ratio of the removal rate between two polished materials, and it affects CMP defects such as dishing and erosion.

1.4.1.1 Dishing and Erosion

Figure 1.6 shows the process steps involved in the fabrication of copper interconnects. Trenches and vias are first etched into the dielectric layer, and then a liner film is deposited to act as a diffusion barrier against copper migration and to promote adhesion, followed by deposition of the copper metal. The next step is to polish the wafer surface by CMP until the copper is left only in the trenches and vias recessed into the dielectric. Because CMP leaves the wafer surface planar on a global scale, this sequence may be easily repeated to add multiple layers of metal. During the CMP process, copper dishing and dielectric erosion may be observed. Copper dishing and SiO_2 erosion occur during the over-polish step which is required to ensure complete copper removal across the entire wafer. This is schematically shown in Figure 1.7. Copper dishing is defined as the difference in height between the center of the copper line, which is the lowest point of the dish, and the point where the dielectric levels off, which is the highest point of the dielectric. Copper dishing occurs because the

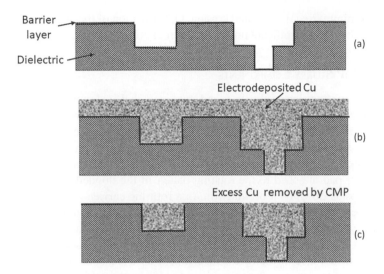

FIGURE 1.6 Process steps involved in the fabrication of copper interconnects. (a) Trenches and vias are etched into dielectric and coated with a diffusion barrier layer, (b) copper is electrodeposited, and (c) excess copper is removed and planarized by CMP.

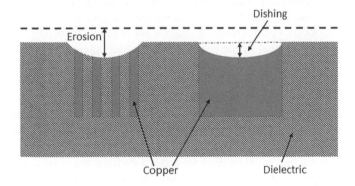

FIGURE 1.7 Schematic presentation of copper dishing and dielectric erosion.

polishing pad bends slightly into the recess and removes copper below the surface of the dielectric. Erosion is defined as the difference in the dielectric thickness before and after the polish step. Erosion is measured at a point where the dielectric is level. Both copper dishing and dielectric erosion are undesirable because they reduce the final thickness of the copper and copper dishing leads to nonplanarity of the surface which results in complications when adding multiple levels of metal. The global surface planarity of a layer is critical for building the next layer for three-dimensional integration technology, which aims to reduce the device size. Dishing and erosion are also critical because they both have a direct impact on the line resistance. It is therefore evident that the development of the CMP process should focus on minimizing dishing and erosion. The optimization of the consumables (pad and slurry) used in CMP has the most impact on minimizing dishing and erosion. Optimization of the slurry chemistry and particles exhibited a significant improvement in dishing and erosion [27].

1.4.2   CMP Slurry Components

The CMP slurry consists of abrasives and chemicals that are homogenously suspended in water. During CMP, the slurry is spread over the pad. The slurry continuously delivers the chemical components and abrasive particles to the entire wafer. The material removal in CMP is the result of an interaction between chemical and mechanical forces. Material removal occurs when the rotating wafer surface is pushed against a soft polymer pad attached to the rotating platen in the flooded slurry. The wafer is mounted upside down on a backing film in a rotating carrier, and a diamond conditioner is used to dress the pad to maintain its surface roughness during polishing. The CMP slurry consists of abrasives and chemicals that are homogenously suspended in water. A chemically reacted layer forms by a chemical reaction between wafer and slurry, which is removed by the mechanical abrasion. The passivation layer is a chemically reacted layer, such as a stabilized oxide on the surface, which protects the final surface from corrosion. The removal mechanism of metal CPM is the repetition of generating a passivation film and its removal by mechanical abrasion [25].

The chemical composition of the slurry and its maintenance is important in metal CMP. The basic chemical components of a metal CMP slurry include an oxidizer, a corrosion inhibitor, a chelating agent, a surfactant, and a pH adjuster [32]. The oxidizer produces an electrochemical dissolution reaction involving the formation of metal ions and removal of the electron(s) from the reacting surface. The corrosion inhibitor helps in controlling and locally inhibiting the dissolution process, thus minimizing dishing and erosion issues. The chelating agent forms complexes with dissolved metal ions, thus providing the desired passivation layer and a controlled dissolution rate of the metal. A surfactant is used to modify the interfacial properties of the slurry by lowering its surface tension and pH adjuster controls and maintains the desired acidity or alkalinity level during the CMP process.

As stated above, the oxidizers generate metal ions from metal via a redox reaction. Oxidation reaction produces metal ions, while the reduction reaction of the oxidizer produces hydroxyl ions. The metal ions and the hydroxyl ions formed at the surface react to form the metal hydroxide passivation layer. Commonly used oxidizers in manufacturing are H_2O_2, $K_2S_2O_8$, and $Fe(NO_3)_3$. Most Cu slurries have alumina abrasive because of its high selectivity toward Cu and the ability to stop on Ta/TaN. Increasingly, silica-based slurries are becoming more common. The silica abrasives have appreciable polish rates for Ta/TaN barrier and need additional components in the slurry to suppress liner erosion. Benzotriazole (BTA) is by far the most widely used inhibitor, even though other triazoles and several organic compounds that are known to adsorb on Cu have been investigated [33].

The addition of complexing agents is essential to keep the Cu ions in solution, and thus reduce surface particulate contamination. Ammonium ions, ethylenediamine, EDTA, and carboxylic acids including amino acids have been used to achieve this goal. Ethylenediamine can be used over a wide range of pH and is used in many slurry formulations.

BTA, a well-known inhibitor, is perhaps the most important and critical component in a CMP slurry. It is known to form an adsorbed film on the copper surface that provides exceptional protection from all forms of corrosion. Numerous studies have focused their attention on characterizing the properties of BTA films formed on a variety of Cu surfaces deposited by PVD, CVD, and electroplated [34–36]. BTA is known to exist in several forms in solution depending on the pH. At pH values around 2–3, the predominant species is the monoprotonated BTAH. It forms diprotonated species with hydrogen on adjacent nitrogens ($BTAH_2^+$) at pH values close to zero. The anionic species BTA^- are formed in alkaline pH. Stewart et al. studied the investigation of the influence of anions on the BTA films formed on copper surfaces, which indicated that films grown in the presence of acetate, sulfate, and perchlorate were much thinner than the films formed in the presence of halides [37,38]. The halide incorporation results in thicker films (>100 nm) because the halide ions coordinate with Cu(I) in the CuI-BTA films stabilizing the Cu(I) and hence the entire film as well [37,38].

Chelating agents promote the production of new metal ions to catalyze oxidation and production of the passivation layer. Acids containing carboxyl [$-C(=O)OH$; succinic acid, acetic acid, oxalic acid] or amine groups ($-NH_2$; ethylene-diamine) or amino acids ($-COOH$ and $-NH_2$; aminobutyric acid, glycine) or hydroxyl ($-OH$; citric acid) are commonly used as complexing agents for metal slurry as well as for post-CMP cleaning solutions. Surfactants have been used in Cu CMP slurries to modify the BTA films or to form a film that is an alternative to BTA films. Depending on the nature of surfactants used, they may increase or decrease the level of particulate contamination of the wafer. Some surfactants may cause aggregation of abrasive particles over time and may lead to increased defects [36]. The surfactant interactions are highly complex and intricate, involving multiple components of the CMP process including abrasives, pad, metal, liner, or dielectric surfaces simultaneously [29].

1.4.3 FINAL REMARKS

The performance of the Cu CMP process is determined by removal rates, selectivity, global and local planarity, surface topography, dishing and erosion, defectivity, and throughput. For low-k and ultra-low-k interconnects, the primary concerns are weak interfaces, delamination, nonuniformity of the cap and dielectric films, and material compatibility with slurry chemistry. With the continuing decrease in interconnect dimensions, many of the requirements and tolerances become extremely stringent. Post-CMP cleaning is a difficult challenge, especially for ultra-low-k interconnects. Innovations in slurry chemistry are essential to meet these requirements. From a processing perspective, CMP has so far accommodated the challenges described above. The CMP process has become a critical part of semiconductor process technology. It is also considered as an enabling process in the fabrication of many other devices including thin-film magnetic heads, and MEMS. Further development efforts will have to address the difficult challenges that many of the materials used in these devices are highly sensitive to the chemical environment and mechanical stress associated with CMP. These challenges are providing unique opportunities for slurry research and planarization process development.

REFERENCES

1. H. H. Uhlig, R. W. Revie, *Corrosion and Corrosion Control*, 3rd edition, John Wiley, New York, (1985).
2. M. Fontana, *Corrosion Engineering*, 3rd edition, McGraw Hill International Edition, New York, (1987).
3. L. L. Shreir, R. A. Jarman, G. T. Burstein eds., *Corrosion*, Butterworth-Heinemann, Oxford, (1994).
4. D. Landolt, *Corrosion and Surface Chemistry of Metals*, EPFL Press (CRC Press), Lausanne, Switzerland, (2007).
5. E. McCafferty, *Introduction to Corrosion Science*, Springer, New York, (2010).
6. K. Khasani, *Case Stud. Eng. Fail. Anal.*, 9, 71–77 (2017).
7. L. Chang, M. Grace Burke, F. Scenini, *Scr. Mater.*, 164, 1–5 (2019).
8. T. Notoya, *Corros. Eng.*, 39, 353–362 (1990); *J. Mater. Sci. Lett.*, 10, 389–391 (1991); Mater. Perform., 32 (5) 53–57 (1993).
9. D. M. Bastidas, I. Cayuela, J. M. Bastidas, *Rev. de Metal.*, 42 (5), 367–381 (2006).
10. M. Pourbaix, *Atlas of Electrochemical Equilibria in Aqueous Solutions*, NACE International, Houston, Houston, (1984).
11. M. Datta, D. Landolt, *Electrochim. Acta*, 45, 2535 (2000).
12. M. Datta, D. Harris, *Electrochim. Acta*, 42, 3007 (1997).
13. G. Bellows, *Chemical Machining*, 2nd edition, pp. 82–102, Machinability Data Center, MDC, Metcut Research Associates Inc., Cincinnati, OH, (1982).
14. D. M. Allen, *The Principles and Practice of Photochemical Machining and Photoetching*, Adam Hilger, Boston, MA, (1986).
15. Advanced Circuits, 4pcb.com/pcb-desgin-specifications/.
16. H. K. Kuiken, R. P. Tijburg, *J. Electrochem. Soc.*, 130, 1722 (1983).
17. R. P. Tijburg, J. G. M. Ligthart, H. K. Kuiken, J. J. Kelly, *J. Electrochem. Soc.*, 150, C440 (2003).
18. M. Georgiadou, R. Alkire, *J. Electrochem. Soc.*, 140, 1340 (1993).
19. M. Georgiadou, R. Alkire, *J. Electrochem. Soc.*, 140, 1348 (1993).
20. L. M. Cook, *J. Non-Crystal. Solids*, 120, 152 (1990).
21. S. R. Runnels, L. M. Eyman, *J. Electrochem. Soc.*, 141, 1698 (1994).
22. C.-W. Liu, B.-T. Dai, W.-T. Tseng, C.-F. Yeh, *J. Electrochem. Soc.*, 143, 716 (1996).
23. J. Warnock, *J. Electrochem. Soc.*, 138, 2398 (1991).
24. F. Preston, *J. Soc. Glass Technol.*, 11, 1, (1927).
25. F. B. Kaufman, D. B. Thompson, R. E. Broadie, M. A. Jaso, W. L. Guthrie, D. J. Pearson, M. B. Small, *J. Electrochem. Soc.*, 138, 3460 (1991).
26. P. B. Zantye, A. Kumar, A. A. K. Sikder, *Mater. Sci. Eng.*, 45, 89 (2004).
27. D. K. Watts, N. Kimura, M. Tsujimura, Chemical Mechanical Planarization: From Scratch to Planar, *in Microelectronic Packaging*, M. Datta, T. Osaka, J. W. Schultze eds., p. 437, CRC Press, Boca Raton (2005).
28. E. Matijevic, S. V. Babu, *J. Colloid Interface Sci.*, 2008 (320), 219 (2008).
29. M. Krishnan, J. W. Nalaskowski, L. M. Cook, *Chem. Rev.*, 110, 178 (2010).
30. D. Lee, H. Lee, H. Jeong, *Int. J. Precis. Eng. Manuf.*, 17 (12), 1751 (2016).
31. A. Mendes, *Intel Next Generation CPU's to Contain 100 Million Transistors Per Square Millimeter, Scientifist*, (2017); M. Lapedus, *Dealing with Resistance in Chips*, (2018).
32. H. Lee, B. Park, H. Jeong, *Microelectron. Eng.*, 85 (4), 689 (2008).
33. S. Govindaswamy, Y. Li, Corrosion Inhibitor for Cu CMP Slurry 249, *in Microelectronic Applications for Chemical Mechanical Planarization*, Y. Li, ed., p. 249, Wiley Interscience John Wiley & Sons Inc., New York (2008).
34. D. J. Tromans, *J. Electrochem. Soc.*, 145, L42 (1998).

35. B. J. Palla, D. O. Shah, *J. Colloid Interface Sci.*, 256 (143), 117 (2002).
36. T. Gopal, J. B. Talbot, *J. Electrochem. Soc.*, 153, G622 (2006).
37. K. L. Stewart, J. J. Keleher, A. A. Gewirth, *J. Electrochem. Soc.*, 155, D625 (2008).
38. K. L. Stewart, J. Zhang, S. Li, P. W. Carter, A. A. Gewirth, *J. Electrochem. Soc.*, 154, D57 (2007).

2 Anodic Behavior of Metals

2.1 INTRODUCTION

Electrochemical processes for shaping, finishing, and surface structuring of metals by controlled anodic dissolution include electrochemical machining (ECM), electropolishing (EP), and electrochemical micromachining. These processes find many applications in the manufacturing industry from the production of large complicated shaped parts for the aerospace and automotive industry to microfabrication of precision parts for microelectromechanical systems. The technical feasibility of these processes depends on the ability of the system to provide high dissolution rates, smooth surfaces, and reliable production of desired surface features. An understanding of the metal-electrolyte interaction including the dependence of anodic reactions and their stoichiometry on anodic potential/current is a prerequisite for optimizing process parameters of such processes.

Anodic reactions occurring at an electrode surface depend on the nature and composition of the electrolyte and the applied potential. A typical presentation of different anodic reactions that may take place [1–3] at increased anodic potential is shown schematically in Figure 2.1. Active dissolution of metal takes place at low potentials where the current density increases with increasing anode potential (region I). In the passive potential region (region II), a protective oxide film forms on the metal, indicated by a decrease in the current density until a very low current is observed. Several different types of anodic reactions may occur in the transpassive potential region. In Figure 2.1, the transpassive region (III), beyond the passive region, is characterized by a renewed increase in the current density, independent of anodic reactions and their mechanisms. Transpassive dissolution may take place at potentials below or above the potential corresponding to oxygen evolution by the decomposition of water. Localized passive film breakdown by pitting (IIIa) and film dissolution due to oxidation or by local acidification (IIIb) involve transpassive metal dissolution at potentials less noble than that necessary for oxygen evolution. On electronically conducting stable anodic films, oxygen evolution takes place on reaching oxygen potential E_{O_2} (IIIc). At still higher potentials, transpassive dissolution takes place at high rates (IIId).

The polarization curves of Figure 2.1 qualitatively describe different possible anodic behaviors of metals and alloys in different electrolytes. In some cases, however, the actual shape of anodic polarization curves may be slightly different from those shown. For example, several metals and alloys may become passivated immediately after they are exposed in a passivating electrolyte. In such systems, the active region is generally absent in the polarization curve, exhibiting only the passive-transpassive transition. In the ECM literature, salt solutions are generally

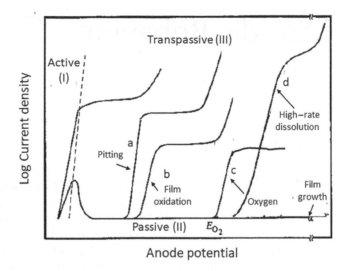

FIGURE 2.1 Schematic presentation of different types of anodic polarization behavior showing active (I), passive (II), and transpassive regions (III). (IIIa) Localized breakdown by pitting, (IIIb) film oxidation, (IIIc) oxygen evolution followed by (IIId) transpassive dissolution at high rates. E_{O_2} is the potential for oxygen evolution.

categorized into two types: passivating electrolytes containing oxidizing anions such as nitrates and chlorates, and nonpassivating electrolytes containing halides such as chlorides and bromides. However, this distinction is mostly applicable only to iron and nickel [9,10]. Other metals such as copper, titanium, and aluminum display qualitatively similar polarization behavior in both types of electrolytes. A limiting current plateau is usually observed in each of the types of polarization curves of Figure 2.1. The importance of limiting current and its influence on ECM and electropolishing performance are discussed later in greater detail. At potentials greater than the limiting current, several new anodic reactions can occur. Anodic polarization of certain metals in concentrated acids at a current beyond the limiting current may lead to the simultaneous production of oxygen. In nonpassivating ECM electrolytes (e.g., in concentrated NaCl), a change in dissolution valence may occur.

From the above discussion, it is clear that the anodic behavior of metals at high potentials involves a variety of processes, the nature, and rate of which may be influenced by electrochemical and hydrodynamic parameters. To acquire an understanding of these processes, experimental investigations must be performed in properly designed electrochemical cells that can provide controlled hydrodynamic conditions at the anode. Rotating disk electrodes have frequently been employed for the study of electropolishing [4–6]. However, they are not suitable for high-rate dissolution studies under ECM conditions, particularly in systems in which metal dissolution is preceded or accompanied by oxygen evolution. Flow channel cells have been used effectively for high-rate dissolution studies, under ECM conditions, of different metals and alloys in passivating as well as nonpassivating electrolytes [7–10].

2.2 ACTIVE DISSOLUTION

An increase in anodic current in the region I of Figure 2.1 (the active region) is due to a dissolution reaction that corresponds to the direct passage of metal atoms into the electrolyte in hydrated or complexed form. The active dissolution involves the removal of atoms from energetically favored kink sites on monoatomic steps situated on closed packed planes [11,12]. For these reasons, the rate of active dissolution at a given potential depends on the crystal planes involved. Since the atoms situated in the proximity of grain boundary and other crystalline defects are energetically more favorable, they are preferentially removed during active dissolution. Studies of the dissolution kinetics and morphology of low-index planes of iron single crystal have demonstrated that a strong correlation exists between the kinetics of the dissolution of an iron electrode and the atomistic structure of its surface. The dissolved surface morphology has been described quantitatively in terms of monoatomic steps and kinks [11,12]. The surface concentration of kinks depends strongly on the electrode potential, while the step density is determined by the pH level of the solution. On randomly oriented crystals or in the presence of a high density of imperfections, the nucleation of monoatomic steps is rapid. Monoatomic steps moving at different velocities lead to the formation of microscopically observable etch patterns. Active dissolution thus leads to surface etching even under extremely high current densities. In the potential region where the metal dissolution rate is not limited by mass transport, the kinetics of active dissolution is generally expressed by the Tafel behavior.

Figure 2.2 shows the anodic polarization behavior of iron and nickel in chloride solution in the active dissolution region. Asakura and Nobe [13] studied the active dissolution of Ferrovac E (99.91%) rotating disk electrodes in a 1 M potassium

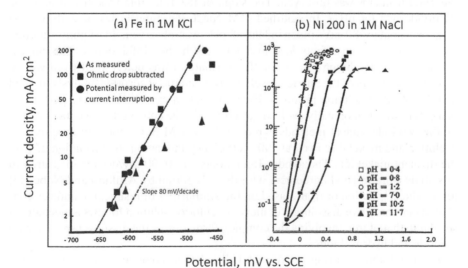

FIGURE 2.2 (a) Steady-state anodic polarization of iron in 1 M KCl. (b) Influence of pH on the anodic polarization of Ni 200 in 1 M NaCl [13,15].

chloride solution. Figure 2.2a shows the potential-current plot under steady-state anodic polarization. While the plot follows the Tafel behavior at low current densities, the deviation from linearity is considerable at high current densities due to the ohmic potential drop. In order to determine the contribution of the ohmic drop, the solution resistance was determined by the pulse technique. In Figure 2.2a, the potential values after subtraction of the ohmic drop from the measured potential are plotted to show a linear relationship even at higher potentials. The electrode potential was also measured by the current-interrupting method. The current was interrupted for 120 μs. The interruption was repeated at intervals of 10 Hz. The open circles in Figure 2.2 represent the potential immediately after interrupting the current. The plots, after correction for the ohmic drop, show Tafel behavior with a slope of about 80 mV/decade up to 200 mA/cm². The absence of passivation even at high current densities and Tafel behavior in the entire range indicated the probable formation of a soluble iron complex with chloride ions. Based on these observations, Asakura and Nobe concluded that the results of their work could be interpreted by assuming the formation of a hydroxochloro-iron complex and a consequent pH change at the electrode surface similar to the mechanism proposed by Bockris et al. on the existence of an iron complex with chloride ions [14].

Datta and Landolt [15] investigated the anodic polarization behavior of Ni 200 in 1 M NaCl electrolyte using a rotating disk electrode. Polarization curves measured in electrolytes of different pH varying between 0.4 and 11.7 are shown in Figure 2.2b. The data shown in Figure 2.2b were measured at a rotation speed of 6,400 rpm. At high current densities, the ohmic drop between the capillary tip and anode surface became the limiting factor. The ohmic contribution to the measured potential data was estimated from the theoretical resistance to a disk under conditions of primary current distribution [16]. Within the current density range studied, the Tafel behavior was observed. The slope of the Tafel line was measured to be 90 mV/decade for pH 7. In acidified 1 M NaCl, the Tafel behavior with a slope of approximately 80 mV, independent of pH, was observed. In slightly alkaline solutions, passivation effects lead to apparently higher Tafel slopes. The current efficiency for dissolution (based on Ni^{2+} formation) was determined from weight loss measurements of the anode as well as from atomic absorption analysis of solid and dissolved dissolution products. Free convection experiments were performed in a cell in which anodically evolved gas could also be collected. The data obtained by the two independent methods agreed well. In 1 M NaCl, there was no oxygen evolution and nickel dissolved at 100% efficiency indicating the formation of only the divalent nickel ion. The anodic Tafel slopes of 80–90 mV/decade measured in this study were in good agreement with those reported by Asakura and Nobe [13] for anodic dissolution of iron. Based on the similarities of results, the authors concluded that the active dissolution of nickel in chloride solution follows a mechanism similar to that of iron involving the complex formation of the dissolved species with chloride and hydroxyl ions.

As stated above, during active metal dissolution atoms are removed from the crystal lattice and pass into solution as hydrated or complexed ions. This process occurs at energetically favorable sites. On low-index surfaces, dissolution takes place from kink sites located at monoatomic steps. Metallic surfaces contain many defects

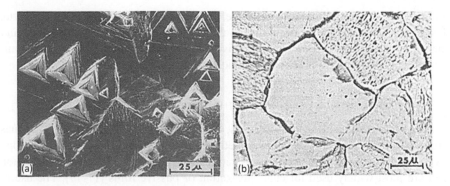

FIGURE 2.3   Surfaces resulting from the active dissolution of nickel (a) and iron (b) in 5 M NaCl. Experiments conducted in a flow channel cell. Flow velocity: 1,000 cm/s; current density: 5 A/cm² [9].

such as vacancies, dislocations, grain boundaries, inclusions, and segregations. Furthermore, they are not atomically flat but have a roughness that depends on the pretreatment. The adsorption of anions or organic molecules at a kink site can block dissolution and thus slow down the propagation of a monoatomic step. The geometry of the facets formed reflects the crystalline orientation of the grains, and thus allows the study of the microstructure of a polycrystalline material. Defects that intersect the surface, modify the crystalline structure, and therefore facilitate localized dissolution. This is shown in Figure 2.3 which shows the surfaces of (i) Nickel 200 (99% min. Ni) and (ii) Armco iron (99.9% Fe) samples. The nickel and iron anodes were dissolved at 5 A/cm² in 5 M NaCl in a flow channel cell [9]. The applied current density corresponded to 50% of the estimated limiting current, hence represented very well within the active dissolution conditions. Figure 2.3a shows the formation of crystallographic etch-pits on the nickel surface after dissolution. Etch pits are generally formed at dislocations that are introduced due to thermomechanical shock or microsegregation of impurities during metal forming. Figure 2.3b shows that the grain boundaries have been preferentially attacked on iron anodes. Grain boundaries are locations where the crystal structure is strongly disturbed or modified and the atoms, therefore, are less solidly bound.

2.3   PASSIVATION

Passive oxide films are formed on a metal surface by its exposure to an oxidizing environment. The oxide films serve to reduce the reaction rate between the metal and the environment. The phenomenon of metal passivation in aqueous electrolytes is illustrated by the anodic polarization curve in Figure 2.1 which shows the active-passive transition region, where the anodic dissolution current passes through a maximum and decreases steeply with increasing anodic polarization. The potential region where the current remains at very low values correspond to the passive state of the metal. The stability of the passive film so formed depends on its composition, thickness, electronic properties, potential, the presence (or absence) of aggressive anions such as halides in the electrolyte, temperature, and pH of the electrolyte. The characterization of the composition, thickness, and structure of such films and the study of their interaction with corrosive environments require the use of a combination of advanced experimental techniques.

2.3.1 EXPERIMENTAL TECHNIQUES FOR THE STUDY OF PASSIVE FILMS

For the characterization of passive films, both ex-situ as well as in-situ techniques are employed. Ex-situ surface analysis methods, such as Auger electron spectroscopy (AES) and X-ray photoelectron spectroscopy (XPS), provide chemical information and are most widely used for characterization of passive films [17–19]. The use of angle-resolved AES and XPS provides information on the depth distribution of different species and their oxidation states [20]. Other methods such as secondary ion mass spectroscopy and ion scattering spectroscopy (ISS) have also been frequently used to obtain important information about passive films [17–19]. All of these techniques require the experiments to be performed in an ultrahigh-vacuum chamber to avoid contamination of the surface by undesired adsorption processes during the analysis.

A critical issue with ex-situ surface analysis is the sample transfer from the electrolyte to the vacuum systems. During the sample transfer, the passive films may undergo structural changes when removed from the electrolyte or when submitted to intense radiation. Different innovative methods for sample transfer without contact with the atmosphere have been developed but complete exclusion of exposure is not easily attainable. However, by developing an experimental protocol that ensures a consistent transfer methodology for all samples complimented by the investigation of reference samples, it is possible to draw valid conclusions from comparisons between a series of measurements. On the other hand, in-situ techniques are available which eliminate the artifacts associated with sample transfer.

In-situ techniques are preferable because the measurements are carried out on passive films as they are formed. Compared to ex-situ techniques, which require a transfer of the sample, the in-situ techniques avoid any chemical modification that might take place during the transfer of the sample from the electrolyte to the instrument used. Coulometry is the most commonly used in-situ method for the determination of the amount of oxide present on a surface. Thus, it is used for determination of thickness and in certain cases the oxidation state of the passive film. The technique consists of measuring the charge required to reduce the passive film by applying a small cathodic current density on the order of $\mu A/cm^2$ to the electrode. Among optical methods, ellipsometry has been frequently used for the characterization of composition, roughness, thickness, crystallinity, and dielectric properties of passive films. Ellipsometry is a contactless, nondestructive method, which measures the change of polarization upon reflection or transmission and compares it to a reference model. Impedance spectroscopy is a widely used electrochemical in-situ method for the characterization of the electrical properties of passive films. The technique can also be used for monitoring changes in the film thickness. Scanning tunneling microscopy is another in-situ method, which provides atomic-scale structural information of passive films.

2.3.2 ANODIC PASSIVE FILMS ON METALS

Anodic passive films on metals and alloys have been studied extensively and several reviews on the topic have appeared [18,21,22]. A comprehensive review of the passivity of different metals of technical interest has been provided by Sato [22]. The passive film generally grows with increasing anodic potential at the rate of 1–3 nm/V. For most transition metals and alloys, the thickness of spontaneously formed passive films is less than several nanometers in the potential region where water is thermodynamically stable. The passive film on iron in aqueous solutions grows up to about 5 nm at potentials prior to the onset of oxygen evolution. If the passivated metals are those such as iron, and nickel on which anodic oxygen evolution takes place, there is virtually no further thickening of the film in aqueous electrolyte. For refractory metals such as aluminum and niobium on which no anodic oxygen evolution proceeds, the film can be grown to thicker than 100 nm by increasing the polarization up to a critical potential at which the film breakdown occurs. The incorporation of electrolyte anions takes place presumably only when the film grows at the oxide-electrolyte interface, and their concentration in the films is greater when the electrolyte concentration ratio [anion]/[OH$^-$] at the film surface is greater [23]. The incorporated anion concentration, therefore, will increase as the anodic current is increased. The incorporated anion concentration is greater in the initial stage than in the final stage of oxide formation, provided the oxide formation and dissolution processes are simultaneously taking place. In general, thin passive oxide films are not crystalline but turn partially crystalline when they become thick for reasons such as internal compressive stresses created in the film during its growth [23]. In the following, passive films on some transition metals and some selected valve metals are briefly described.

2.3.2.1 Passive Films on Ni, Fe, and Their Alloys

2.3.2.1.1 Nickel

Passivating oxide films on nickel have been studied by many authors and literature reviews on the subject are available [24–26]. When immersed in alkaline solution, nickel spontaneously passivates and becomes covered with a film of Ni(OH)$_2$ [27,28]. Upon anodic polarization, depending on the pH and the potential, higher oxidation states may be formed such as NiO, Ni$_3$O$_4$, Ni$_2$O$_3$, and NiO$_2$ of different structures and degrees of hydration [29]. Evidence for the existence of divalent and trivalent nickel oxide on passive nickel has been produced by a number of authors [28,30–32]. For example, using XPS, Dickinson et al. [30] concluded that the film present on nickel in the passive potential region is composed of NiO and Ni(OH)$_2$, the former being the passivating species. Similar conclusions have been drawn by Arvia et al. [33]. At higher potentials, Ni(OH)$_2$ may be oxidized to β-NiOOH and a series of solid solutions of the two species may be formed [34,35]. X-ray diffraction studies [31,32] and ellipsometric studies indicate that Ni(OH)$_2$ and β-NiOOH are usually present in the potential region preceding oxygen evolution. Very conflicting opinions, however, exist regarding the existence of Ni$_3$O$_4$, and NiO$_2$. Okuyawa and Haruyama [36] claim that Ni$_3$O$_4$ forms at intermediate potentials, whereas Davies and Barker [28] could not find any evidence for the existence of a definite oxide corresponding to Ni$_3$O$_4$. Although β-NiOOH is the highest oxidation state of Ni observed by X-ray diffraction studies [31], this does not necessarily exclude the existence of higher oxides such as NiO_2 since these species may be amorphous to X-rays. Several authors have found the final oxidation state of nickel oxide to be $NiO_{1.7-1.9}$ [28,34] which was explained by assuming that the oxide observed is a solid solution of NiO_2 and Ni_2O_3.

Anodic film thickness on passive nickel has been investigated by a number of authors [37–40] using different experimental techniques but reported results do not always agree. Nuclear microanalysis experiments by Siejka et al. [38] and coulometric determination of film thickness by MacDougall and Cohen [39] in neutral and acid solutions yielded film thickness values not higher than 1.6 nm. Sato and Kudo [40] investigated the film thickness on nickel in an alkaline solution using ellipsometry and coulometry. For the passive potential region, film thickness measured by ellipsometry was found to increase from 0.4 to 1.0 nm with increasing potential while coulometry yielded higher values. Lu and Srinivasan [41], using ellipsometry, reported film thickness on nickel in the oxygen evolution region in 1 M KOH to be as high as 70 nm.

2.3.2.1.2 Iron

Anodic polarization of iron in sulfuric acid leads to the formation of passive films with an average film thickness of less than a monolayer at the passivation potential. The passive film thickness, however, increases with increasing potential with values up to several nanometers at potentials below the oxygen evolution region. The composition of anodic passive films on iron has been investigated by a number of workers using electrochemical, chemical, and surface analytical techniques [42–46]. Nagayama and Cohen [43,44] proposed a sandwich model consisting of an inner layer of Fe_3O_4, and an outer layer of y-Fe_2O_3. The existence of these oxides has been confirmed by electron diffraction [43–45]. According to Vetter [46], the potential drop in a passive film is across the outer layer, since Fe_3O_4 is a good electronic conductor. Bloom and Goldenberg have reported that y-Fe_2O_3 is stable only if it contains a minimum concentration of hydrogen ions [47]. Indeed, there are indications in the literature that the outer layer contains hydrogen in the form of hydroxyl ions or water [47,48].

Sato et al. [49] investigated the formation of anodic passive oxide film on iron in boric acid and sodium borate solution at a pH range of 5.45–10.45. The chemical composition of the films was investigated by cathodic reduction, chemical analysis, and ellipsometry. The cathodic reduction using a borate solution of pH 6.35 containing arsenic trioxide as inhibitor estimated the iron in the film to be all iron (III), indicating that no magnetite layer was present. Oxygen in the film was estimated from the ellipsometric thickness to be in excess of the stoichiometric ferric oxide, suggesting the presence of bound water. The average composition of the film was presented as $Fe_2O_3 \cdot 0.4H_2O$, in which hydrogen may be replaced partly with iron-ion vacancy. According to the authors, the oxide film on iron may be divided into two layers, an inner layer of anhydrous ferric oxide which thickens with an increase of potential, and an outer layer of hydrous ferric oxide whose thickness depends on the solution environment. It is across the inner layer in which most of the overpotential develops. The thickness of the film formed in the oxygen-evolution region ceases to increase and becomes potential-independent above a critical potential where the film loses its passivating property and the transpassive dissolution of iron occurs.

In another publication, Sato et al. [50] investigated the passivating films on iron in phosphate and borate solutions at pH values ranging from 1.85 to 11.50 by ellipsometric measurements and chemical analyses. In neutral solutions, a duplex passive film formed which consisted of an inner oxide barrier layer and an outer hydroxide deposit layer (Figure 2.4). The barrier layer was found to be potential-dependent and its thickness increased linearly with the potential. At the steady-state, a high electric field 5.6–6 V/cm (18 Å/V) developed in the layer. The outer hydroxide deposit layer depended on the solution composition rather than on the potential. In acid solution, the deposited layer dissolved away and only the barrier layer remained stable on the passive surface (Figure 2.5). The barrier layer in neutral solution was found to be ferric oxide and the deposited layer was found to be hydrous ferric oxide whose water content increased with increasing solution pH.

2.3.2.1.3   FeNi Alloys
Seo and Sato [51] used AES depth profiling with Ar^+ ion sputter-etching to determine the thickness and composition of anodic oxide films formed on 55Fe45Ni alloy in boric acid-sodium borate solutions. Their results showed that the thickness of anodic oxide film on Fe-Ni alloy increased linearly with anodic potential in the passive region. Depth profiling of the anodic oxide film revealed that a significant part of the Fe component in the film was in the divalent state. Enrichment of the Ni component at the oxide-substrate interface and depletion of Ni component in the film was measured as a function of pH and anodic potential. The authors interpreted the results

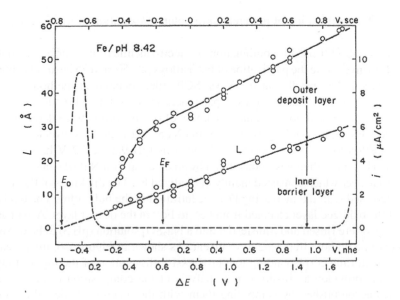

FIGURE 2.4 Thickness L of the inner barrier layer and the outer deposit layer of the anodic passivating film formed on iron at various potentials for 1 hour in sodium borate solution at pH 8.42. The broken line indicates the stationary anodic polarization curve. E_F is Flade potential [50].

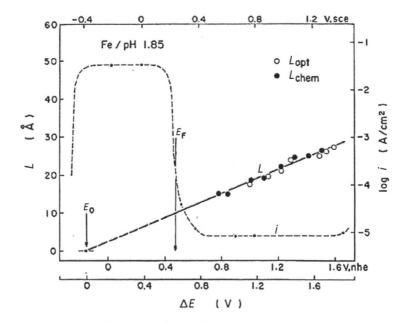

FIGURE 2.5 Thickness L of the anodic passivating film formed on iron at various potentials for 1 hour in phosphoric acid solution at pH 1.85. The film consists of only the barrier layer. The broken line indicates the stationary anodic polarization curve [50].

in terms of a combination of preferential dissolution/oxidation of the Fe component in the film.

Basile et al. [52] used a combination of electrochemical and surface analytical methods to investigate the passivation of FeNi alloys (25, 50, and 75 Ni at.%) in borate-boric acid solution, pH 9.2, at 0.2 V vs. SCE. Electrochemical methods included measurement of polarization curves, capacitance, and dissolution kinetics. The depth composition, chemical bonding, and thickness of passive layers were determined by AES and XPS depth profiling. Both AES and XPS depth analyses revealed that passivated FeNi alloys in borate-boric acid solutions (pH 9.2) at 0.2 V have a duplex structure (1–2.5 nm thickness) constituted by an external thin $Ni(OH)_2$ layer covering an internal layer which consisted mainly of oxidized Fe. The thickness of the internal layer increased with the increasing Fe concentration in the bulk alloy. The oxidation state of the oxidized layer changed from Fe^{3+} to Fe^{2+} in the deeper layer. After passivation, films were rinsed and transferred into a Na_2SO_4 solution (pH 3) to be dissolved at a slow rate without imposing any current. As shown in Figure 2.6, the potentials evolved with time corresponding to a progressive dissolution of the topmost layers. These data provided an in-situ depth analysis of the composition of passive film. This was accomplished by comparing them with the potential scale calibration of standard films of FeOOH, Fe_2O_3, NiO, or $Ni(OH)_2$. In passive nickel films, where $Ni(OH)_2$ covers NiO layers, the potential traces exhibited a fast potential decrease down to 0.2 V due to the rapid dissolution of the upper hydroxide layer (Figure 2.7), followed by a large plateau at −0.24 V assigned to the NiO dissolution reaction.

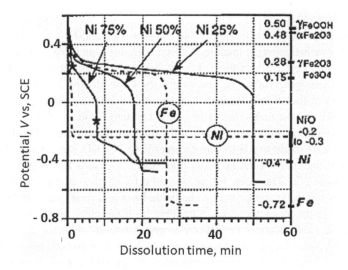

FIGURE 2.6 Dissolution kinetics of passivated Fe, Ni, and FeNi alloys in Na₂SO₄ solution, pH 3 [52].

In alloys, iron oxides, hydroxides, and Ni(OH)₂ dissolution occur at potentials higher than +0.1 V, as indicated in Figure 2.7. Iron-rich alloys containing larger oxidized Fe amounts display an abrupt potential decay which reveals the existence of step-like interfaces between the film and the alloy. In contrast, in the Ni-rich alloy (75 at.%) and to a lesser extent in the 50 at.% alloy, a second stage is displayed between −0.25 and −0.45 V that can be attributed to NiO dissolution.

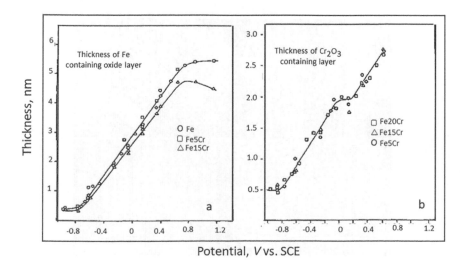

FIGURE 2.7 (a) Fe-containing oxide thickness of Fe and Fe-Cr alloys and (b) Cr₂O₃ thickness as a function of the anode potential. Passivation time: 300 seconds in 1 M NaOH [53].

2.3.2.1.4 Fe-Cr/FeCrNi Alloys

Haupt et al. [53] studied the oxide film formation on Fe and Fe-Cr alloys using XPS and ISS. They investigated the thickness of iron-containing and Cr_2O_3-containing oxide layers on Fe and Fe-Cr alloys formed by polarizing for 300 seconds in 1 M NaOH at different anode potentials. The results are shown in Figure 2.7. In order to avoid contamination of the sample surfaces by exposure to air, they developed a specimen preparation and transfer methodology in a closed system. Oxide thickness was deduced from Fe $2p_{3/2}$ signal after deconvolution of the integrated peak areas. For this purpose, a simple two-layer oxide structure with an inner FeO, $Fe(OH)_2$ and an outer Fe_2O_3 were assumed. The thickness of the Fe containing oxide layer as a function of potential is shown in Figure 2.7a. The oxide formation on Fe-Cr alloys is complicated because of the presence of both Fe and Cr within the oxide layer and their concentration changes continuously. In this case, therefore, a multiple layer structure, as in the case of pure iron, is an oversimplification. Fe (III) oxide grows linearly with the electrode potential up to 0.8 V, however, the thickness decreases in the transpassive potentials >0.8 V. The iron oxide thickness is almost independent of the bulk composition as shown in Figure 2.8a for Fe and two Fe-Cr alloys. The thickness of Cr_2O_3, as deduced from the Cr $2p_{3/2}$ signal, increases linearly with the electrode potential (Figure 2.7b). The small step at nearly 0 V is attributed to the release of Cr^{3+} by its oxidation to soluble CrO_4^-, which becomes thermodynamically possible at potential >−0.2 V. Angular-dependent XPS studies suggested that the passive film on Fe-Cr alloys contained an outer Cr-depleted and an inner Cr-enriched part. These results were also confirmed by ISS depth profiling.

Olsson and Landolt [54] reviewed the literature on the characterization and understanding of passive films on stainless steel. According to them, substantial progress has been made in ex-situ surface analytical techniques, while new in-situ methods for the study of passive films with atomic resolution have been introduced. These advances have provided real-time information on film chemistry and growth.

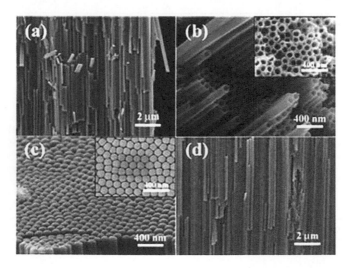

FIGURE 2.8 FE-SEM cross-sectional (a and d), top (b) and bottom (c) views of TiO_2 nanotubes produced by anodizing in a mixture of ethylene glycol +2 wt.% water+0.25 wt.% NH_4F [66].

Different literature results indicated that the thickness of passive films on stainless steels grows linearly with the applied potential. Based on the investigation of passive films on Fe-Cr alloys in acidic sulfate and sodium hydroxide solutions [55,56], it is concluded that the passive films on these alloys constitute two components – an oxide part and a hydroxide part. In alkaline solution, considerably thicker films are obtained since film dissolution is retarded. In acidic solution, the passive film is due mainly to the oxide part, the hydroxide part of the film is approximately independent of potential [57]. The composition and chemistry of the film depends on the potential. For Fe-Cr alloys, enrichment of chromium occurs in the film in the low passive region, but in the high passive region, the chromium content decreases since the stability of iron surpasses that of chromium. Chromium, in turn, may oxidize to its soluble hexavalent state. At high passive potentials, Fe has a stabilizing action on the trivalent Cr, making Fe-Cr alloys stable over a wider potential range than pure Cr.

2.3.2.2 Anodic Oxide Films on Valve Metals

Valve metals, such as Al, Ti, Ta, Zr, Hf, Nb, and W can form thick oxide films that can withstand high voltages in selected electrolytes. In the following, the formation of oxide films on valve metals is briefly described focusing mainly on Al and Ti.

2.3.2.2.1 Anodic Oxide Films on Aluminum

The formation of oxide films on aluminum and its alloys has been studied extensively because of their significance in commercial applications [58]. The type of anodic oxide film that can be produced on aluminum depends upon several factors, the most important of which is the nature of the electrolyte. Barrier-type films are produced in several electrolytes including aqueous solutions of boric acid, ammonium borate or ammonium tartrate, ammonium tetraborate in ethylene glycol, and several organic electrolytes including citric, malic, and glycolic acids in the pH range of 5–7. In acidic solutions, these electrolytes do not form nonporous barrier-type films. Porous-type films are produced in electrolytes in which the anodically formed oxide film is slightly soluble. Examples of such electrolytes include sulfuric, phosphoric, chromic, and oxalic acids at almost any concentration. The thickness of barrier-type film is controlled by voltage and electrolyte temperature, whereas porous-type film thickness depends on the current density and time. The maximum film thickness attainable for barrier-type films is restricted to a voltage below the oxide breakdown voltage value, i.e., 500–700 V, which corresponds to 700–1,000 nm. The thickness of porous films, being time-dependent, can grow to many times higher than the upper limit placed upon barrier-type films. For porous films, apart from the current density and time, the electrolyte temperature is also an important criterion in determining the film thickness. At low temperatures (0°C–5°C), the porous film formed is thick, compact, and hard and is the process is known as hard anodizing. At high temperatures (60°C–75°C), the porous film is thin, soft, and nonprotective; under such conditions the process of electropolishing is achieved, i.e., where the oxide film is dissolved by the electrolyte almost as soon as it is formed.

The most important commercial use of the barrier-type films of aluminum is in the field of dielectric capacitors. The porous-type films used commercially possess excellent corrosion and abrasion resistance and, owing to high porosity, form a good base for paints or dyes. The industrial uses of porous anodic oxide films are restricted to those films which are sealed. This process of sealing of the porous films involves immersing the films in hot water, or in aqueous solutions of certain salts above 90°C.

2.3.2.2.2   Anodic Films on Titanium

Titanium is a highly corrosion-resistant light metal owing to the high chemical stability of a thin passive film formed on titanium in various corrosive environments. Because of high biocompatibility, titanium is also of interest as a biomaterial. Coloring of titanium occurs as a consequence of the thickening of the surface oxide film by anodic polarization. Such oxide films, showing the interference color, have been utilized for decoration of titanium as well as corrosion protection. The structure of the oxide films formed on titanium by anodic polarization is strongly dependent on the formation potential. The oxide films formed on titanium are amorphous below 5 V, but an amorphous-to-crystalline transition takes place at 5–20 V. During the growth of the amorphous oxide, the film growth proceeds at high current efficiency close to 100%, while the film growth involves oxygen gas generation after the transition to crystalline oxide [59]. The transition is also dependent upon the grain orientation of the polycrystalline titanium substrate. Alloying of titanium with aluminum, molybdenum, tungsten, and zirconium suppresses the amorphous-to-crystalline transition.

2.3.2.2.3   Comparison of Anodic Films on Ti, Zr, Nb, and Ta

Anodic oxide films on Ti, Zr, Nb, and Ta in 0.5 M H_2SO_4 and 0.5 M H_3PO_4 was studied by Diamanti et al. [60]. Cell voltages ranging from 10 to 100 V were applied on each metal in order to record the growth kinetics of these four metal oxides in these electrolytes. Oxide thickness was measured by spectrophotometry. The position of interference maxima and minima were used to determine the oxide thickness by applying Bragg's law [61]. The oxide thickness increased linearly as a function of the applied cell voltage. They found their results to be consistent with the reported trends of anodizing ratios between 1.5 and 2.1 nm/V [62–65]. Oxidation in dilute phosphoric acid produced slightly thinner oxides compared with dilute sulfuric acid. They concluded that all the valve metals examined presented similar behavior in diluted acids, with the formation of a barrier oxide film of similar thickness as a function of the applied voltage. XRD analyses showed that the titanium oxides produced in phosphoric acid are amorphous, while titanium oxides produced in sulfuric acid are crystalline which produces favorable conditions for sparking. Among the four oxides, titanium oxide films have found applications in different fields. The electrochemically formed titanium oxide films can be tailored by modulating process parameters to obtain a variety of thicknesses, morphologies, structures, and compositions of the oxide. Lee et al. described the fabrication of crystallographically preferred oriented TiO_2 anatase nanotube arrays (p-NTAs) and the characterization of their photovoltaic properties [66]. Figure 2.8 shows the scanning electron microscope images of the titanium samples anodized in a mixture of water (2%–5%), ethylene glycol, and NH_4F (0.25 wt.%) at an applied voltage of 50 V. Figure 2.8 shows that an array of highly ordered nanotubes are formed that stand vertically on the Ti substrate. The nanotubes grew to approximately 11.5 and 11.8 mm in length. Die-sensitized solar cells constructed by employing the nanotube array showed excellent photovoltaic current density and high energy conversion efficiency [66]. In addition to the demonstrated application of titanium nanotubes in photovoltaic cells, the engineered properties of anodized titanium films have been put to several other applications which include water splitting, pollution control, and biomedical industry.

2.4 TRANSPASSIVITY

The stability of passive films and the reactions taking place in the transpassive potential region depend on the anode potential and the nature of anions present in the electrolyte. As shown in Figure 2.1, the transpassive potential region consists of several reactions such as localized passive film breakdown leading to pitting (IIIa), film oxidation and electropolishing (IIIb), oxygen evolution, and transpassive dissolution beyond oxygen evolution (IIIc).

2.4.1 PITTING

In the presence of aggressive anions (such as chloride ions) many metals exhibit localized film breakdown and dissolution by pitting when their potential exceeds the critical pitting potential (IIIa). Pitting phenomena, which are of great technical importance in corrosion, have been studied extensively and the topic has been reviewed [67–70]. Under high-field conditions in the film and at the film-solution interface, aggressive anions may penetrate through defects in the passive film and may provoke its localized destruction, leading to metal dissolution from the bare surface. Pitting is autocatalytic in nature. Once a pit is initiated the conditions developed are such that further pit growth is promoted. The pit environment becomes enriched in metal cations and an anionic species such as chloride ions. Chloride is a relatively small anion with a high diffusivity, which migrates into the pit to maintain charge neutrality by balancing the charge associated with the cation concentration. The pH in the pit is lower owing to cation hydrolysis. The acidic chloride environment thus generated in pits is aggressive to most metals and tends to propagate the pit growth. As the pit current density increases, the ionic concentration in the pit solution increases, often reaching supersaturation conditions. A solid salt film may form on the pit surface. Under these conditions, the pit growth rate is limited by mass transport out of the pit. Salt films are not required for pit stability, but they enhance stability by providing a buffer of ionic species that can dissolve into the pit. Under mass-transport-limited growth, pits will be hemispherical with polished surfaces. In the absence of a salt film (at lower potentials), pits may be crystallographically etched or they may be irregularly shaped.

The metal composition can have strong effects on the tendency for a metal to pit. In an alloy such as stainless steel, Cr concentration plays a dominant role in conferring passivity to the alloy. The pitting potential increases dramatically as the Cr content increases above the critical 13% value needed to create stainless steel. Increasing concentration of Ni, which stabilizes the austenitic phase, moderately improves the pitting resistance of Fe-Cr. Small increases in certain minor alloying elements, such as Mo in stainless steels, can greatly reduce pitting susceptibility. Pits almost always initiate at some chemical or physical heterogeneity at the surface, such as inclusions, flaws, dislocations, second-phase particles, solute-segregated grain boundaries, or mechanical damage. Most metals and alloys have some or all such defects, and pits will tend to form at the most susceptible sites first. Pits in stainless steel are often associated with inclusions such as MnS.

2.4.2   FILM OXIDATION AND ELECTROPOLISHING

In some cases, transpassive dissolution may involve film oxidation to soluble species. A well-known example is the transpassive dissolution of chromium in sulfuric acid [71]. In the passive potential region, the chromium surface is covered with a thin oxide film of trivalent chromium which protects the metal. In the transpassive potential region, the trivalent oxide film is transformed into an oxide of hexavalent chromium, which is highly soluble, namely

$$Cr_2O_3 + 4H_2O \rightarrow Cr_2O_7^{2-} + 8H^+ + 6e^-. \tag{2.1}$$

Chromium, therefore, goes into solution as chromate during the transpassive dissolution of chromium. The transpassive dissolution of chromium in concentrated phosphoric-sulfuric acids also involves the formation of soluble chromate. Anodic dissolution of many metals and alloys in concentrated acids lead to electropolishing. As an example, the anodic polarization behavior of 420 (Fe13Cr) stainless steel in phosphoric acid at 60°C is shown in Figure 2.9 [5]. The polarization curves show typical

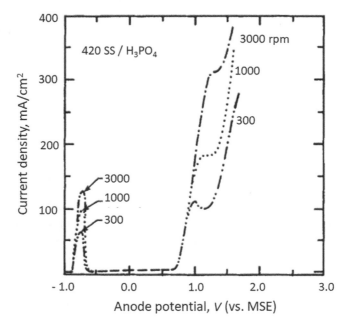

FIGURE 2.9   Anodic polarization behavior of 420 stainless steel in concentrated phosphoric acid (85%) electrolyte. Temperature 60°C; scan rate: 5 mv/s [5].

active-passive-transpassive transition behavior. In the active region, a current peak is observed at −0.8 V. The magnitude of the peak current increased with increasing rotation speed, indicating that mass transport processes play a role in the metal dissolution reaction under these conditions. In the passive potential region from −0.7 to 0.75 V, the anodic currents dropped to very low values, independent of the rotation speed. A sharp rise in the anodic current beyond 0.75 V is due to transpassive metal dissolution. A well-defined, rotation-speed-dependent limiting current is observed at potentials above 1 V. Under these conditions electropolishing is observed. It is well established that electropolishing is achieved only under conditions in which the metal dissolution is mass transport-controlled and the formation of a precipitated salt layer at the surface is possible [5–7,9,10]. The presence of a salt layer suppresses the influence of crystallographic orientation and surface defects on the dissolution process, thus leading to microfinishing. At potentials higher than 1.25 V, a renewed increase in the current is accompanied by oxygen evolution. A more detailed discussion of the fundamental aspects of electropolishing and its application in the industry will be presented in separate chapters.

2.4.3 OXYGEN EVOLUTION AND HIGH-RATE TRANSPASSIVE DISSOLUTION

As noted earlier, the thickness of passive films normally increases with the applied potential. Most passive films are semiconducting, but their electronic properties can vary greatly. For chemically stable, nonconducting films, it can reach several tens of nanometers at anodization potentials of 100 V or more. On the other hand, on films exhibiting electronic conductivity, electron-transfer reactions can take place at the film-electrolyte interface, and therefore above a certain potential, water is oxidized into oxygen. A mainly electronic current then crosses the interface and film growth ceases. Oxygen evolution takes place by solution oxidation according to the following reaction:

$$2H_2O \rightarrow O_2 + 4H^+ + 4e^- \tag{2.2}$$

The potential required for oxidation of water generally is on the order of 1.2–2 V and the thickness of electron-conducting films, therefore, does not exceed a few nm. The passive films formed on nickel and iron in passivating ECM electrolytes such as nitrate, chlorate, and sulfate solution are typical examples of this type of behavior. In such systems, an increase in potentials higher than that of oxygen evolution leads to metal dissolution. At still higher potentials, the rate of metal dissolution increases with increasing potential while the rate of oxygen evolution decreases. Lohrengel et al. used a capillary-based droplet cell to analyze reaction products including anodically evolved oxygen during the transpassive dissolution of many metallic and ceramic materials [72–76]. A continuous electrolyte flow was realized in the droplet cell by a computer-controlled gear pump with flow rates up to 10 m/s to guarantee the removal of all products. With typical spot diameters of about 100 mm current densities >100 A/cm² was reached at absolute currents of 1 A. Oxygen was quantified by fluorescence quenching of Ru complexes [75,76]. Contribution of the oxygen evolution reaction in the total reaction product at a current density of 30 A/cm² depended on the material [72]. They reported that negligible oxygen evolution was observed (<5% of charge) with Al, CrC, Mg, Mo₂C, NbC, Ni, TaC, TiC, VC, WC, and ZrC; around 10% of the charge contributed to the oxygen evolution with Ag, Al_3Fe, Al_3Mg_2, $Al_{12}Mg_{17}$, Mo, Mo_2C, and VC; around 20% with Al_2CuMg, Al_4Mn, Co, Cr, Cr_3C_2, Cu, Fe, and ZrC; around 40% with Al_2Cu, Fe_3C, $Fe_{64}Ni_{36}$ (Invar1), Mn, and Zr; and almost 100% with Au and Pt. No oxygen evolution and no machining took place with Nb, NbC, Ta, TaC, Ti, V, W, and WC.

It is interesting to note that high-rate anodic dissolution in the active region as well as in the transpassive region eventually exhibits a limiting current at high potentials indicating a change in the reaction stoichiometry. As will be discussed in the following chapter, the limiting current depends on the hydrodynamic conditions at the dissolving anode surface and is mass transport controlled. A detailed description of reaction stoichiometry for different metal-electrolyte systems under different polarization conditions is presented in the following.

2.5   ANODIC REACTION STOICHIOMETRY

A complete analysis of all reaction products is required in order to obtain information on the anodic reactions taking place at high potentials. This, however, is not easily accomplished, since it may require the collection of solid, dissolved, and gaseous products during anodic dissolution studies. Furthermore, separation of the anodic products from the cathodic reaction products is not easily possible either. In ECM and electropolishing, a knowledge of anodic reactions taking place at high potentials is derived primarily from weight loss measurements and applying the Faraday law [5–10]. The weight loss, ΔW, is related to the metal dissolution stoichiometry by

$$\Delta W = \frac{ItM}{nF} \tag{2.3}$$

where I is the applied current, t is the time, F is the Faraday constant, and n is the valence of metal dissolution which corresponds to the number of electrons removed from dissolving metal atoms by anodic oxidation. The use of weight loss as a measure of dissolution valence is strictly applicable to anodic reactions involving metal dissolution only. In the presence of other reactions occurring simultaneously at the anode, such as oxygen evolution, weight loss measurements have frequently been used to determine an "apparent dissolution valence" [5–10]. Under these conditions, knowledge of current efficiency is required to calculate n using the above equation. The value of n provides information on the overall reaction stoichiometry. Its measurement is, therefore, a prerequisite for any mechanistic studies. From a practical point of view, a knowledge of the value of n is useful for the calculation of metal removal rate. During the high-rate dissolution of copper, the value of n undergoes a noticeable change with the transition from active to transpassive dissolution. The anodic polarization of copper does not exhibit the conventional passivation behavior, as described earlier. Transpassive dissolution here refers to dissolution at current densities higher than the limiting current for salt film formation. In nitrate and sulfate solutions, the active dissolution of copper proceeds with a dissolution valence of 2. Transpassive dissolution leads to the simultaneous production of monovalent copper reaching a limiting value of 1.6 at high current densities [7,8]. The active dissolution of copper in chloride solution proceeds with a dissolution valence of 1. This is due to the stabilization of cuprous ions by complex formation. In the transpassive mode, the simultaneous formation of cupric ions increases the apparent dissolution valence, going through a maximum of 1.4 before reaching a limiting value of 1.2 at high current densities. No oxygen evolution has been reported during copper dissolution in these electrolytes [8].

High-rate dissolution of iron in a chloride solution at low current densities involves iron dissolution in its divalent state [9,77]. However, at current densities higher than the limiting current density, simultaneous production of Fe^{2+} and Fe^{3+} takes place [9]. This has been confirmed by the complete analysis of reaction products which have shown no evidence of anodic oxygen evolution during the high-rate dissolution of iron in concentrated chloride solutions [77]. High-rate anodic dissolution of nickel in chloride solutions involves the formation of divalent nickel ions independently of the current density level [9,15]. Anodic reaction stoichiometry during the high-rate dissolution of iron and nickel in passivating ECM solutions (sodium nitrate and sodium chlorate) has been studied extensively [10,77–84]. In these studies, oxygen evolution has been found to be the predominant anodic reaction at relatively low current densities in the transpassive potential region. At higher current densities, the relative rate of metal dissolution increases with increasing current density. This is shown in Figure 2.10, which contains apparent dissolution valence data for iron dissolution in chloride, nitrate, and chlorate electrolytes. In ECM, the distribution of metal

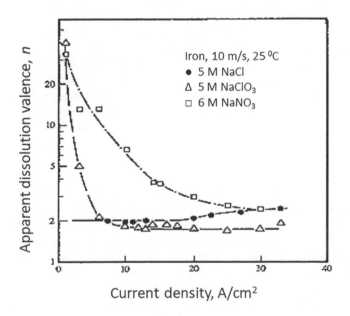

FIGURE 2.10 Apparent dissolution valence data for iron in neutral salt solutions as a function of current density. Data were obtained in a flow channel cell at 10 m/s. In 5 M NaCl, simultaneous production of Fe(II) and Fe(III) takes place at current densities higher than the limiting current. In nitrate and chlorate electrolytes, oxygen evolution is the dominant anodic reaction at low current densities, while high-rate metal dissolution takes place at high current densities [9,10].

dissolution rate on the workpiece determines the final shape in relation to the tool. The machining performance is, therefore, influenced significantly by the dependence on the current density of the anodic reactions. Passivating metal-electrolyte systems are known to give better ECM precision because of their ability to form oxide films and evolve oxygen in the stray-current region [85]. During the high-rate dissolution of iron in the chlorate solution, the apparent dissolution valence was found to be less than 2 at high current densities. Similar results have been obtained for the high-rate dissolution of nickel in chlorate solution [10]. Such anomalous behavior may result from an intergranular attack, leading to the removal of metallic particles from the anode, or they may be the result of chemical reactions occurring at the anode surface. X-ray diffraction analysis of dissolution products has not shown any evidence of the presence of metallic particles [15]. The low dissolution valence values have been attributed to chlorate reduction at active sites on the anode by the corrosion reactions:

$$2Fe + ClO_3^- + 6H^+ \rightarrow 2Fe^{3+} + Cl^- + 3H_2O \tag{2.4}$$

and

$$6Fe^{2+} + ClO_3^- + 6H^+ \rightarrow 6Fe^{3+} + Cl^- + 3H_2O. \tag{2.5}$$

Solution analysis, indeed, has shown an accumulation of Cl^-, providing conclusive evidence of chlorate reduction on the active surface produced during the high-rate anodic dissolution of iron and nickel, and thus also providing an explanation for the experimentally obtained anomalous dissolution valence [10].

Anodic dissolution of Cr in ECM electrolytes takes place in the transpassive potential region, yielding Cr^{6+}; the dissolution valence is independent of current density and the nature of the electrolyte anion [86,87]. Molybdenum also yields Mo^{6+} during anodic dissolution, which takes place exclusively in the transpassive potential region [88–90]. The high-rate anodic dissolution of titanium takes place at sufficiently high anode potentials through the breakdown of passive films. In chloride and bromide solutions, the dissolution valence is close to 4, similar to that found under pitting corrosion conditions [91].

During electropolishing of nickel in concentrated sulfuric acid, the transpassive dissolution occurs at potentials well below that required for oxygen evolution and the associated metal dissolution valence is 2. At higher current densities, oxygen evolution occurs, and the sum of the current efficiencies for metal dissolution based on Ni^{2+} formation and oxygen evolution is 100%. The transpassive dissolution of Cr in concentrated phosphoric acid and a mixture of concentrated phosphoric and sulfuric acids yields a dissolution valence of 6, independent of anode potential [92]. On the other hand, anodic reaction stoichiometry for iron dissolution in phosphoric acid depends on the anode potential [6]. In the active region, iron dissolves as Fe^{2+}. At higher potentials Fe^{2+} and Fe^{3+} are formed, the measured value of n is 2.5. At potentials corresponding to the limiting current, iron dissolution leads to Fe^{3+} formation. At higher potentials, a renewed increase in current density is associated with oxygen evolution, as evidenced by apparent valence values of n in excess of 3 under these conditions. In a mixture of concentrated phosphoric and sulfuric acids, anodic polarization leads to passivation of iron; anodic dissolution of iron takes place exclusively in the transpassive potential region. At potentials below the current plateau, the simultaneous production of Fe^{2+} and Fe^{3+} occurs, the latter becoming increasingly important until only the Fe^{3+} is formed at the limiting current plateau. At potentials higher than the limiting current, apparent dissolution valence values greater than 3 are due to simultaneous oxygen evolution [6]. During the high-rate dissolution of copper in concentrated phosphoric acids, copper dissolves as Cu^{2+} in the active region and in the current plateau region. At potentials higher than those of the current plateau region, simultaneous oxygen evolution occurs, yielding a value of n greater than 2. The transpassive dissolution of titanium in perchloric acid leads to an anomalous dissolution valence of 1.8; this has been explained by the reduction of perchlorate ions at the anode [91].

REFERENCES

1. T. P. Hoar, *Corros. Sci.*, 7, 341 (1967).
2. D. Landolt, Transpassivity, *in Passivity of Metals*, R. Frankenthal, J. Kruger eds., p. 488, The Electrochemical Society, Inc., Pennington, NJ, (1978).
3. M. Datta, *IBM J. Res. Dev.*, 37 (2), 207 (1993).
4. L. Ponto, M. Datta, D. Landolt, *Surf. Coating. Tech.*, 30, 265 (1987).
5. M. Datta, D. Vercruysse, *J. Electrochem. Soc.*, 137, 3016 (1990).
6. M. Datta, L. F. Vega, L. T. Romankiw, P. Duby, *Electrochim. Acta*, 37 (13), 2475 (1992).
7. D. Landolt, R. H. Muller, C. W. Tobias, *J. Electrochem. Soc.*, 116, 1384 (1969).
8. K. Kinoshita, D. Landolt, R. H. Muller, C. W. Tobias, *J. Electrochem. Soc.*, 117, 1246 (1970).
9. M. Datta, D. Landolt, *Electrochim. Acta*, 25, 1255 (1980).
10. M. Datta, D. Landolt, *Electrochim. Acta*, 25, 1263 (1980).
11. W. Allgaier, K. E. Heusler, *J. Appl. Electrochem.*, 9, 155 (1979).
12. W. J. Lorenz, K. E. Heusler, Anodic Dissolution of Iron Group Metals, *in Corrosion Mechanisms*, F. Mansfeld ed., pp. 1–83, Marcel Dekker, Inc., New York, (1987).
13. S. Asakura, K. Nobe, *J. Electrochem. Soc.*, 118, 13 (1971).
14. J. O. M. Bockris, D. Drazic, A. R. Despic, *Electrochim. Acta*, 4, 325 (1961).
15. M. Datta, D. Landolt, *Corros. Sci.*, 13, 187 (1973).
16. J. Newman, *J. Electrochem. Soc.*, 113, 501 (1966).
17. D. Briggs, M. P. Seah, *Practical Surface Analysis*, John Wiley, New York, (1983).
18. M. Fremont ed., *Passivity of Metals and Semiconductors*, Elsevier, Amsterdam, (1983).
19. H.-H. Strehblow, *Electrochim. Acta*, 212, 630 (2016).
20. M. Datta, H. J. Mathieu, D. Landolt, *Appl. Surf. Sci.*, 18, 299–314 (1984).
21. R. Frankenthal, J. Kruger eds., *Passivity of Metals*, Electrochemical Soc., Princeton, NJ, (1978).
22. N. Sato, *Corros. Sci.*, 3 (I), 1–19 (1990).
23. J. S. L. Leach, B. R. Pearson, *Corros. Sci.*, 28, 43 (1988).
24. L. Young, *Anodic Oxide Films*, Academic Press, New York, (1961).
25. A. J. Arvia, P. Posadas, Passivity of Nickel, *in The Electrochemistry of the Elements*, A. J. Bard ed., vol. III, p. 212, Marcel Dekker, New York, (1975).
26. E. Sikora, D. D. Macdonald, *Electrochim. Acta*, 48, 69 (2002).
27. J. L. Weininger, M. W. Breiter, *J. Electrochem. Soc.*, 110, 484 (1963).
28. D. E. Davies, W. Barker, *Corrosion*, 20, 47t (1964).

29. M. Pourbaix, *Atlas of Electrochemical Equilibria in Aqueous Solutions*, NACE International, Houston, TX, (1984).
30. T. Dickinson, A. F. Povey, P. M. A. Sherwood, *J. Chem. Soc. Faraday Trans.*, I (73), 327 (1977).
31. S. U. Falk, *J. Electrochem. Soc.*, 107, 661 (1960).
32. A. J. Salkind, P. F. Bruins, *J. Electrochem. Soc.*, 109, 356 (1962).
33. R. S. Scbrebler-Gutzman, J. R. Vilche, A. J. Arvia, *Corros. Sci.*, 18. 765 (1978).
34. G. W. D. Briggs, W. F. K. Wynne Jones, *Electrochim. Acta*, 7, 241 (1962).
35. G. W. D. Brings, M. Fleischmann, *Trans. Faraday Soc.*, 62, 3217 (1966).
36. M. Okuyawa, S. Haruyama, *Electrochim. Acta*, 14, 1 (1974).
37. K. Kudo, N. Sato, *Corros. Eng.*, 21, 24 (1972).
38. J. Siejka, C. Cherki, J. Yahalom, *Electrochim. Acta*, 17, 61 (1972).
39. B. MacDougall, M. Cohen, *J. Electrochem. Soc.*, 123, 1783 (1976).
40. N. Sato, K. Kudo, *Electrochim. Acta*, 19, 461 (1974).
41. P. W. T. Lu, S. Srinivasan, *J. Electrochem. Soc.*, 125, 1416 (1978).
42. M. Cohen, *in Passivity of Metals*, R. P. Frankenthal, J. Kruger eds., p. 521, The Electrochemical Society, Inc., Pennington, NJ, (1978).
43. M. Nagayama, M. Cohen, *J. Electrochem. Soc.*, 109, 781 (1962).
44. M. Nagayama, M. Cohen, *J. Electrochem. Soc.*, 110, 670 (1963).
45. C. L. Foley, J. Kruger, C. J. Bechtoldt, *J. Electrochem. Soc.*, 114, 994 (1967).
46. K. J. Vetter, *J. Electrochem. Soc.*, 110, 597 (1963).
47. M. C. Bloom, M. Goldenberg, *Corros. Sci.*, 5, 623 (1965).
48. N. Sato, K. Kudo, T. Noda, *Corros. Sci.*, 10, 785 (1970).
49. N. Sato, K. Kudo, T. Noda, *Electrochim. Acta*, 16, 1909 (1971).
50. N. Sato, K. Kudo, T. Noda, *Z. Phys. Chem. N.F.*, 98, 271 (1975).
51. M. Seo, N. Sato, *Corros. Sci.*, 18, 577 (1978).
52. F. Basile, J. Bergner, C. Bombart, B. Rondot, P. Le Guevel, G. Lorang, *Surf. Interface Anal.*, 30, 154 (2000).
53. S. Haupt, C. Calinski, U. Collisi, H. W. Hoppe, H.-D. Speckmann, H.-H. Strehblow, *Surf. Interface Anal.*, 9, 357 (1986).
54. C.-O. A. Olsson, D. Landolt, *Electrochim. Acta*, 48, 1093 (2003).
55. S. Haupt, H.-H. Strehblow, *Corros. Sci.*, 37, 43 (1995).
56. H.-W. Hoppe, S. Haupt, H.-H. Strehblow, *Surf. Interface Anal.*, 21, 514 (1994).
57. C.-O. A. Olsson, D. Hamm, D. Landolt, *J. Electrochem. Soc.*, 147, 2563 (2000).
58. J. W. Diggle, T. C. Downie, C. W. Goulding, *Chem. Rev.*, 69 (3), 365 (1969).
59. C. K. Dyer, J. S. L. Leach, *J. Electrochem. Soc.*, 125, 1032 (1978).
60. M. V. Diamanti, P. Garbagnoli, B. Del Curto, M. P. Pedeferri, *Curr. Nanosci.*, 2015 (11), 307 (2015).
61. M. V. Diamanti, B. Del Curto, M. Pedeferri, *Color Res. Appl.*, 33, 221 (2008).
62. S. Van Gils, P. Mast, E. Stinjns, H. Terryn, *Surf. Coat. Technol.*, 185, 303 (2004).
63. Y. T. Sul, C. B. Johansson, Y. Jeong, Y. Albrektsson, *Med. Eng. Phys.*, 23, 329 (2001).
64. D. Velten, V. Biehl, F. Aubertin, B. Valeske, W. Possart, J. Breme, *J. Biomed. Mater. Res.*, 59, 18 (2001).
65. M. V. Diamanti, B. Del Curto, V. Masconale, C. Passaro, M. P. Pedeferri, *Color Res. Appl.*, 37, 384 (2012).
66. S. Lee, I. J. Park, D. H. Kim, W. M. Seong, D. W. Kim, G. S. Han, J. Y. Kim, H. S. Jung, K. S. Hong, *Energy Environ. Sci.*, 5, 7989 (2012).
67. Z. Szklarska-Smialowska, *Pitting Corrosion of Metals*, National Association of Corrosion Engineers, Houston, TX, (1986).
68. G. S. Frankel, *J. Electrochem. Soc.*, 145, 2186 (1998).
69. G. S. Frankel, T. Li, J. R. Scully, *J. Electrochem. Soc.*, 164, C180 (2017).
70. R. E. Melchers, *Corros Mater. Degrad.*, 1, 42 (2018).

71 W. J. Plieth, K. J. Vetter, *Ber. Bunsenges.*, 74, 1077 (1969).
72. M. M. Lohrengel, K. P. Rataj, T. Munninghoff, *Electrochim. Acta*, 201, 348 (2016).
73. M. M. Lohrengel, I. Klüppel, C. Rosenkranz, H. Bettermann, J. W. Schultze, *Electrochim. Acta*, 48, 3203 (2003).
74. M. M. Lohrengel, C. Rosenkranz, *Corros. Sci.*, 47 785 (2005).
75. C. Hammer, B. Walther, H. Karabulut, M. M. Lohrengel, *J. Solid State Electrochem.*, 15, 1885 (2011).
76. K. P. Rataj, C. Hammer, B. Walther, M. M. Lohrengel, *Electrochim. Acta*, 90, 12 (2013).
77. K. W. Mao, *J. Electrochem. Soc.*, 118, 1870 (1971).
78. D. T. Chin, *J. Electrochem. Soc.*, 118, 174 (1971).
79. M. A. LaBoda, A. J. Chartrand, J. P. Hoare, C. R. Wiese, K. W. Mao, *J. Electrochem. Soc.*, 120, 643 (1973).
80. K. W. Mao, D. T. Chin, *J. Electrochem. Soc.*, 121, 191 (1974).
81. M. Datta, D. Landolt, *J. Electrochem. Soc.*, 122, 1466 (1975).
82. J. P. Hoare, C. R. Wiese, *Corros. Sci.*, 15, 435 (1975).
83. M. Datta, D. Landolt, *J. Appl. Electrochem.*, 7, 247 (1977).
84. M. Datta, D. Landolt, *J. Electrochem. Soc.*, 124, 483 (1977).
85. M. Datta and D. Landolt, *Proceedings of the 2nd International Symposium on Industrial and Oriented Basic Electrochemistry*, SAEST, Madras, India, p. 4.3.1, (1980).
86. O. A. Arzhintar, A. I. Dikusar, V. I. Petrenko, Y. N. Petrov, *Electronnaya Orabot. Mat.*, 6, 9 (1974).
87. E. Rosset, M. Datta, D. Landolt, *J. Appl. Electrochem.*, 20, 69 (1990).
88. J. W. Johnson, C. H. Chi, C. K. Chen, W. J. James, *Corrosion*, 26, 238 (1970).
89. M. N. Hull, *J. Electroanal. Chem.*, 38, 143 (1972).
90. M. Datta, *IBM J. Res. Dev.*, 37 (2), 207 (1998).
91. J. B. Mathieu, D. Landolt, *J. Electrocehm. Soc.*, 125, 1044 (1978).
92. L. Ponto, D. Landolt, *J. Appl. Electrochem.*, 17, 205 (1987).

3 Transpassive Films and Their Breakdown under ECM Conditions

3.1 INTRODUCTION

Investigations of anodic dissolution of nickel, iron, and its alloys in passivating electrochemical machining (ECM) electrolytes have been conducted by several authors [1–22]. As discussed in Chapter 2, passive films formed on nickel and iron are electronically conducting such that oxygen evolution takes place above oxygen potential. When the applied anodic potential or current density in the transpassive potential region is further increased, the relative importance of oxygen evolution decreases, and that of transpassive metal dissolution increases. Eventually, the metal dissolves at high current efficiency yielding generally the same valence state as under the active dissolution conditions. The described behavior indicates that the passive film which normally protects the anode becomes nonprotective as current density is increased. The processes responsible for the change in film properties or the surface morphology under high-rate transpassive dissolution conditions have been investigated [1–16]. Earlier studies in this area were reported by General Motors researchers [1–4,6–10], who investigated the anodic films formed on iron and nickel in ECM electrolytes. They performed potentiostatic polarization experiments in conjunction with coulometry in a stationary electrochemical cell. According to Mao [4], the oxide film formed on steel in chlorate solution dissolves under conditions of low pH at the electrode resulting from oxygen evolution. However, in a later work, Mao et al. [6] explained the dissolution mechanism of steel in different passivating electrolytes in terms of an ion-exchange model. According to the model, anions adsorbed on the film are exchanged with the oxygen ions of the oxide matrix which eventually breaks down and dissolves resulting in film thinning leading to high-rate dissolution. Chin and Mao [7] reported that oxide film on steel is ruptured by an intergranular attack and high-rate dissolution takes place from the underlying bare metal surface. In a series of publications, Datta and Landolt [11–18] reported the results of their investigation on the transpassive films and their breakdown during high-rate dissolution of nickel and iron in nitrate electrolytes under controlled flow conditions in a flow channel cell. Their studies involved the determination of the experimental variables which govern the metal dissolution efficiency, microscopic investigation of surface attack by applying short anodic current pulses, and the use of surface analytical (Auger Electron Spectroscopy (AES)/X-ray photoelectron spectroscopy (XPS)) techniques to study the transpassive films. These studies provided an insight into the phenomena of the transpassive film breakdown leading to high-rate dissolution under ECM condition.

51

3.2   FACTORS INFLUENCING TRANSPASSIVE DISSOLUTION

As stated above, transpassive metal dissolution sets in at potentials above the oxygen potential. At such potentials, simultaneous oxygen evolution and metal dissolution take place, the latter being an increasing function of the applied current density. Datta and Landolt investigated the influence of electrolyte concentration, pH, and temperature on the current efficiency for metal dissolution by using high-rate dissolution of nickel in nitrate electrolytes as a model system [12]. The current efficiency for nickel dissolution was determined from weight loss experiments and computed based on Ni^{2+} formation [11]. The results are summarized in Figure 3.1 which shows the nickel dissolution efficiency data as a function of current density for the three different parameters. The electrolyte flow velocity was kept constant at 1 m/s in these experiments. Nickel dissolution efficiency increased with an increase in bulk nitrate concentration (Figure 3.1a). The experimental data shows that a clear relationship exists among anion concentration, current density (anode potential), and metal dissolution efficiency indicating that transpassive dissolution being favored by high nitrate ion concentration and high anode potential. With regards to the influence of pH, Figure 3.1b shows that in the pH range of 0.5–12, metal dissolution efficiency is relatively little influenced by the pH of the electrolyte. This behavior is explained by considering that the local pH at the anode changes due to oxygen evolution which results in H^+ production. An estimation of H^+ ion buildup at the anode indicates that the local pH can indeed be neutral or acidic even in a slightly alkaline bulk solution (6 M $NaNO_3$ + 0.1 M NaOH). These results lead to the conclusion that in ECM practice, where neutral electrolytes are commonly used, small changes in electrolyte pH should not affect the machining performance. In a strongly alkaline solution, however, the local pH at the anode remains alkaline. Metal dissolution is, therefore, inhibited, i.e., oxygen evolution takes place over a wide range of current density as seen in Figure 3.1b which shows that in strongly alkaline solutions (pH 14) the

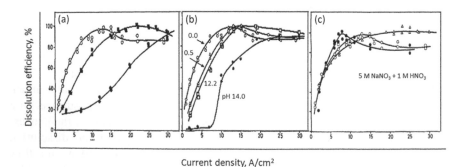

Current density, A/cm²

FIGURE 3.1   Influence of electrolyte concentration (a), pH (b), and temperature (c) on current efficiency for nickel dissolution in 6 M nitrate solution. (a): (O) 5 M $NaNO_3$ + 1 M HNO_3, (■) 3 M $NaNO_3$ + 1 M HNO_3, (●) 1 M $NaNO_3$ + 1 M HNO_3; (b) (O) 5 M $NaNO_3$ + 1 M HNO_3 (pH = 0), (Δ) 5.9 M $NaNO_3$ + 0.1 M HNO_3 (pH = 0.5), (□) 6 M $NaNO_3$ + 0.1 M NaOH (pH = 12.2), (●) 6 M $NaNO_3$ + 1 M NaOH (pH = 14); and (c): (●) 5°C, (O) 25°C, (Δ) 40°C [12].

transpassive nickel dissolution sets in at much higher current densities. The data in Figure 3.1c shows that at low current densities where metal dissolution efficiency is an increasing function of current density, the electrolyte temperature has little influence on metal dissolution efficiency.

The above data present some analogy to the well-known film breakdown phenomena leading to pitting in chloride solution. For example, the critical pitting potential of stainless steel has been reported to be strongly dependent on chloride ion concentration [23,24]. The pitting potential has been reported to be independent of pH in the range of 1–7 but for pH values higher than 7, the OH⁻ ions act as pit inhibitor [20]. The presence of NaOH inhibits the passivity breakdown of steel in NaCl solution [25,26]. Except at very low temperatures, the critical pitting potential is little influenced by increasing temperature [20].

In the transpassive dissolution experiments described above, the nickel dissolution efficiency can be taken as a measure of the stability of the oxide film, i.e., oxygen evolution takes place on stable films whereas increasing metal dissolution efficiency indicates film breakdown. The obtained results for metal dissolution efficiency as a function of nitrate ion concentration, pH, and temperature show a dependency that is similar to that for pitting. These results, therefore, suggest that a pitting-type mechanism is responsible for transpassive film breakdown that leads to high-rate dissolution in passivating ECM electrolytes.

3.3 MICROSCOPIC INVESTIGATION OF TRANSPASSIVE FILM BREAKDOWN

Microscopic studies of surfaces resulting from dissolution by applying very short current pulses validated the above hypothesis [12]. Experiments at small dissolution times were performed in 5 M $NaNO_3$ + 1 M HNO_3 solution. High-purity polycrystalline nickel (99.95%) anodes were used in a flow channel cell at a constant flow velocity of 1,000 cm/s. On the application of a very short current pulse, localized breakdown sites resembling etch pits could be seen. Initiation and growth of these dissolution sites were studied as a function of dissolution time and current density. Figure 3.2 shows the effect of dissolution time on the surface attack at a current density of 3 A/cm². Figure 3.2a corresponds to a charge of 1 C/cm². It shows that the initiation of dissolution occurs at discrete sites that are randomly distributed. With increasing dissolution time (hence increasing charge), grain boundaries become severely attacked and the dissolution sites grow giving rise to characteristic dissolution patterns that are oriented in definite directions. For example, three grains and their boundaries are visible in Figure 3.2b–d. The dissolution lines are pointed toward a particular direction which is different in different grains. The directional growth of the dissolution sites must, therefore, be related to grain orientation. The influence of current density on the evolution of surface attack is shown in Figure 3.3. These experiments were conducted at a constant charge of 5 C/cm². Keeping in mind that increasing current density corresponds to increasing dissolution efficiency, it is observed that the number of dissolution sites increases and the dissolution sites grow bigger with increasing current density until the whole surface becomes densely covered with small etch pits. These pits finally merge giving rise to a surface that

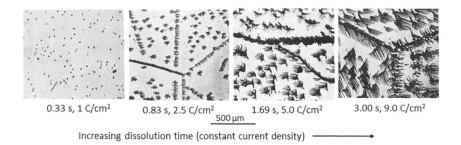

0.33 s, 1 C/cm²   0.83 s, 2.5 C/cm²   1.69 s, 5.0 C/cm²   3.00 s, 9.0 C/cm²

500 µm

Increasing dissolution time (constant current density) ⟶

FIGURE 3.2 Effect of dissolution time on surface attack of polycrystalline nickel. Electrolyte: 5 M NaNO$_3$ + 1 M HNO$_3$; current density: 3 A/cm² [12].

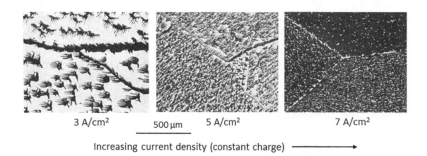

3 A/cm²   500 µm   5 A/cm²   7 A/cm²

Increasing current density (constant charge) ⟶

FIGURE 3.3 Effect of current density on surface attack of polycrystalline nickel. Electrolyte: 5 M NaNO$_3$ + 1 M HNO$_3$; constant charge: 5 C/cm² [12].

10 µm   (a)   2 µm   (b)

FIGURE 3.4 (a and b) SEM photographs showing dissolution pits on a polycrystalline nickel anode after a dissolution time of 0.85 seconds; current density, 3 A/cm²; electrolyte: 5 M NaNO$_3$ + 1 M HNO$_3$ [12].

appears black to the naked eye. It is interesting to note that under a given dissolution condition, the number of dissolution sites and their size is different in different grains indicating that different crystals in a polycrystalline nickel anode dissolve at different rates. Dissolution efficiency should, therefore, be related to the crystal orientation.

The scanning electron microscope (SEM) micrographs presented in Figure 3.4 demonstrate that initiation of dissolution is highly localized, i.e., dissolution pits with sharp boundaries are apparent in an otherwise intact passive film on which probably oxygen is evolved. The pits are randomly distributed, and their shape is also random. At higher magnification (Figure 3.4b), submicroscopic steps are visible within the pit indicating that dissolution proceeds from preferred crystallographic planes which are similar to active dissolution. With increasing dissolution time, the randomly shaped pits grow in certain specific directions giving rise to parallel ditches. These data show that the growth direction is different for different grains but within a single grain, growth is usually in the same direction.

Experiments with high-purity [111] oriented single-crystal nickel (99.95%) provided further evidence of the influence of crystallographic orientation on the dissolution process [12]. Figure 3.5 shows SEM photographs of a surface resulting from dissolution at 3 A/cm². Dissolution proceeds in three directions oriented approximately 120° to each other corresponding to the [111] projections. This gives rise to Y-shaped dissolution sites, all oriented in parallel. This behavior further confirms that the process of dissolution is related to grain orientation and probably proceeds from close-packed planes. Some undercutting of the film is observed.

The above data and discussions demonstrate that the transpassive dissolution of nickel in nitrate solution is initiated by a local breakdown of the passive film and, therefore, presents some analogies to the pitting behavior of passive metals in aggressive media under corrosion conditions. The process of nucleation and growth of corrosion pits and the electrochemical conditions within the pits have been widely studied [27–38]. Hoar et al. [28] proposed that preferential nucleation occurs on those parts of the film overlying grain boundaries and dislocations and suggested a mechanical mechanism for anion penetration into the passive film. Smialowska and coworkers [29,30] studying the pit morphology on stainless steel showed that pits may also nucleate at inclusions in the metal. According to Vetter and Strehblow [31,32], Sato [33], and Galvele [34], the breakdown of the anodic oxide film is controlled not only by the electrochemical reactivity of the aggressive ions but also by mechanical and dielectric properties of the film itself. Macdonald [35,36] described a Point Defect Model, according to which transport of cation vacancies through the

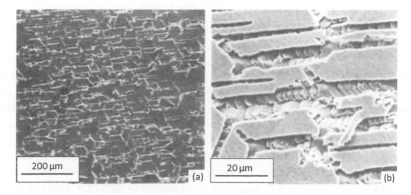

FIGURE 3.5 (a and b) SEM photographs of surface attack resulting from dissolution of a [111] oriented single-crystal nickel anode; current density: 3 A/cm², dissolution time: 2.66 seconds, electrolyte: 5 M $NaNO_3$ + 1 M HNO_3 [12].

oxide lattice and their condensation at the metal/film interface form a vacancy that leads to breakdown. Frankel et al. [37,38] provided passive film breakdown and pit growth stability perspective for more aggressive and less aggressive pitting conditions. In a more aggressive situation, passive film breakdown is easy and frequent so that pit stability considerations determine whether pitting occurs. In less aggressive situations, when high potentials lead to film breakdown, pit growth is likely to be rapid [38]. Similar factors may play a role in the breakdown processes leading to local breakdown and high-rate dissolution under ECM conditions.

3.4   AES/XPS STUDIES OF TRANSPASSIVE FILMS ON Ni AND Fe

The discussions presented above conclusively demonstrate that high-rate transpassive dissolution of nickel in sodium nitrate electrolyte is initiated by a local breakdown of the passive film similar to the conventional pitting phenomena. A delicate balance between depassivation and repassivation causes a change from anodic oxygen evolution to high-rate metal dissolution. Datta and Landolt investigated the anodic films formed on nickel and iron under high-rate transpassive dissolution conditions and their role in governing the metal dissolution efficiency in passivating ECM electrolytes [14,17,18]. Experimental investigations were aimed at finding a correlation among anodic polarization behavior, dissolution current efficiency, and transpassive film thickness over a wide range of current densities up to 30 A/cm^2 under well-controlled hydrodynamic conditions. AES and XPS were employed for characterizing the transpassive films.

3.4.1   TRANSPASSIVE FILMS ON NICKEL

Single-crystal nickel (99.95%, orientation [111]) anodes were used in a flow channel cell through which the electrolyte was pumped at a flow velocity of 1,000 cm/s [14,17,18]. Alkaline solutions were employed to avoid self-activation of the anode after interruption of the polarizing current. For AES analysis, the samples after anodic polarization were removed from the ceil, washed with distilled water, and dried. They were then introduced into the AES vacuum chamber. AES depth profile analysis provided information about the film thickness and incorporation of electrolyte anions in the film. Figure 3.6 shows typical Auger depth profiles of anodic films formed at two different current densities, 0.01 and 0.33 A/cm^2, in 6 M $NaNO_3$ + 0.1 M NaOH. The amplitude of the oxygen peak decreases whereas that of the nickel peak increases as the oxide film is removed by sputtering. Carbon present as an impurity on the surface disappears rapidly with sputtering, The nitrogen peak amplitude monitored with a magnification of five, remains close to zero for the film formed at a current density of 0.01 A/cm^2 but it exhibits a finite value which increases with increasing sputter time for the film formed at 0.33 A/cm^2. A maximum in the nitrogen peak amplitude is reached at the oxide-metal interface. The sputtering time corresponding to a decrease of the oxygen peak amplitude to 50% of its maximum value in the film was measured and apparent film thickness was calculated from these values assuming a constant sputter rate of 0.6 nm/min. This sputter rate corresponded

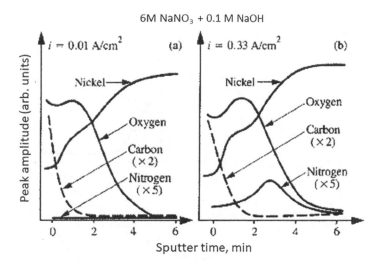

6M NaNO$_3$ + 0.1 M NaOH

FIGURE 3.6 AES concentration depth profiles of oxide films on nickel formed in 6 M NaNO$_3$ + 0.1 M NaOH at two different anodic current densities: (a) $i = 0.01$ A/cm^2, (b) $i = 0.33$ A/cm^2 [14].

to that measured experimentally with a NiO single-crystal sample using identical experimental conditions.

Figure 3.7 shows the transpassive film thickness derived from AES analysis together with the steady-state anode potentials measured during galvanostatic polarization and the current efficiency for metal dissolution. Current efficiency for metal dissolution was determined in separate experiments by weight toss measurements assuming divalent nickel formation [11,15,16]. At anodic current densities below 1 A/cm^2, the current efficiency for metal dissolution is very low indicating that oxygen evolution is the dominant reaction. Anodic film thickness increased in this region and the current-voltage curve exhibited a Tafel behavior with a slope of 250 mV. At 0.8 A/cm^2, a change in the slope of the current-voltage curve is observed. Separate experiments showed that the current density corresponding to this change is dependent on flow rate and OH$^-$ concentration. The apparent change in the slope is due to reaching the limiting current for OH$^-$ discharge at the anode. At still higher current densities, under conditions where apparent film thickness becomes independent of current density, metal dissolution sets in, and the current efficiency for nickel dissolution increases with increasing current density. Qualitatively similar behavior was obtained in an electrolyte with higher pH (6 M NaNO$_3$ + 1 M NaOH) and in an electrolyte of lower nitrate concentration (1 M NaNO$_3$ + 0.1 M NaOH). In all these electrolytes, the film thickness reaches a constant value at current densities where transpassive dissolution takes place.

Figure 3.7b shows that the transpassive film thickness data goes through a maximum before reaching a steady-state at higher current densities. Also shown in the figure is the normalized nitrogen peak height which is the ratio of the maximum nitrogen peak amplitude at the metal-oxide interface to the steady-state nickel peak

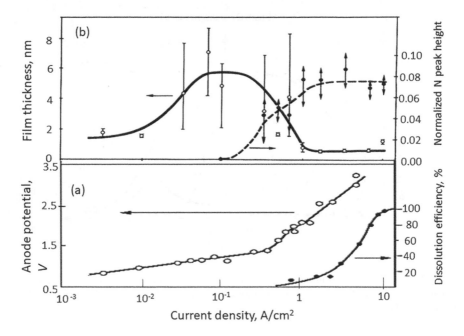

FIGURE 3.7 (a) Anodic polarization curve and current efficiency for nickel dissolution in 6 M NaNO$_3$ + 0.1 M NaOH; (b) oxide film thickness derived from AES depth profiles data and normalized N peak at the interface of oxide films on nickel as a function of different anodic current densities [14].

amplitude in the metal matrix. The normalized nitrogen value is zero at low current densities below the apparent film thickness maximum. Above the maximum, the value increases with current density then reach a constant value at current densities where apparent film thickness also becomes constant. Using elemental sensitivity factors [39] the nitrogen concentration corresponding to the measured steady-state value of its normalized value at the oxide-metal interface is estimated to be 8 at.%. AES Film thickness data obtained for transpassive films formed in 1 M NaNO$_3$ + 0.1 M NaOH and in 6 M NaNO$_3$ + 1 M NaOH showed a qualitatively similar behavior as described above [14].

In another work, Datta et al. [18] employed XPS and AES depth profiling and angle-resolved XPS/AES to characterize the thickness and chemical nature of transpassive films formed at 3 A/cm^2 in 6 M NaNO$_3$ + 1 M NaOH electrolyte. The data obtained for transpassive films were compared with those obtained for air-formed films. Figure 3.8 shows the XPS depth profile data for a nickel surface after transpassive dissolution and for an air-formed film. The amplitude of the oxygen peak decreased whereas that of nickel increased as the oxide film was removed by sputtering. Carbon present as impurity disappeared rapidly. For transpassive films, the nitrogen peak (monitored with a magnification of 40) increased with increasing sputter time reaching a maximum at the metal-oxide interface. Similar results were obtained by AES depth profiling.

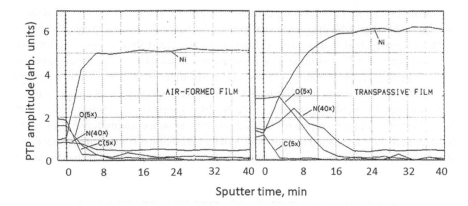

FIGURE 3.8 XPS depth profiles of different elements in the transpassive and air-formed films. N amplitude is 40×, C amplitude is 5×, and O amplitude is 5×. The transpassive dissolution was performed at 3 A/cm² in 6 M $NaNO_3$ + 0.1 M NaOH [18].

From the measured XPS and AES depth profile data, the sputtering time corresponding to a decrease of the oxygen peak amplitude to its 50% value in the film was determined. The apparent film thickness was then calculated using the sputter rate value determined for a NiO single crystal under similar conditions (0.10 nm/min for XPS and 1.86 nm/min for AES and measurements). The transpassive film thickness was thus estimated to be 0.9 nm from XPS and 1.1 nm from AES depth profiling. For air-formed films the corresponding thickness values were 0.3 nm by XPS and 0.4 nm by AES depth profiling (Table 3.1). Separate XPS and AES depth profiling were performed on nickel samples that were subjected to only cathodic treatment in the flow channel cell. The results obtained were similar to those of air formed films: 0.3 nm by XPS and 0.4 nm by AES and no nitrogen was observed at the metal-oxide interface. These data indicate that prior exposure to the electrolyte in the absence of anodic polarization does not significantly alter the thickness of the film that is formed spontaneously in air.

TABLE 3.1
Apparent Oxide Film Thickness on Nickel Determined by Different Methods

		Film Thickness (nm)	
Analytical Technique		**Transpassive**	**Air Formed**
Sputter profiling	AES	1.1	0.4
	XPS	0.9	0.3
Angle resolved	AES	1.0	0.5
	XPS	2.6	0.7

Source: Adapted from Ref. [18].

Information about the film composition is obtained from the elemental XPS spectra. Figure 3.9 shows Ni $2p_{3/2}$ and O 1s spectra for three differently treated samples: (a, d) transpassively dissolved, (b, e) exposure to the electrolyte at open circuit, and (c, f) air-exposed nickel surface. For transpassive films, the Ni $2p_{3/2}$ spectrum (3a) shows a peak at approximately 855 eV and a shoulder at 852 eV binding energy attributed to oxidized and metallic nickel, respectively [40,41]. On the other hand, a dominant metallic peak is observed in the Ni $2p_{3/2}$ spectrum of (Figure 3.9b and c). The O 1s spectrum for transpassive films shows a peak at approximately 531 eV and a shoulder at 529 eV attributed to hydroxide (OH^-) and oxide (O^{2-}) species respectively [40,42]. Only the OH^- peak is dominant in the O 1s spectra of Figure 3.9e and f.

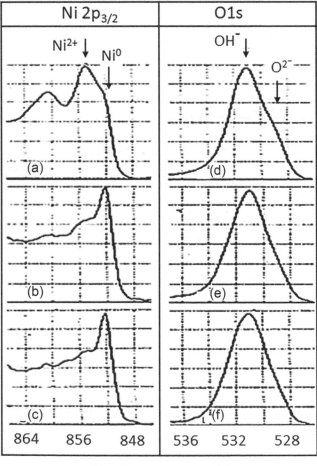

FIGURE 3.9 Ni $2p_{3/2}$ and O 1s spectra for differently treated nickel samples: (a and d) transpassively dissolved, (b and e) exposure to electrolyte without polarization, (c and f) mechanically polished without further treatment [18].

The above data suggest that the chemical nature of air-formed films is little influenced by exposure to the electrolyte.

For studying the chemical composition of thin films, angle-resolved XPS is used as a nondestructive depth profiling technique. The principle of angle-resolved XPS and AES lie in the fact that the analysis information depth is determined by the effective escape depth, Λ, of the emitted electrons which is given by:

$$\Lambda = \lambda \cos\varphi \qquad (3.1)$$

where λ is the mean free path of Auger electrons or photoelectrons in the solid and φ is the take-off angle with respect to the surface normal. It follows from Equation 3.1 that by varying the take-off angle, φ, the depth of information can be varied which allows one to study the composition as a function of the depth of very thin films without the need for ion sputtering.

Figure 3.10 shows the Ni $2p_{3/2}$ spectra for transpassive and air-formed films at different take-off angles. For a transpassive film at $\varphi = 0°$, corresponding to deep inside the film, two peaks corresponding to metallic and oxidized nickel are observed. The intensity of the metallic nickel peak decreased with an increasing take-off angle. At $\varphi = 57°$, the metallic nickel peak is present only as a weak shoulder since at this angle most of the information is derived from the surface film. On the air-formed film, the metallic nickel peak was dominant even at $\varphi = 57°$. In addition, the shake-up satellite signal is more dominant in the transpassive film. All these data indicate that the air-formed films are thinner, in qualitative agreement with the sputter profile AES/XPS data. Figure 3.11 shows deconvoluted Ni $2p_{3/2}$ spectrum (after background subtraction) of a transpassive film at a take-off angle, $\varphi = 0°$ (Figure 3.10a) and deconvoluted O 1s spectrum (after background subtraction) of a transpassive film at a take-off angle, $\varphi = 57°$ (Figure 3.10f). The deconvolution of XPS spectra was carried out after background subtraction using a curve fitting routine. The Ni $2p_{3/2}$ spectrum from 849 to 867 eV was resolved into four peaks as shown in Figure 3.11a. The two peaks at 852.1 ± 0.2 and 854.7 ± 0.4 eV correspond to the photoelectron signals of metallic and oxidized nickel, respectively [40,41]. The peaks at 857.7 ± 0.9 and 860.7 ± 0.6 eV are shake-up satellite signals of metallic and oxidized nickel, respectively. The intensities of oxidized nickel and that of metallic nickel were determined from the peak area which included the contributions from shake-up satellite peaks [43]. From the measured peak intensity ratio, the film thickness was calculated at different take-off angles. The calculated thickness values for transpassive and air-formed films on nickel remained constant as a function of take-off angle. The transpassive and the air-formed film thickness were determined to be 2.6 and 0.7 nm, respectively. Similar results were obtained by angle-resolved AES studies which yielded a thickness value of 1 nm for transpassive film and 0.5 nm for air-formed films. These data confirm that transpassive films on nickel are thicker than air-formed films, in agreement with AES and XPS sputter depth profile data.

The O 1s spectrum measured on air-formed and on transpassive film was deconvoluted as shown in Figure 3.11b, attributing the peaks at 528.8 ± 0.3, 530.9 ± 0.1, and 532.7 ± 0.6 eV to O^{2-}, OH^-, and H_2O, respectively [40,41,44]. The relative intensity of the signals corresponding to these species was determined as a function of take-off

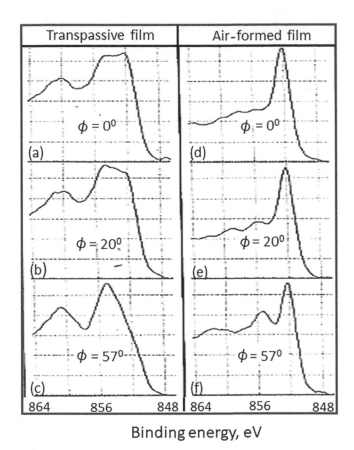

Binding energy, eV

FIGURE 3.10 Ni $2p_{3/2}$ spectra of (a–c) transpassive and (d–f) air-formed films on nickel at different take off angles: (a and d) $\varphi = 0°$, (b and e) $\varphi = 20°$, (c and f) $\varphi = 57°$. Take-off angle is measured with respect to the surface normal [18].

angle to get information on their depth distribution. In all cases, a systematic difference between air-formed and transpassive films existed. For transpassive films, the relative signal intensities for OH$^-$, O^{2-}, and H$_2$O were approximately 0.6, 0.3, and 0.1, and for air-formed films 0.9, 0, and 0.1, respectively. The data suggested that the transpassive films contain a significant fraction of O^{2-} which is mostly absent in air-formed films.

XPS studies provided information on the chemical nature of the nitrogen maximum observed at the metal-oxide interface. Table 3.2 lists the binding energies of the N 1s electrons for different compounds and compares them with the values obtained for transpassive films and air-formed films on nickel [18,41,45,46]. It is concluded that the nitrogen in the transpassive film corresponds to a reduced form, the observed N 1s binding being close to the value for nitride.

The signal intensity of the N 1s spectrum, determined as a function of take-off angle, indicated that both air-formed and transpassive films contain nearly an equal quantity of nitrogen at the surface ($\varphi = 57°$). The N 1s intensity in the air-formed

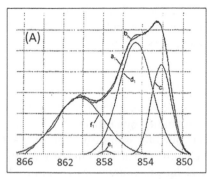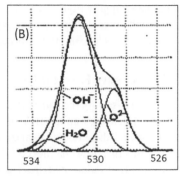

Biding energy, eV

FIGURE 3.11　(A) Deconvoluted Ni $2p_{3/2}$ spectrum (after background subtraction) of a transpassive film at a take-off angle, $\varphi = 0°$: (a) raw data; (b) fitted curve; (c) metallic nickel signal; (d) oxidized nickel signal; (e) metallic nickel shake up signal; (f) oxidized nickel shake-up signal. (B) Deconvoluted O 1s spectrum (after background subtraction) of a transpassive film at a take-off angle, $\varphi = 57°$ [18].

TABLE 3.2
Binding Energies of N 1s Electrons in Different Compounds

Compound	Binding Energies of N 1s Electrons in Different Compounds (eV)			
	[45]	[46]	[41]	[18]
$NaNO_3$	407.3	407.0		407.1
$NaNO_2$	403.3	403.3		403.6
NH_4NO_3	406.0			
NH_4Cl				401.8
Iron nitride				398.0
NaN_3	398.6			
BN			397.8	
WN			397.3	
Transpassive film				397.5
Air-formed film				398.7

film decreased with a decreasing take-off angle because nitrogen is present mainly as surface contamination. On the transpassive film, the most intensive nitrogen signal was observed at low take-off angle, suggesting that nitrogen is present in the interior of the film, in agreement with the sputter profiling data.

　　The above data show that the thickness of transpassive films on nickel formed in nitrate solution is two to three times higher than that of air-formed films. Compared to air-formed films the proportion of O^{2-} and OH^- in the transpassive film was different and the nitrogen present at the film-metal interface is in a chemically reduced form. The data suggest that transpassive films under the conditions of high anodic potentials (high current densities) are permeable to electrolyte anions through pores or other defects thus leading to high-rate metal dissolution.

3.4.2 TRANSPASSIVE FILMS ON IRON

Anodic films on iron formed in the transpassive potential region in nitrate electrolytes at current densities up to 40 A/cm² was studied by Datta et al. [17] using AES. Figure 3.12 shows the anodic polarization curve for iron in 6 M NaNO₃ + 0.1 M NaOH at 2.5 m/s in a flow channel cell. The figure also includes the current efficiency for iron dissolution calculated from weight loss measurements assuming divalent iron formation [16]. At current densities below 8 A/cm², current efficiency is low because oxygen evolution is the main reaction. The current-potential curve exhibits Tafel behavior, with a slope of −100 mV up to 0.2 A/cm², where the limiting current for OH⁻ discharge is reached. Measured potentials at higher current densities are strongly influenced by ohmic drops, which explains the shape of the current-voltage curve above 2 A/cm². Figure 3.13 shows Auger depth profiles of films formed at 0.042 and 12.5 A/cm². The important difference between the two films is in their nitrogen profiles. For the films formed at 0.042 A/cm², the nitrogen peak amplitude decreases rapidly with sputtering. Similar behavior was observed for air-formed films. The observed surface nitrogen is due to adsorption from the air. For the film formed at 12.5 A/cm², however, the nitrogen peak amplitude increased with time reaching a maximum at the oxide-metal interface. Using elemental sensitivity factors [39], the atomic concentration of nitrogen at the metal-oxide interface was estimated from AES depth profiles. Figure 3.14 shows the transpassive film thickness and the nitrogen concentration at the interface as a function of current density. As seen in

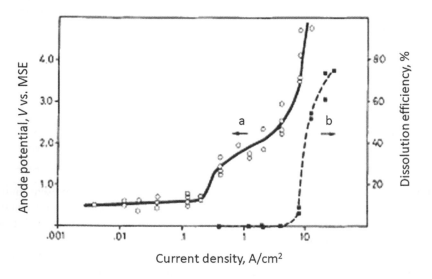

FIGURE 3.12 Anodic potential (a) and current efficiency for iron dissolution (b) as a function of current density. Electrolyte: 5 M NaNO₃ + 0.1 M NaOH [17].

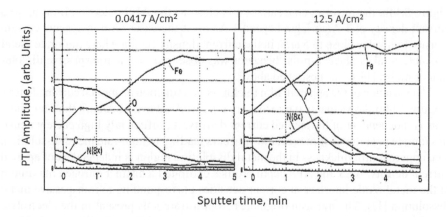

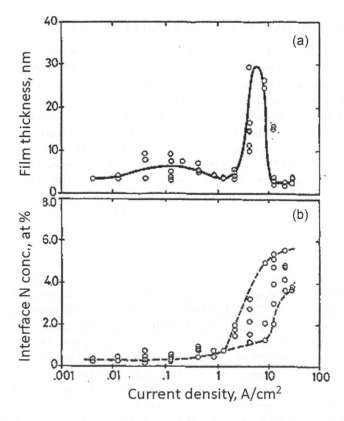

FIGURE 3.13 AES depth profiles of transpassive oxide films formed on iron at two different anodic current densities. Electrolyte: 5 M NaNO$_3$ + 0.1 M NaOH [17].

FIGURE 3.14 (a) Apparent film thickness on iron and (b) nitrogen concentration at the metal-film interface determined from AES depth profile data as a function of current density. Electrolyte: 5 M NaNO$_3$ + 0.1 M NaOH [17].

Figure 3.14b, the nitrogen concentration is close to the AES detection limit up to a current density of 1 A/cm^2, but beyond 1 A/cm^2 it increases with increasing current density to a value close to 5 at.% in the current density region where high-rate metal dissolution takes place. Similar behavior was observed for the transpassive dissolution of nickel [14,18] as described above.

The variation of the apparent film thickness as a function of applied current density (Figure 3.14a) shows that two maxima are observed. Above 12 A/cm^2, the film thickness is independent of the applied current density. At sufficiently high current densities (anode potentials), film breakdown occurs leading to high-rate metal dissolution. The apparent film thickness in the metal dissolution region is independent of applied current density. AES and XPS depth profiles exhibited a distinct nitrogen maximum at the metal-oxide interface at current densities corresponding to transpassive metal dissolution [17]. The nitrogen originates from nitrate ions present in the electrolyte which penetrates through defects in the film as observed for nickel [12,14,18]. These results further confirm that transpassive dissolution is governed by the kinetics of depassivation and repassivation phenomena, which determine the average surface fraction available for oxygen evolution and metal dissolution, respectively.

3.5   TRANSPASSIVE DISSOLUTION MECHANISM

A generalized picture of the transpassive dissolution mechanism emerges from the data presented above. The different data presented above clearly demonstrate that the transpassive dissolution of nickel and iron in nitrate solution is initiated by a local breakdown of the passive film and, therefore, presents some analogies to the pitting behavior of passive metals in aggressive media under corrosion conditions.

The film thickness studies demonstrate that at high current densities where transpassive metal dissolution occurs, oxide films are extremely thin, observed values lying between 0.5 and 2.5 nm. On the other hand, in the oxygen evolution region film thickness is usually higher, exhibiting a maximum at intermediate current densities. This behavior is evidenced by both, AES and coulometric data [14,17,18]. No direct relation exists between the position of the film thickness maximum on the current axis and the onset of transpassive metal dissolution or the limiting current for OH$^-$ discharge. This suggests that film thinning due to local acidification of the anodic diffusion layer cannot be responsible for the observed transition from predominant oxygen evolution to metal dissolution at high current density. Depth profiling data show that nitrogen is present in the anodic films formed at current densities above the thickness maximum. To explain this observation one could postulate that nitrate ion is incorporated into the film lattice according to the ion exchange mechanism [6]. Indeed, the data show apparent film thickness to decrease with increasing nitrogen content in apparent agreement with the ion exchange model for film dissolution. On the other hand, the maximum value for the nitrogen amplitude was always observed at the metal-oxide interface rather than at the oxide-solution interface. This is in contradiction with the ion exchange mechanism which requires the maximum nitrate ion concentration to lie at the film solution interface. It is therefore believed that the observed nitrogen concentration maximum at the film-metal interface is due to nitrate ion penetration through flaws in the film and local accumulation at the metal surface rather than to a solid-state diffusion mechanism. The anion in the film-metal interface has been found to be in the reduced state further presenting similarities to the pitting behavior under corrosion conditions. Similar behavior has been found for transpassive dissolution of titanium in perchloric acid solution where much thicker films are formed [47].

From ECM considerations, the most important aspect of the transpassive behavior of metals is the functional dependence of current efficiency on current density. The slope of the current efficiency vs. current density curves depends on the electrolyte concentration and pH. On the other hand, in the current density region of increasing metal dissolution efficiency, the transpassive film thickness stays constant. Therefore, it is concluded that metal dissolution efficiency in the transpassive potential region is independent of film thickness. This behavior can be rationalized by taking into account observations indicating that metal dissolution is initiated at discrete sites and proceeds by a pitting-type mechanism. Such a mechanism involves two parallel paths for the anodic current: direct metal dissolution through pores and pits, and electron transfer through the oxide film with oxygen evolution. The observed increase in metal dissolution efficiency with current density may thus be explained by an increase in the effective surface area available for dissolution.

REFERENCES

1. M. A. LaBoda, M. L. McMillan, *Electrochem. Technol.*, 5, 340 (1967).
2. J. P. Hoare, M. A. LaBoda, M. L. McMillan, A. J. Wallace, *J. Electrochem. Soc.*, 116, 199 (1969).
3. P. A. Brook, Q. Iqbal, *J. Electrochem. Soc.*, 116, 1458 (1969).
4. K. W. Mao, *J. Electrochem. Soc.*, 118, 1870 (1971); 118, 1876 (1971).
5. M. Datta, D. Landolt, *Corros. Sci.*, 13, 187 (1973).
6. K. W. Mao, M. A. LaBoda, J. P. Hoare, *J. Electrochem. Soc.*, 119, 419 (1972).
7. D. T. Chin, K. W. Mao, *J. Appl. Electrochem.*, 4, 155 (1974).
8. K. W. Mao, D. T. Chin, *J. Electrochem. Soc.*, 121, 191 (1974).
9. M. A. LaBoda, A. J. Chartrand, J. P. Hoare, C. R. Weise, K. W. Mao, *J. Electrochem. Soc.*, 120, 643 (1973).
10. J. P. Hoare, C. R. Weise, *Corros. Sci.*, 15, 435 (1975).
11. M. Datta, D. Landolt, *J. Electrochem. Soc.*, 122, 1466 (1975).
12. M. Datta, D. Landolt, *J. Electrochem. Soc.*, 124, 483 (1977).
13. M. Datta, D. Landolt, *J. Appl. Electrochem.*, 7, 247 (1977).
14. M. Datta, H. J. Mathieu, D. Landolt, *Electrochim. Acta*, 24, 843 (1979).
15. M. Datta, D. Landolt, *Electrochim. Acta*, 25, 1255 (1980).
16. M. Datta, D. Landolt, *Electrochim. Acta*, 25, 1263 (1980).
17. M. Datta, H. J. Mathieu, D. Landolt, *J. Electrochem. Soc.*, 131, 2484 (1984).
18. M. Datta, H. J. Mathieu, D. Landolt, *Appl. Surf. Sci.*, 18, 299 (1984).
19. M. Datta, *IBM J. Res. Develop.*, 37 (2) 207 (1993).
20. M. M. Lohrengel, I. Kluppel, C. Rosenkranz, H. Bettermann, J. W. Schultze, *Electrochim. Acta*, 48, 3203 (2003).
21. M. M. Lohrengel, C. Rosenkranz, I. Klüppel, A. Moehring, H. Bettermann, B. Van den Bossche, J. Deconinck, *Electrochim. Acta*, 49, 2863 (2004).
22. M. M. Lohrengel, K. P. Rataj, T. Münninghoff, *Electrochim. Acta*, 201, 348 (2016).
23. H. P. Leckie, H. H. Uhlig, *J. Electrochem. Soc.*, 113, 1262 (1966).
24. A. Broli, H. Holtan, T. B. Andreassen, *Werkst. Korros.*, 27, 497 (1976).

25. K. Venu, K. Balakrishnan, K. S. Rajagopalan, *Corros. Sci.*, 5, 59 (1965).
26. K. S. Rajagopalan, K. Venu, *Corros. Sci.*, 8, 557 (1968).
27. H. H. Strehblow, J. Wenners, *Z. Phys. Chem. N.F.*, 98S, 199 (1975).
28. T. P. Hoar, D. C. Mears, G. P. Rothwell, *Corros. Sci.*, 5, 279 (1965).
29. A. Szummer, Z. Szklarska-Smialowska, M. Janik-Czachor, *Corros. Sci.*, 8, 833 (1968).
30. Z. Szklarska-Smialowska, *Corros. Sci.*, 11, 209 (1971).
31. K. J. Vetter, H. H. Strehblow, *Bet. Bunsenges. Phys. Chem.*, 74, 449 (1970).
32. K. J. Vetter, H. H. Strehblow. Pitting Corrosion in an Early Stage and Its Theoretical Implications, *in Localized Corrosion*, R. W. Staehle, B. F. Brown, J. Kruger, A. Agarwal eds., p. 240, NACE-3, Houston, (1974).
33. N. Sato, *Electrochim. Acta*, 16, 1683 (1971).
34. J. R. Galvele, *J. Electrochem. Soc.*, 123, 464 (1976).
35. D. D. Macdonald, *J. Electrochem. Soc.*, 139, 3434 (1992).
36. D. D. Macdonald, *Pure Appl. Chem.*, 71, 951 (1999).
37. G. S. Frankel, *J. Electrochem. Soc.*, 145, 2186 (1998).
38. G. S. Frankel, T. Li, J. R. Scully, *J. Electrochem. Soc.*, 164, C180 (2017).
39. L. E. Davis, N. C. MacDonald, P. W. Palmberg, G. E. Riach, R. E. Weber, *Handbook of Auger Electron Spectroscopy*, 2nd edition, p. 11, Physical Electronics Industries, Inc., Eden Prairie, MN, (1976).
40. P. Marcus, J. Oudar, I. Olefjord, J. Micros. Spectrosc. Electron., 4, 63 (1979).
41. J. F. Moulder, *Handbook of X-Ray Photoelectron Spectroscopy*, Physical Electronics Division, Perkin-Elmer Corp., (1992).
42. N. S. McIntyre, D. G. Zetaruk, *Anal. Chem.*, 49, 1521 (1977).
43. T. A. Carlson, *Surf. Interface Anal.*, 4, 125 (1982).
44. K. Konno, M. Nagayama, *in Passivity of Metals*, R. P. Frankenthal, J. Kruger eds., p. 585, Electrochemical Society, Princeton, (1978).
45. B. Folkesson, *Acta Chem. Stand.*, 27, 287 (1973).
46. H. Schultheiss, E. Fluck, *J. Inorg. Nucl. Chem.*, 37, 2109 (1975).
47. J. B. Mathieu, H. J. Mathieu, D. Landolt, *J. Electrochem. Soc.*, 125, 1039 (1978).

4 Mass Transport and Current Distribution

4.1 INTRODUCTION

The technical feasibility of an electrochemical metal shaping process is determined by its ability to provide the desired metal removal rate, shape profile, and surface finish. These criteria are dependent on the prevailing mass transport, current distribution, and surface film properties at the dissolving workpiece surface. In most applications, mass transport and current distribution are intimately related. Mass transport conditions influence the macroscopic and microscopic current distribution on the workpiece and hence determine the shape evolution and surface finish of the dissolving anode. An understanding of the underlying principles of mass transport and current distribution is crucial for the successful implementation of electrochemical metal shaping processes.

The basic principles governing mass transport and current distribution in electrochemical systems have been described and several reviews are available in the literature [1–7]. In the following, a brief description of mass transport and current distribution aspects related to high-rate anodic dissolution processes is presented. Experimental tools for investigating high-rate anodic dissolution processes under controlled hydrodynamic conditions are also described.

4.2 MASS TRANSPORT

The rate of an electrochemical reaction is measured by the Faradaic current that flows in the electrochemical cell. The current flow is dependent on the rate of material movement (mass transport) and on the rate at which electrons can transfer across the interface (charge transfer). Mass transport stems from differences in electrical potential, chemical potential (concentration gradient), and movement of the electrolyte.

During an electrochemical reaction, the material can be transported by three different modes:

1. Diffusion,
2. Migration, and
3. Convection.

Diffusion is the movement of a species from a region of high concentration to a region of lower concentration. The rate of diffusion is dependent on the concentration gradient, and the diffusion coefficient, which has a specific value for a given solution species at a given temperature.

Migration is the movement of a charged body (ion) under the influence of an electric field. Anions are attracted to the positively charged anode and are repelled by a negatively charged cathode.

Convection is the forced movement of solution species by mechanical means such as stirring, vibration, or pumping. The rate at which a solution is moved can generally be controlled.

The current passing through the electrochemical cell is given by the molar flux density of the reacting species. Mass transfer to the working electrode surface is described by the Nernst-Planck equation, which is written for a one-dimensional mass transfer along the *x*-axis as [8]:

$$N(x) = -D\frac{\partial C(x)}{\partial(x)} - \frac{nF}{RT}DC\frac{\partial \Phi(x)}{\partial(x)} - Cu(x)$$

(4.1)

$$\text{diffusion} \qquad \text{migration} \qquad \text{convection}$$

Here $N(x)$ is the molar flux density at a distance x from the surface, D is the diffusion coefficient, $\partial C(x)/\partial x$ is the concentration gradient at distance x, $\partial \phi(x)/\partial x$ is the potential gradient, n is the charge, C is the concentration of the species, and $u(x)$ is the flow velocity along the axis. The sign associated with molar flux density identifies the direction of motion: positive toward and negative away from the plane.

To study the contribution of pure diffusion, it is necessary to suppress the contributions from migration and convection. The contribution of migration can be minimized by the addition of an excess of nonelectroactive ions (supporting electrolyte), at a concentration that is significantly higher than that of the electroactive species. The contribution of convection can be eliminated by preventing stirring, vibration, and electrolyte circulation in the electrochemical cell. Also, natural convection, which is due to thermal gradients, is a possible contribution to mass transfer which can also be prevented by selecting an isothermal environment where natural convection is negligible. Under conditions where both migration and convection terms are negligible in Equation 4.1, diffusion can be described by Fick's first law:

$$N(x) = -D\frac{\partial C(x)}{\partial(x)}$$

(4.2)

In Equation 4.2, the negative sign signifies that the net molar flux density is from a region of high concentration to one of low concentration. Changes in concentration with time caused by diffusion is predicted by Fick's second law, which for a one-dimensional system is expressed as:

$$\frac{\partial C(x)}{\partial t} = D\frac{\partial^2 C}{\partial x^2}.$$

(4.3)

Fick's Second Law expresses non-steady-state diffusion, which deals with the concentration C being a function of both time t and position x. In Equation 4.3, the diffusion coefficient is a physical constant that is dependent on the size of the molecule, temperature, and pressure. Diffusion coefficients can be calculated by Equation 4.4:

$$D = mk_B T \qquad (4.4)$$

where m is the mobility, k_B is Boltzmann's constant, and T is the temperature.

4.2.1 CONVECTIVE DIFFUSION

Let us consider a molecule undergoing mass transport in a convective system. Since chemical species have a finite diffusivity, it is normal to consider a situation where both the diffusive and convective contributions to the mass transport are included. During high-rate anodic dissolution processes under conditions of high electrolyte flow, mass transport processes can be described by convective diffusion [9–12]. To apply Equation 4.1 to convective mass transport problems, we take a simplified approach of linear convective diffusion [9–14], originally proposed by Nernst. According to this concept, the electrolyte is divided into two zones. Near the electrode surface, a stagnant diffusion layer is assumed where convection is absent while in the bulk solution due to perfect mixing no concentration gradients exist. The thickness of the stagnant diffusion layer depends on prevailing convection conditions and can be estimated from dimensionless mass transport correlations. The described approach permits one to treat convective mass transport problems without the need to know the velocity profile near the electrode. In these applications, the contribution of migration to transport in the diffusion layer can be neglected permitting a further simplification of the problem.

The Nernst diffusion layer concept for a mass transport-controlled anodic dissolution process is shown in Figure 4.1. Inside the diffusion layer, δ, a concentration gradient exists, and the transport occurs exclusively by linear diffusion. Outside the diffusion layer, transport occurs by convection, where the electrolyte concentration

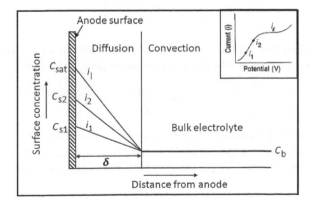

FIGURE 4.1 Nernst diffusion layer concept describing mass transport at the dissolving anode surface. Inset: A typical anodic polarization curve exhibiting limiting current.

is assumed to be constant. The thickness of the anodic diffusion layer depends on hydrodynamic conditions and is given by:

$$\delta = \frac{L}{\text{Sh}} \tag{4.5}$$

where L is a characteristic length and Sh the Sherwood number that represents non-dimensional mass transport rate. The Sherwood number can also be regarded as the normalized or nondimensional diffusion layer thickness, the value of which depends on the electrolyte properties and hydrodynamic conditions. An exhaustive list of derived expressions describing $\text{Sh} = f(\text{Re, Sc})$ is available in the literature. Here $\text{Re} = D_h u/v$ is the Reynolds number that characterizes the flow, and $\text{Sc}(= v/D)$ is the Schmidt number which is the ratio of the viscous diffusion rate and the molecular diffusion rate; D_h is the hydraulic diameter, u is the flow velocity, and v is the kinematic viscosity and D is the diffusion coefficient.

For flow channel cell under laminar conditions, mass transport rates are given by [9,11,15],

$$\text{Sh} = 1.85 \left(\frac{\text{Re Sc } D_h}{L} \right)^{1/3} \tag{4.6}$$

and for turbulent flow conditions [9,11,15],

$$\text{Sh} = 0.22 \ \text{Re}^{7/8} \text{Sc}^{1/4} \tag{4.7}$$

For a rotating disk electrode (RDE) [8],

$$\text{Sh} = 0.62 \ \text{Re}^{1/2} \text{Sc}^{1/3}. \tag{4.8}$$

For an unsubmerged circular impinging jet, the mass transport in the impinging region is given by [16]:

$$\text{Sh} = 0.9 \ \text{Re}^{1/2} \text{Sc}^{1/2} \left(\frac{h}{d} \right)^{-0.09}. \tag{4.9}$$

The effect of fluid flow on the convective mass transport in a photoresist patterned work is expressed by the nondimensional Peclet number, Pe, which is defined as:

$$\text{Pe} = \frac{Lu}{D} \tag{4.10}$$

The influence of the Peclet number on the average mass transport rate in a cavity has been correlated based on experimentally measured average mass transfer coefficient during etching of patterned copper samples [17]. The following empirical correlation was obtained:

$$\text{Sh} = 0.3 \left(\frac{L}{H} \right)^{0.83} \text{Pe}^{0.33} \tag{4.11}$$

where L is the width and H is the height of the cavity. Equation 4.11 agreed well with the average mass transport rates calculated by solving the equations for Stokes flow using the finite element method (FEM) and by a combination of the boundary integral method and Lighthill boundary layer analysis [17].

4.2.2 ANODIC LIMITING CURRENT

Investigation of the anodic dissolution behavior of many electrochemical machining (ECM) and electropolishing systems have indicated that polarization of a workpiece at high anodic potentials leads to a limiting current plateau for metal dissolution reaction [9–12,18–20]. These studies have confirmed that the limiting current density is controlled by convective mass transport. For an anodic reaction that is controlled by convective mass transport, the anodic current density, i, is given by

$$i = nFD\frac{C_s - C_b}{\delta} \tag{4.12}$$

where n is the valence of metal dissolution, F is the Faraday constant, C_s is the surface concentration, and C_b is the bulk concentration.

An increase in current density leads to an increase in the rate of metal ion production at the anode so that the surface concentration of the dissolved metal ions increases accordingly. When the metal ion concentration at the surface reaches or exceeds the saturation limit, precipitation of a thin salt film occurs. The polarization curve under these conditions exhibits a limiting current plateau.[1] At the limiting current, i_l, Equation 4.12 becomes

$$i_l = nFD\frac{C_{sat}}{\delta} \tag{4.13}$$

where C_{sat} is the saturation concentration of the precipitating salt at the anode surface. The bulk electrolyte is assumed to contain a negligibly small concentration of the metal; hence, C_b value is zero.

It must be noted that in many metal dissolution systems, the limiting current plateau does not correspond to the maximum machining rate. It merely signifies the limiting rate of a dissolution reaction for a given oxidation state [13,14]. The maximum rate of machining is limited by the ability of the system to eliminate reaction products and electrolyte heating in the gap. Nevertheless, the limiting current has a profound influence on the machining performance since it influences the dissolution efficiency, surface finish, and current distribution.

---

[1] It must be noted that in the above discussion, the transport of dissolved metal ions from the anode surface into the bulk is considered to be rate-limiting. This mechanism is valid for most of the anodic dissolution processes in ECM electrolytes involving salt solutions. On the other hand, in some electropolishing systems involving concentrated acids (with or without organic solvents), acceptor limited transport may lead to the limiting current. In such cases, transport rate toward the anode of the acceptor species such as complexing ions or water is rate-limiting. At the limiting current, the surface concentration of the acceptor species drops to zero. These aspects will be discussed in more details in Chapters 5 and 7.

As mentioned above, the limiting current corresponds to the precipitation of a salt layer on the dissolving surface. The formation of salt films on the anode influences the surface morphology of a dissolved workpiece [9–14,18,21]. Different studies have conclusively demonstrated that two distinctly different surface morphologies result from dissolution as shown in Figure 4.2. At current densities lower than the limiting current density, surface etching is observed which, depending on the metal-electrolyte combination, reveals crystallographic steps and etch pits, preferred grain boundary attack or finely dispersed microstructure. Anodic dissolution under these conditions leads to extremely rough surfaces. On the other hand, electropolished surfaces are obtained from dissolution at or above the limiting current density. Under these conditions, the formation of salt films at the surface suppresses the influence of crystallographic orientation and surface defects on the dissolution process, thus leading to microfinished surfaces. The presence of salt films may increase the anode potential to such high values that dissolution reactions involving higher oxidation state, or the onset of oxygen evolution may become possible [13]. The presence of salt films has also been found to significantly affect the current distribution during electrochemical micromachining of photoresist masked anodes [21–24].

From the above discussion, it is evident that the metal-shaping operations should be conducted at current densities equal to or higher than the limiting current density to obtain microfinished surfaces. In through-mask microfabrication by electrochemical dissolution, the anodic surface films formed at the limiting current density improves the shape profile uniformity. The operating current density in the through-mask electroetching process should be high enough to yield the desired directionality of dissolution. On the other hand, the application of high current density in the processing of thin films leads to an uncontrollable fast process and issues related to loss of electrical contact. To alleviate these problems in thin-film processing, it is desirable to choose conditions that provide a low rate of metal removal and hence a low operating current density. These two opposing requirements are best met by using pulsating current. Extremely high peak currents (voltages) applicable in pulsating current provides directionality of dissolution and permits the metal dissolution to take place with high efficiency and produces smooth surfaces [25,26].

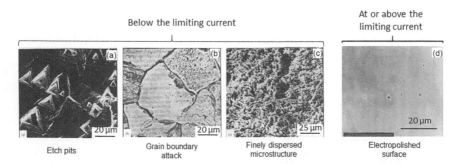

FIGURE 4.2 Typical surfaces resulting from anodic dissolution. Below the limiting current: (a) Ni in 5 M NaCl, 5 A/cm², 10 m/s, (b) Fe in 5 M NaCl, 5 A/cm², 10 m/s, (c) Ni in 5 M NaNO₃ + 1 M HNO₃, 3 A/cm², 4 m/s. At or above the limiting current: (d) 304 Stainless steel in 6 M NaNO₃, 28 A/cm², 2.5 m/s [9,11,21].

4.2.3 Mass Transport in Pulsating Current Dissolution

The use of pulsating current allows one to apply a high instantaneous current density without the necessity of a high electrolyte flow rate since each current pulse is followed by an off-time such that the system relaxes between two consecutive pulses [25,26]. One of the attractive features of using pulsating current is that three parameters, namely peak current density, pulse-on time, and pulse-off time, can be varied independently while in conventional direct current operation only average current can be chosen. In principle, either current or voltage pulses of any shape can be applied. For simplicity, the present discussion will be restricted to rectangular constant current pulses separated by intervals of zero current. Figure 4.3 shows schematically the current pulses to be considered, where i_p is the peak current density, t_p is the pulse-on time, t'_p is the pulse-off time, and t_{pp} is the pulse period. The ratio t_p/t_{pp} is called the duty cycle. The process then takes place at an average current density given by

$$i_a = i_p \left(\frac{t_p}{t_{pp}} \right)$$

(4.14)

To describe mass transport in pulsating current operation, let us recollect the mass transport situation under direct current operation. The anodic current density for a steady-state metal dissolution process is described by Equation 4.12, and a limiting current is observed when the surface concentration of the dissolved metal ions reaches saturation concentration thus leading to the precipitation of a salt film at the anode. The anodic limiting current, i_l, is described by Equation 4.13. It has been shown previously that i_l is an important quantity in metal dissolution processes since anode potential, reaction stoichiometry, and surface finish may differ drastically depending on whether the dissolution is carried out below or above the limiting current [9–14,18].

Mass transport in pulsed dissolution *is a combination of steady-state and non-steady-state diffusion processes* which can be described by a duplex diffusion layer

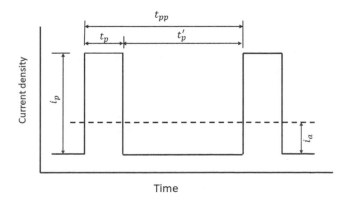

FIGURE 4.3 Schematic diagram of current pulses. i_p: peak current density, t_p: pulse-on time, t'_p: pulse-off time, t_{pp}: pulse period, i_a: average current density.

model originally proposed by Ibl [27] for pulse plating. According to this model, a time-independent stagnant diffusion layer (Nernst diffusion layer) of thickness δ is assumed to exist at the electrode, the value of δ being determined by hydrodynamic conditions. As shown in Figure 4.4, the diffusion layer is divided into two parts, a pulsating diffusion layer of thickness δ_p, within which the dissolved metal ion concentration is a periodic function of time and an outer diffusion layer of thickness $\delta'_p = \delta - \delta_p$, in which the concentration of dissolved metal ion is time-invariant. The proposed duplex diffusion layer model is a gross simplification of real behavior. A more exact description of mass transport could be obtained by solving the relevant non-steady-state diffusion equations [28]. However, as shown by Ibl [27], the approximate model gives results that are close to those obtained by more elaborate calculations and it has the advantage of yielding simple expressions and of providing a good physical insight into the mass transport processes occurring. To calculate the thickness of the pulsating diffusion layer, δ_p, one considers the mass balance of dissolved metal ions over the diffusion layer for the pulse-on time, t_p. Considering a metal dissolution process with 100% current efficiency, the amount of metal dissolved per unit area during time t_p is $i_p t_p / nF$. The amount of metal ion lost into solution by diffusion through the outer diffusion layer during this time is $i_a t_p / nF$, where i_a is the average current density. The difference between the two terms is the amount of metal ions accumulated in the pulsating diffusion layer, i.e., the amount of metal ions that will diffuse out into the bulk during the pulse-off time. Assuming linear concentration profiles, the amount of metal ions accumulated in the pulsating diffusion layer is $(C_s - C_n)\delta_p / 2$, where C_n is the concentration at the δ_p boundary as shown in Figure 4.4. The mass balance thus reads:

$$\frac{i_p t_p}{nF} = \frac{i_a t_p}{nF} + (C_s - C_n)\frac{\delta_p}{2} \tag{4.15}$$

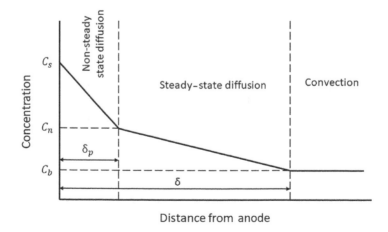

FIGURE 4.4 Schematic diagram of the duplex diffusion layer model for mass transport during anodic pulse dissolution [25].

The peak current density, i_p, is given by:

$$i_p = nFD\frac{C_s - C_n}{\delta_p}$$

(4.16)

From Equations 4.14–4.16 one gets the thickness of the pulsating diffusion layer, δ_p, at the end of a pulse:

$$\delta_p = \sqrt{\left[2Dt_p\left(1 - \frac{t_p}{t_{pp}}\right)\right]}.$$

(4.17)

At low duty cycles, Equation 4.17 reduces to Equation 4.18:

$$\delta_p = \sqrt{2Dt_p}$$

(4.18)

For short current pulses, t_p (which leads to $\delta_p \ll \delta$). Much higher instantaneous current densities can be applied in pulsed dissolution than under steady-state conditions without reaching the limiting current. Defining the limiting pulse current density, i_{pl}, to be the current density at which $C_s = C_{sat}$ at the end of a pulse,

$$i_{pl} = nFD\frac{C_{sat} - C_n}{\delta_p}$$

(4.19)

From Equations 4.13 and 4.19, the ratio of the pulse-limiting current and the steady-state limiting current i_{pl}/i_l is given as:

$$\frac{i_{pl}}{i_l} = \frac{\delta}{\delta_p}\left(1 - \frac{C_n}{C_{sat}}\right)$$

(4.20)

The ratio C_n/C_{sat} is obtained by considering that the amount of metal dissolved during pulse-on time t_p is equal to that diffusing through the outer part of the diffusion layer ($\delta - \delta_p$) during the pulse period t_{pp}:

$$\frac{i_p t_p}{nF} = D\frac{C_n - C_b}{\delta - \delta_p}t_{pp}$$

(4.21)

Setting $C_b = 0$ and with Equation 4.16 one obtains Equation 4.22:

$$\frac{C_n}{C_s} = \frac{t_p/t_{pp}\left((\delta/\delta_p) - 1\right)}{1 + (t_p/t_{pp})\left((\delta/\delta_p) - 1\right)}$$

(4.22)

Setting $C_s = C_{sat}$ at the limiting current density, i_{pl}, one obtains from Equations 4.20 and 4.22 after rearranging:

$$i_{pl} = i_l\left[\frac{\delta_p}{\delta}\left(1 - \frac{t_p}{t_{pp}}\right) + \frac{t_p}{t_{pp}}\right]^{-1}$$

(4.23)

Equation 4.23 allows one to adjust the different pulse parameters in such a way as to work above or below the limiting pulse current density. By using small pulse times one can apply high instantaneous current densities even at relatively low electrolyte flow rates, but the duty cycle must be kept small to allow the removal of reaction products and dissipation of the generated heat. The pulse dissolution thus provides directionality and good surface finish and is particularly useful in microfabrication by electrochemical micromachining.

4.3   CURRENT DISTRIBUTION

The current distribution at the anode depends on the geometry, anodic reaction kinetics, electrolyte conductivity, and hydrodynamic conditions. In the absence of concentration gradients, the potential distribution in the electrolyte obeys Laplace's Equation 4.24,

$$\nabla^2 \phi = 0 \qquad (4.24)$$

and the current density is given by Ohms law, Equation 4.25.

$$i = -\kappa \nabla \phi \qquad (4.25)$$

The local current density at the electrode, in this case, is proportional to the potential gradient normal to the surface. To calculate the potential distribution in the electrolyte and the corresponding current distribution on the electrode the functional relationship between the potential just outside the double layer and that in the metal must be known. Three conditions can be distinguished: primary, secondary, and tertiary current distribution.

Primary current distribution: If the electrode potential does not vary appreciably with current density, the so-called *primary* current distribution prevails which depends only on the geometry of the electrochemical system. Since the kinetic resistances at the electrodes are neglected, the electrodes are treated as equipotential surfaces. At the anode, the potential, ϕ, is given by:

$$\phi = V_a, \qquad (4.26)$$

and at the cathode:

$$\phi = 0 \qquad (4.27)$$

The gradient of the potential at all lines of symmetry and insulators is zero:

$$\frac{\partial \phi}{\partial n} = 0. \qquad (4.28)$$

Secondary current distribution: The secondary current distribution accounts for the effect of the electrode kinetics in addition to solution resistance. The uniformity of the current density on the electrode depends on the relative magnitude of the polarization resistance, $d\eta/di$, and the ohmic resistance in the electrolyte. This relationship is expressed by the dimensionless Wagner number:

$$W_a = \frac{d\eta/di}{\rho_e L} \tag{4.29}$$

where ρ_e is the electrolyte resistivity and L is the characteristic length of the electrochemical system. The secondary current distribution is always more uniform than the primary current distribution and for $W_a \gg 1$, it becomes largely independent of cell geometry.

Tertiary current distribution: In the presence of significant mass transport effects, the tertiary current distribution prevails, which depends on both the potential distribution in the bulk and on the local rate of mass transport. At the limiting current, the current distribution becomes entirely governed by mass transport. Under these conditions, potential gradients and electrode polarization can be neglected and the current distribution can be obtained by solving the distribution of the concentration of the metal ions. The mass transport limited condition is simulated in the diffusion layer in which the concentration of the metal ions is governed by Laplace's equation:

$$\nabla^2 C = 0 \tag{4.30}$$

For salt film formation at the anode, the concentration of metal ions at the surface is the saturation concentration:

$$C = C_{sat} \tag{4.31}$$

In the bulk electrolyte, the concentration of metal ions is negligible:

$$C = 0 \tag{4.32}$$

The concentration gradient at all lines of symmetry and insulators is zero:

$$\frac{\partial C}{\partial n} = 0 \tag{4.33}$$

For the computation of shape evolution in ECM and through-mask micromachining, the current distribution at the anode is combined with Faraday's law according to which the rate, r, at which the anodic surface recedes is determined as:

$$r = i \frac{M\theta}{nF\rho} \tag{4.34}$$

where i is the current density, M is the molecular weight of the metal, n is the metal dissolution valence, ρ is the density of the metal film, F is Faraday constant, and θ is the current efficiency of metal dissolution. For the simulation of shape evolution, the current distribution is solved for an initial anode profile. The surface of the anode is then moved proportionately to the current distribution using Faraday's law. This process is repeated at several time steps to predict the shape evolution at the anode.

A variety of numerical techniques are available for solving fluid flow, mass transport, and current distribution problems. Finite difference methods have been extensively used for this purpose due to their ease of implementation and readily available routines and algorithms [1,29]. However, finite difference methods have difficulty in handling shape evolution and irregular geometry. FEM can handle irregular geometry and can be used to solve current distribution and shape evolution problems. Boundary element methods are more suitable for solving the current distribution and shape evolution under primary current distribution or under tertiary current distribution. Unlike the finite difference or FEMs, which are domain-based, the boundary element methods can be used to solve the current/flux distribution at the electrode without solving the potential or concentration distribution in the interelectrode gap. This eliminates the need to generate a new mesh over the domain, each time the position of the electrode is to be moved to account for the shape changes. In the following, a brief review of the literature on simulation of the current distribution and shape evolution for selected applications is presented.

The transport mechanisms responsible for anodic leveling play a crucial role in electrochemical micromachining because they control the shapes and the surface finish that can be achieved. Anodic leveling was numerically simulated by several authors. Finite element and boundary element methods have been used for simulation of shape evolution during anodic leveling of model surface profiles [30–33]. The results of these simulation techniques agreed well with the experimental results of anodic leveling of model profiles [30,32]. Under conditions well below the limiting current, the shape evolution was modeled assuming primary current distribution conditions. Nonuniformity of current density along the profile surface led to a higher rate of metal dissolution at the peaks than in the valleys. The surface roughness, defined as the profile height to profile amplitude ratio, decreased as the anodic leveling proceeded. Figure 4.5 shows a comparison of the numerical simulation with the experimental results for the dissolution of triangular copper profiles in 2 M NaNO$_3$ [30,33]. An agreement of results was obtained at short dissolution times. However, at long dissolution times, the surface roughness was found to be higher than that predicted by the model. This was attributed to observed crystallographic etching under these conditions that yielded rough surfaces. At or above the limiting current, the shape evolution has been modeled assuming tertiary current distribution conditions. For the simulation under tertiary current distribution, two distinct types of profiles, based on the ratio of the profile height to the anodic diffusion layer, were investigated [32,33]. The microprofiles were defined to be smaller than the diffusion layer thickness whereas the macroprofiles were larger than the diffusion layer thickness. The simulation results were compared with the results obtained from experiments that were performed with triangular Ni profiles in NaCl solution [33]. The predictions for the rate of leveling for microprofiles compared very well with the experimental data at both long and short times. Because of the absence of crystallographic etching at the limiting current, experimental data agreed well with the predicted values even at long dissolution times.

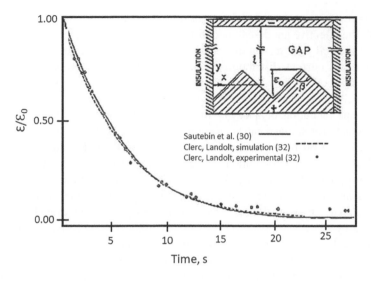

FIGURE 4.5 Variation of dimensionless profile height as a function of dissolution time. $\mathcal{E}$: profile height at time > 0, $\mathcal{E}_0$: initial profile height [30,32].

For the simulation of shaping in ECM, solutions to the above current distribution equations have been computed in several ways including analytical, graphical analog, and computational methods [34]. Out of these methods, computational methods undoubtedly give the most practical results. However, according to McGeough [34], the complex nature of the shaping problem in ECM renders difficulties in the direct application of these theoretical methods to practical problems. As a result, computational methods are taken as guidelines to develop an empirical design of tool shapes. A combination of computational methods and empirical design is widely practiced for obtaining a required workpiece configuration [34].

Computational techniques have been frequently and successfully used to compute current distribution in through-mask electrochemical microfabrication where additional factors play an important role. For example, on photoresist patterned electrodes, the macroscopic current distribution depends on the arrangement and density of the features [35–38]. In through-mask electroetching, the current distribution in the individual features evolves with time due to the shape change of the electrode. Three different scales are distinguished for current distribution and mass transport on photoresist patterned electrodes: the workpiece scale, the pattern scale, and the feature scale [39]. At the workpiece scale, the current distribution depends primarily on cell geometry and the presence or absence of auxiliary electrodes. It can be calculated in the same way as in other electrochemical reactors. On the pattern scale, the current distribution depends strongly on the spacing of the features and their geometry. West et al. [35] analyzed the primary current distribution on the feature-scale on a photoresist covered anode containing features at different distances from each other. The current density on features spaced farther apart was found to be higher leading to an increased etch rate of these features. The results confirmed the previous experimental data of Rosset et al. [21,40].

To improve the uniformity of current distribution on the pattern scale one can diminish the effective anode-cathode distance by optimizing the electrode arrangement and the cell design. Another approach is to work under mass transport controlled tertiary current distribution conditions. On the feature-scale, the current distribution is governed by the concentration field of the reacting species and/or the potential field in the feature.

The shape evolution of a cavity in isotropic electrochemical micromachining through a photoresist mask has been studied theoretically by West et al. [36] for the limiting cases of primary current distribution and uniform dissolution rate, respectively. With progressing dissolution, a given cavity evolves from an initially flat shape into a hemispherical shape. The simulation yielded the variation of the etch factor which is a measure of the undercutting as a function of the dimensionless charge of dissolution. The shape evolution in two-sided through foil electroetching was also considered by these authors. During electroetching of photoresist patterned metal films/sheets, the slope of the cavity wall increases with increasing dissolution time resulting in an increased extent of undercutting of the photoresist. The numerical results permit the estimation of the undercutting required to achieve a given wall steepness. Shenoy and Datta [38] theoretically modeled the shape evolution of relatively large cavities which can lead to island formation. They proposed to protect the center of a feature with a dummy mask which eventually is eliminated by undercutting. Shenoy and Datta also investigated the effect of the resist wall angle [37]. Madore et al. [41] presented a numerical simulation of the shape evolution of cavities in titanium, a valve metal, and found good correspondence with measured cavity shapes produced in a methanol-based electrolyte. Some of the aspects of shape evolution in through-mask electrochemical micromachining will be discussed in detail in another chapter.

4.4   EXPERIMENTAL TOOLS FOR THE INVESTIGATION OF HIGH-RATE ANODIC DISSOLUTION PROCESSES

High-rate anodic dissolution processes under ECM conditions involve high current densities ranging up to 150 A/cm^2. The reaction products that are produced at a fast speed at the electrode surfaces must be removed and fresh electrolyte supplied for the process to continue at a steady state. High electrolyte flow is therefore required not only to remove the reaction products as soon as they are formed but also to dissipate the heat generated at such high current densities. In many metal-electrolyte systems, the dissolution process is controlled by mass transport which affects the anodic overvoltage, anodic reaction stoichiometry, and the surface finish. Therefore, it is evident that the experimental investigation of high-rate anodic dissolution processes must be conducted using electrochemical cells with provisions for high electrolyte flow under controlled hydrodynamic conditions. While different electrolyte agitation systems that are applicable include RDE, channel flow, electrolytic jet, slotted jet, and multinozzle systems [9,15,19,42–44], only RDE and channel flow systems are commonly employed as experimental tools for the investigation of mass transport-controlled electrochemical processes. In the following, RDE and flow channel systems are described. These tools provide controlled hydrodynamic conditions at the reacting surface.

4.4.1 ROTATING DISK ELECTRODE

The RDE is a commonly used hydrodynamic analytical tool for electrochemical studies. The uniqueness of the RDE system is that the diffusion layer thickness does not depend on the radius thus the mass transport rate is uniform all over the surface. The RDE consists of a disk made of the working electrode material which is set into an insulating surround generally PTFE or a ceramic material. The RDE is rotated about its vertical axis typically between 100 and 10,000 rpm. The principle of operation of an RDE is shown in Figure 4.6. The electrolyte is drawn towards the disk by its rotation while the tangential forces project the electrolyte outward toward the edge of the disk.

Levich [45] presented an analytical solution for the flux to the RDE under laminar conditions (Re < 2,300). According to his theory, the anodic limiting current is given by what is known as Levich equation:

$$i_l = 0.62 \ nF \ C_{sat} \ D^{2/3} \ v^{-1/6} \ \omega^{1/2}, \tag{4.35}$$

where D is the diffusion coefficient, v is the kinematic viscosity, and ω is the angular velocity in rad/s.

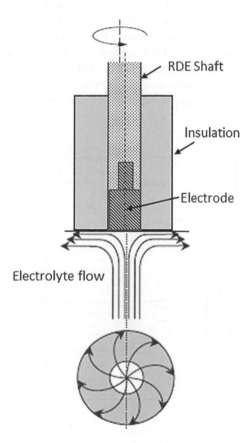

FIGURE 4.6 RDE assembly showing the electrolyte flow near the disk surface [8].

The limiting current density corresponds to the maximum reaction rate allowed by mass transport under a given hydrodynamic condition. In an anodic dissolution reaction, the limiting rate corresponds to the condition when the concentration of the anodic reaction product reaches its saturation concentration, C_{sat}, at the electrode given by Equation 4.13. By combining Equations 4.13 and 4.35, we obtain an expression for the thickness of the Nernst diffusion layer for an RDE:

$$\delta = 1.61\ D^{1/3}\ v^{1/6}\ \omega^{-1/2}. \tag{4.36}$$

It is interesting to note that in Equation 4.36, the diffusion layer thickness is independent of the disk radius. Furthermore, since the value of δ depends on the diffusion coefficient and the kinematic viscosity, every reacting species in an electrochemical system has its own specific Nernst diffusion layer.

An experimental RDE cell is shown in Figure 4.7. For anodic dissolution experiments involving high currents, the commercially available RDE systems may need to be modified to include a rugged shaft and an improved electrical contact to withstand high currents. The electrochemical cell in Figure 4.7 consists of a working electrode attached to the RDE shaft, a counter electrode separated by a frit to separate and minimize the mixing of cathodic products into the electrolyte and a reference electrode with a lugging capillary that is placed close to the working anode. The complete system includes a thermostated water bath to keep the electrolyte at a desired constant temperature. In some cases, purging the electrolyte with an inert gas to eliminate possible interference from dissolved oxygen in the electrolyte may be included (not shown). Depending on the nature of experiments and the information being sought,

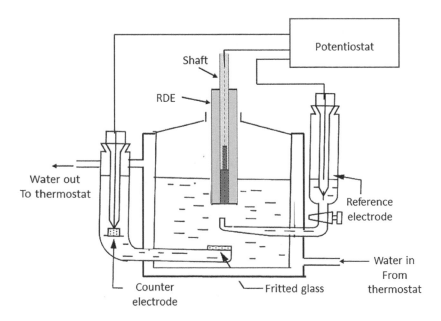

FIGURE 4.7 An experimental set-up for anodic dissolution studies using an RDE system.

some of these items may be considered as optional. One of the drawbacks of the RDE system is that in passivating metal-electrolyte systems where metal dissolution is accompanied by gas evolution, the coalescence of small gas bubbles can lead to the formation of an adherent big bubble on the anode surface, thereby blocking the passage of electric current and stopping the electrochemical reaction. On the other hand, RDE systems have been useful in studying mass transport-controlled processes in nonpassivating systems, which will be discussed later in a separate chapter.

4.4.2 FLOW CHANNEL CELL

A flow channel cell is particularly suited for the investigation of high-rate anodic dissolution of different metal-electrolyte systems including those where metal dissolution is accompanied by anodic gas evolution. The high linear velocity of electrolyte flow removes all reaction products from the reacting surface and prevents the accumulation of gas bubbles. Figure 4.8 shows a flow channel cell system which incorporates the provision for measuring anode potential. It consists of a piston pump and a rectangular flow channel cell in which linear electrolyte velocities of up to 2,000 cm/s are reached with fully developed velocity profiles at the electrodes. Such a pumping system provides precise electrolyte flow velocities without fluctuations which is essential for experimental investigation of mass transport-controlled processes. The pumping unit consists of a PVC cylinder mounted vertically and the

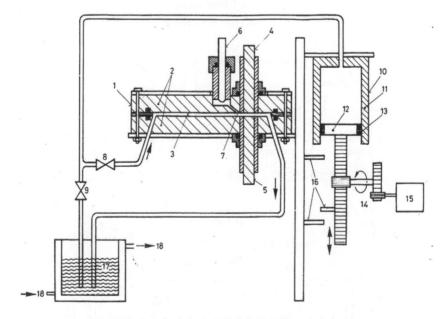

FIGURE 4.8 Schematic diagram of the flow channel cell apparatus: 1. electrochemical cell, 2. PVC blocks, 3. electrolyte flow channel, 4. anode, 5. cathode, 6. reference electrode, 7. capillary, 8, 9. valves, 10. piston pump, 11. PVC cylinder, 12. piston, 13. 0-rings with Teflon slipper seals, 14. variable-speed motor drive unit, 15. motor, 16. microswitch, 17. electrolyte reservoir, 18. thermostat [11].

pumping action results from the movement of a PVC piston. Two Viton O-rings with Teflon slipper seals prevent electrolyte leakage. A nylon diaphragm inserted into the fitting of the drainage outlet serves as a safety device against high pressures of more than 15 atm. The operation of the system is discontinuous. After the electrolyte is drawn from the reservoir into the pump cylinder, the direction of the motor is reversed. The advancing piston pushes the electrolyte through the electrochemical cell back into the reservoir. The automatic shut-off of the motor at the end of each filling or displacement stroke of the pump is controlled by a microswitch. The details of the electrochemical cell are shown in Figure 4.9. It consists of two rectangular PVC blocks separated by a 0.5 mm thick Teflon spacer and backed by two stainless steel plates. The Teflon spacer is cut out in the center (3 mm wide) to provide a flow path for the electrolyte past the electrodes. 0-rings are used to prevent leakage between the Teflon spacer and PVC blocks.

Establishing fully developed velocity profiles is important for achieving well-controlled mass transport rates at the electrodes. This requires providing sufficiently long hydrodynamic length upstream from the electrodes. Under laminar conditions $(\mathrm{Re} < 2{,}300)$, the hydrodynamic entrance length, L_e, is given by [46]:

$$L_e = 0.035 \, D_h R_e. \tag{4.37}$$

For a rectangular flow channel, the hydraulic diameter, D_h, is defined by:

$$D_h = \frac{4 \times \text{Cross section}}{\text{Wetted perimeter}}. \tag{4.38}$$

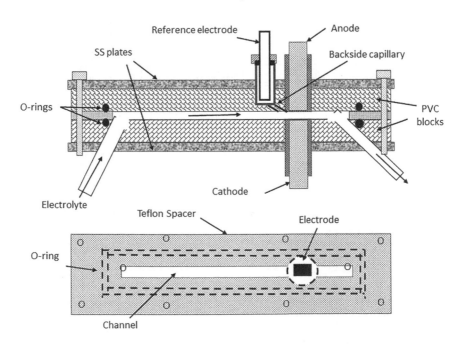

FIGURE 4.9 Details of the flow channel cell components.

No equivalent formula to Equation 4.37 exists for turbulent flow, although experiments have shown that the entrance length is often 50–100 times the hydraulic diameter [47]. On the other hand, the inclusion of the hydrodynamic length in the flow channel cell increases manifold the inlet pressure necessary to maintain a given linear velocity at the electrode. The pressure drop in a flow channel is given by [46]:

$$\Delta P = \frac{2\rho u^2 f L_c}{D_h}.$$ (4.39)

where ρ is the density of the electrolyte, u is the linear velocity, and f is the Fanning friction factor, which is defined as [46]:

$$f = \frac{16}{\mathrm{Re}} \text{ for } \mathrm{Re} < 2.3 \times 10^3,$$ (4.40)

and

$$f = \frac{0.0792}{\mathrm{Re}^{1/4}} \text{ for } 2.3 \times 10^3 \langle \mathrm{Re} \rangle 10^{5*}.$$ (4.41)

(*Note: Equation 4.41 applies to hydraulically smooth surfaces. For rough surfaces involving turbulent flow, charts are available [46] relating friction factor and Reynolds number for different relative surface roughness.)

For the microchannel created by the Teflon spacer described above, a hydrodynamic entrance length of 7 cm upstream from the electrodes served for establishing fully developed velocity profiles at the electrodes. Rectangular shaped anodes and cathodes cast in epoxy are positioned flush with the cell wall. Anodes are made of the material to be investigated and cathodes are made of an inert metal such as stainless steel or platinum. The reactive surface of all electrodes is 1 cm long and 0.3 cm wide.

The mass transfer characteristics of the apparatus were determined by limiting current measurements of the reduction of ferricyanide in a solution containing 0.05 mol/L potassium ferricyanide, 0.1 mol/L potassium ferrocyanide, and 2 mol/L sodium hydroxide [15]. Figure 4.10 shows the current potential curves measured at a scan rate of 100 mV/s. The curves exhibit a well-defined limiting current plateau over a wide range of potential. The limiting current values are dependent on the electrolyte flow indicating that the ferricyanide reduction is a mass transport-controlled process. The mass transport rate, Sh, was calculated from the measured limiting currents using the following equation:

$$\mathrm{Sh} = \frac{i_l D_h}{nFC_b D}.$$ (4.42)

Using $D = 5.06 \times 10^{-6}$ cm^2/s, $\eta = 1.467 \times 10^{-2}$ g/cm s, and $\rho = 1.107$ g/cm^3, values of Sh were calculated at different flow velocities (Reynolds number) and compared with the literature values for channel flow under laminar and turbulent flow conditions given by Equations 4.6 and 4.7, respectively. Figure 4.11 shows a comparison of the Sh

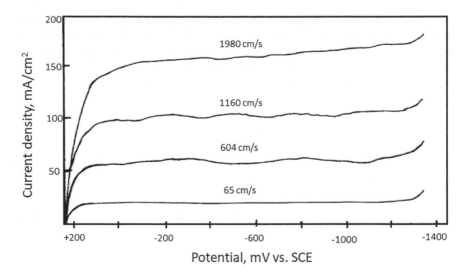

FIGURE 4.10 Current potential curves showing limiting current densities for cathodic reduction of ferricyanide at different flow rates in the flow channel cell. Potential scan rate = 100 mV/s. (Reproduced from Ref. [15] with the permission of AIP Publishing.)

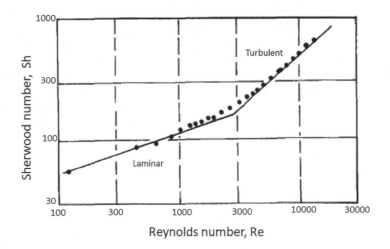

FIGURE 4.11 Mass transfer characteristics of the flow channel cell derived from measured limiting currents for ferricyanide reduction. Solid lines represent mass transfer rates predicted for laminar and turbulent flow by the correlations given in the text. (Reproduced from Ref. [15] with the permission of AIP Publishing.)

values calculated from the measured limiting currents and the values taken from the literature [48,49]. The measured values compare very well with the literature values confirming the usefulness of the system for mass transport studies. The flow channel cell has been extensively used in the investigation of high-rate anodic dissolution processes which will be discussed in the following chapters.

APPENDIX

DIMENSIONLESS NUMBERS USED IN THIS CHAPTER AND THEIR SIGNIFICANCE

The **Sherwood number, Sh**, is the dimensionless version of the mass transfer coefficient. It is the ratio of the convective mass transfer to the rate of diffusive mass transport, and is expressed as:

$$\text{Sh} = \frac{KL}{D} = \frac{L}{\delta} \tag{4.43}$$

where K is the mass transfer coefficient, D is the diffusion coefficient, L is a characteristic length, δ is the diffusion layer thickness. For an anodic dissolution process with 100% dissolution efficiency, $\text{Sh} = i_l L / nFC_{sat}D$, where i_l is the limiting current density, n is the valence of dissolution, F is the Faraday constant, and C_{sat} is the saturation concentration of the dissolving species.

The dimensionless **Reynolds number, Re,** is defined as the ratio of inertial forces to viscous forces and can be expressed as:

$$\text{Re} = \frac{\rho u L}{\mu} = \frac{uL}{v} \tag{4.44}$$

where ρ is the density of the fluid, u is the flow velocity, μ is the dynamic viscosity of fluid, and v is the kinematic viscosity ($=\mu/\rho$). For a given fluid, Re predicts whether the fluid flow would be in laminar or turbulent regimes, referring to its flow velocity. If the viscous forces which provide resistance to flow, are dominant, the flow is laminar. The flow is turbulent if the inertial forces are the dominant cause of the fluid movement, the flow is turbulent. Otherwise, if the viscous forces, defined as the resistance to flow, are dominant – the flow is laminar. The transition between the regimes is an important issue that is driven by both fluid and flow properties. For a fluid flow in a pipe or a closed channel, the critical Reynolds numbers for different flow types are: Laminar regime up to $\text{Re} = 2,300$, Transition regime $- 2,300 < \text{Re} < 4,000$, and for Turbulent regime $- \text{Re} > 4,000$.

The dimensionless **Schmidt number, Sc**, is defined as the ratio of momentum diffusivity or viscosity and mass diffusivity. The Sc is expressed by:

$$\text{Sc} = \frac{\text{Viscous diffusion rate}}{\text{mass diffusion rate}} = \frac{v}{D} = \frac{\mu}{\rho D} \tag{4.45}$$

where v is the momentum diffusivity (kinematic viscosity), D is the mass diffusivity, μ is the dynamic viscosity, and ρ is the density. The Schmidt number physically relates to the relative thickness of the hydrodynamic layer and mass-transfer boundary layer. The Sc corresponds to the Prandtl number in heat transfer. A Schmidt number equal to one indicates that momentum and mass transfer by diffusion are comparable, and velocity and concentration boundary layers almost coincide with each other.

The dimensionless **Peclet number (Pe)** is the ratio of the contributions to mass transport by convection to that by diffusion:

$$Pe = \frac{N_{conv}}{N_{diff}} = \frac{Lu}{D} = ReSc \qquad (4.46)$$

where L is a characteristic length scale, u is the velocity, and D is the diffusion coefficient. The Peclet number for mass transport is comparable to the Reynolds number for momentum transport. When the Peclet number is greater than one, the effects of convection exceed those of diffusion in determining the overall mass flux. This is normally the case for systems larger than the micrometer scale. However, at small scales, diffusion contributes much more effectively to mass transfer, so that mixing can be achieved without stirring. Most forms of forced convection increase the local magnitude of mass transfer by diffusion.

REFERENCES

1. J. Newman, *Electrochemical Systems*, 2nd edition, Prentice-Hall, Englewood Cliff, NJ, (1991).
2. N. Ibl, Fundamentals of Transport Phenomena in Electrolytic Systems, *in Comprehensive Treatise of Electrochemistry*, P. Horsman, B.E. Conway, E. Yeager, eds., vol. 6, p. 1, Plenum Press, New York, (1983).
3. N. Ibl, O. Dossenbach, Convective Mass Transport, *in Comprehensive Treatise of Electrochemistry*, P. Horsman, B.E. Conway, E. Yeager eds., vol. 6, p. 133, Plenum Press, New York, (1983).
4. J.R. Selman, C. W. Tobias, Mass Transfer Measurements by the Limiting Current Technique, *in Advances in Chemical Engineering*, T. A. Drew ed., vol. 10, p. 211, Academic Press, New York, (1978).
5. N. Ibl, Current Distribution, *in Comprehensive Treatise of Electrochemistry*, P. Horsman, B. E. Conway, E. Yeager eds., vol. 6, p. 239, Plenum Press, New York, (1982).
6. A. C. West, J. Newman, Determination of Current Distributions Governed by Laplace's Equation, in *Modern Aspects of Electrochemistry No 23*, B. E. Conway, J. O'M Bockris, R. E. White eds., p. 101, Plenum Press, New York, (1992).
7. M. Datta, D. Landolt, *Electrochim. Acta*, 45, 2535 (2000).
8. A. J. Bard, L. R. Faulkner, *Electrochemical Methods; Fundamentals and Applications*, John Wiley & Sons, Inc., New York, (1980).
9. M. Datta, D. Landolt, *J. Electrochem. Soc.*, 122, 1466 (1975).
10. M. Datta, D. Landolt, *J. Appl. Electrochem.*, 7, 247 (1977).
11. M. Datta, D. Landolt, *Electrochim. Acta*, 25, 1255 (1980).
12. M. Datta, D. Landolt, *Electrochim. Acta*, 25, 1263 (1980).
13. M. Datta, *IBM J. Res. Dev.*, 37 (2), 207 (1993).
14. M. Datta, Micromachining by Electrochemical Dissolution, *in Micromachining of Engineering Materials*, J. A. McGeough, ed., Marcel Dekker, Inc., New York, (2002).
15. D. Landolt, *Rev. Sci. Instrum.*, 43, 592 (1972).
16. D. T. Chin, K.-L. Hsueh, *Electrochim. Acta*, 31, 561 (1986).
17. R. C. Alkire, H. Deligianni, J.-B. Ju, *J. Electrochem. Soc.*, 137, 818 (1990).
18. D. Landolt, R. H. Muller, C. W. Tobias, *J. Electrochem. Soc.*, 116, 1384 (1968).
19. H. C. Kuo, D. Landolt, *Electrochim. Acta.*, 20, 393 (1975).
20. C. Clerc, M. Datta, D. Landolt, *Electrochim. Acta*, 29, 787 (1984).
21. E. Rosset, M. Datta, D. Landolt, *J. Appl. Electrochem.*, 20, 69 (1990).

22. R. C. Alkire, P. B. Reiser, *J. Electrochem. Soc.*, 131, 2797 (1984).
23. R. C. Alkire, H. Deligianni, *J. Electrochem. Soc.*, 135, 1093 (1988).
24. M. Datta, *J. Electrochem. Soc.*, 142, 3801 (1995).
25. M. Datta, D. Landolt, *Electrochim. Acta*, 26, 899 (1981).
26. M. Datta, D. Landolt, *Electrochim. Acta*, 27, 385 (1982).
27. N. Ibl, *Surf. Tech.*, 10, 81 (1980).
28. H. Y. Cheh, *J. Electrochem. Soc.*, 118, 551 (1971).
29. B. Carnahan, H. A. Luther, J. O. Wilkes, *Applied Numerical Methods*, John Wiley & Sons, Inc., New York, (1969).
30. R. Sautebin, H. Froidevaux, D. Landolt, *J. Electrochem. Soc.*, 127, 1096 (1980).
31. R. Sautebin, D. Landolt, *J. Electrochem. Soc.*, 129, 946 (1982).
32. C. Clerc, D. Landolt, *Electrochim. Acta*, 29, 787 (1984).
33. C. Clerc, M. Datta, D. Landolt, *Electrochim. Acta*, 29, 1477 (1984).
34. J. A. McGeough, *Advanced Methods of Machining*, Chapman and Hall, London, (1988).
35. A. C. West, M. Matlosz, D. Landolt, *J. Electrochem. Soc.*, 138, 728 (1991).
36. A. C. West, C. Madore, M. Matlosz, D. Landolt, *J. Electrochem. Soc.*, 139, 499 (1992).
37. R. V. Shenoy, M. Datta, *J. Electrochem. Soc.*, 143, 544 (1996).
38. R. V. Shenoy, M. Datta, L. T. Romankiw, *J. Electrochem. Soc.*, 143, 2305 (1996).
39. J. O. Dukovic, *IBM J. Res. Dev.*, 37 (2), 125 (1993).
40. E. Rosset, M. Datta, D. Landolt, *Plat. Surf. Finish*, 72, 60 (1985).
41. C. Madore, O. Piotrowski, D. Landolt, *J. Electrochem. Soc.*, 146, 2526 (1999).
42. M. Datta, D. Landolt, *Corros. Sci.*, 13, 187 (1973).
43. M. Datta, L. T. Romankiw, D. R. Vigliotti, R. J. von Gutfeld, *J. Electrochem. Soc.*, 136, 2251 (1989).
44. M. Datta, *IBM J. Res. Dev.*, 42 (5), 655 (1998).
45. V. G. Levich, *Physicochemical Hydrodynamics*, Prentice Hall, Englewood Cliffs, NJ, (1962).
46. R. B. Bird, W. E. Stewart, E. N. Lightfoot, *Transport Phenomena*, 2nd edition, John Wiley & Sons, Inc., New York, (2007).
47. J. A. McGeough, *Principles of Electrochemical Machining*, Chapman and Hall, London, (1974).
48. J. Newman, *Ind. Eng. Chem.*, 60, 12 (1968).
49. P. Van Shaw, I. P. Reiss, T. J. Hanratty, *Am. Inst. Chem. Eng.*, 9, 362 (1963).

5 High-Rate Anodic Dissolution of Fe, Ni, Cr, and Their Alloys

5.1 INTRODUCTION

High-rate anodic dissolution processes, as employed in electrochemical machining (ECM), involve high current densities ranging up to 150 A/cm², high electrolyte velocity (5–50 m/s), and a narrow electrode spacing (0.1–1 mm). A high electrolyte flow velocity is required to remove reaction products as they are formed and to dissipate the heat generated. Close spacing is essential to reproduce the contours of the cathode onto the anode workpiece. Experimental investigation of such processes poses several challenges that are not normally encountered in conventional electrolysis. Measurement of anode potentials at such high current densities is rendered difficult since they are dominated by the IR drop in the solution. It is evident that for an understanding of high-rate anodic dissolution processes, experimental investigations must be conducted in properly designed electrochemical cells which take into consideration the diverse conditions encountered in ECM.

Early works on high-rate anodic dissolution related to the understanding of ECM processes were conducted in the 1960s and 1970s by Hoare et al. in the Electrochemistry Department of General Motors Corporation [1–7], and by Landolt et al. in the Chemical Engineering Department of University of California at Berkeley [8–11].

Hoare et al. [3] investigated the differences between different electrolytes in ECM. From the anodic polarization curves obtained on steel rotating disk anodes and stationary iron anodes in concentrated $NaCl$, $NaClO_3$, $NaNO_3$, and $Na_2Cr_2O_7$ electrolytes, they concluded that the metal removal reaction is activation controlled and that $NaClO_3$ electrolytes are reduced to $NaCl$ with extended periods of ECM. $NaCl$ electrolytes do not form protective films on ferrous metals, whereas $Na_2Cr_2O_7$ form too highly protecting films of $\gamma\text{-}Fe_2O_3$. The $NaClO_3$ electrolyte behaved in an intermediate way and exhibited a sharp transition from the passive to the transpassive state. $NaClO_3$ electrolyte provided better dimensional control.

Mao [4] studied the electrode reactions involved in ECM of mild steel in $NaClO_3$ in a closed cell system at constant current. At the anode, in addition to the metal removal, significant amounts of oxygen were produced and the total anodic current efficiency for Fe^{2+} and O_2 formation was found to be close to 100%. The reduction of $NaClO_3$ to $NaCl$ during ECM was caused by the chemical reaction between $NaClO_3$ and Fe^{2+}. Based on the experimental results, the presence of porous anodic oxide films on the anode was postulated. In a related study, Mao [5] investigated the

93

anodic reactions during ECM of mild steel in NaCl, NaClO$_4$, and NaNO$_3$ solutions in a closed-cell system. The anodic processes were found to be different in different solutions. In NaCl, the anodic reaction was essentially the iron dissolution. In NaNO$_3$, however, most of the current was consumed for oxygen generation due to the presence of an electronically conductive oxide film. A small amount of oxygen (less than approximately 10% of the current) was observed in the NaClO$_4$ system. The cathodic reaction in both NaCl and NaClO$_3$ solutions was hydrogen evolution, whereas, in the case of NaNO$_3$, little or no hydrogen was observed – the current was mainly consumed in the reduction of nitrate ions.

Chin and Wallace [6] examined the relationship between dissolution current efficiency and dimensional control during the ECM of steel in NaCl, NaClO$_3$, and NaNO$_3$ electrolytes. The current efficiency for iron dissolution was measured in a flow cell under controlled anodic current densities ranging from 1 to 100 A/cm^2. In NaCl electrolyte, steel dissolved with 100% efficiency while in NaClO$_3$ and NaNO$_3$, the current efficiency increased with increasing current density. Comparison of laboratory results with those obtained under actual ECM conditions indicated that the current efficiency vs. current density curve is a useful tool to determine optimal ECM operating conditions for achieving the best possible combination of dimensional control and surface finish.

Mao and Chin [7] investigated the anodic behavior of mild steel in NaClO$_3$ at high current densities under ECM conditions. Based on the results, they concluded that the transpassive dissolution of mild steel in NaClO$_3$ takes place in three reaction stages depending on current density (applied potential). In the first stage, at low current densities, the mild steel surface is covered with an electronically conductive oxide film, and the current is consumed in oxygen evolution. In the second stage, at higher current densities, the oxide film is broken apart and the metal starts to dissolve into the electrolyte. The breakdown of the oxide film is started by the formation of localized pits in the film. In the third stage at still higher current densities, the oxide film has completely disappeared, and the high-rate metal dissolution takes place with a current efficiency greater than 100%. The metal surface after dissolution in this final stage is found to be highly polished probably due to the presence of a precipitated salt layer on the anode. However, they found no effect of electrolyte flow rate on surface finish rendering the proposed mechanism of surface polishing due to salt film formation questionable.

Landolt et al. [8] emphasized the importance of using a properly designed electrochemical cell that allows simulating electrochemical conditions while providing controlled hydrodynamic conditions at the anode surface. They built an experimental flow system with two capillaries placed closed to the anode and the cathode for potential measurements. The flow cell provided fully developed velocity profiles at the electrodes. They chose to work with copper because of its reasonably well understood electrochemical and metallurgical properties. The anodic dissolution of copper at current densities of 10–150 A/cm^2 was studied in neutral KNO$_3$, and K$_2$SO$_4$ solutions. The copper dissolved in an active or a transpassive mode which differs in overvoltage by 10–20 V. At a certain current density, the value of which depended on the electrolyte flow rate, a steep rise in the anode potential was observed. The observed steep rise in anode potential was attributed to surface films formed at the anode. The onset of "passivation" was assumed to be due to the limiting mass transfer of dissolved metal from the electrode surface. Etched, dull surfaces resulted from active dissolution, while pitted, bright surfaces resulted from passive dissolution. Similar behavior was observed in sulfate electrolytes.

Kinoshita et al. [9] investigated the influence of current density, flow rate, and electrolyte composition on the stoichiometry of anodic copper dissolution under conditions comparable to those of ECM. Weight loss measurements, X-ray diffraction, and chemical analysis were used for the characterization of the dissolution reaction. They observed that a sharp change in dissolution mechanism coincided with the transition from active to transpassive dissolution. For active dissolution, an apparent valence of the dissolution process of Cu^{2+} was found in sulfate and nitrate electrolytes, while in chloride electrolytes the dissolution valence was measured to be Cu^{+}. For transpassive dissolution, mixed valences lying between 1 and 2 were found in all electrolytes. The mixed valences were interpreted to result from the simultaneous production of monovalent and divalent reaction products, some of which in solid form.

Landolt et al. [10] studied the surface textures resulting from the anodic dissolution of polycrystalline and single crystal copper at 50 A/cm^2 in 2 M KNO$_3$. Electrolyte flow velocities were changed to obtain dissolution in the active (2,500 cm/s), and transpassive (400 cm/s) modes. Results obtained by optical and scanning electron microscopy showed that the surface topography resulting from active dissolution depended on crystal orientation. Flow streaks appeared on surfaces dissolved under transpassive conditions. Transpassive pitting was observed in polycrystalline samples only.

Datta and Landolt [12–18] employed electrochemical and surface analytical techniques to investigate the role of mass transport and surface films in influencing the high-rate dissolution of iron and nickel. They conducted a systematic study of the active, and transpassive dissolution behavior of these metals in a flow channel cell using nonpassivating (NaCl) and passivating (NaNO$_3$ and NaClO$_3$) electrolytes that are normally used in ECM. These studies were aimed at obtaining information on the relationship between anodic polarization behavior, metal dissolution efficiency, and the surface finish. Their studies included a detailed investigation of the nature of oxide films present under transpassive dissolution conditions employing in-situ electrochemical and ex-situ Auger electron spectroscopy and X-ray photoelectron spectroscopy techniques. These topics have been discussed previously in Chapter 3. In this chapter, we will discuss the high-rate dissolution behavior of iron, nickel, and FeCr alloys focusing on the influence of mass transport on anodic polarization behavior and dissolution stoichiometry and the conditions leading to salt films on the anode surface that govern the surface finish of the dissolving anode.

5.2   IRON AND NICKEL IN CHLORIDE ELECTROLYTES

Datta and Landolt [12–15] investigated the high-rate anodic dissolution of iron and nickel in 5 M NaCl electrolyte in a flow channel cell. The special features of the cell included the establishment of fully developed velocity profiles at the reacting surfaces and provision for measuring anode potential by incorporating a backside capillary close to the anode and connected to a reference electrode. Anodes were made of nickel 200 (99.0% min.) or Armco iron (99.9%) or of nickel 200 (99.0% min.). A platinum cathode was used. Since the experiments involved the use of high currents ranging up to 35 A/cm², the measured potentials were dominated by a large IR drop between the capillary and the anode. Therefore, the galvanostatic experiments were conducted involving the control of current rather than the potential.

Typical potential transients obtained on the application of constant current pulses to iron anodes are shown in Figure 5.1 for a constant flow rate of 100 cm/s. At low current densities, the anode potential reached a steady value almost immediately (<10 ms) and stayed constant throughout the pulse duration (Figure 5.1a). Beyond a certain current density, depending on the flow rate, potential jumped to an initial value then increased to a maximum before reaching its steady-state value (Figure 5.1b). Analogous potential transients were observed with nickel anodes. In Figure 5.2, the initial and peak values of the potential transients for iron dissolution are plotted as a function of current density for different flow rates. Initial values of the transients fall on a straight line indicating that the ohmic contribution of the electrolyte is predominant. Peak values lie higher and exhibit a different slope. This indicates that above a flow rate-dependent critical current density, additional resistance is present at the anode. Similar current-voltage curves showing flow rate-dependent limiting current were obtained for nickel dissolution in 5 M NaCl. The current density at a deviation from the linear relationship occurs is dependent on the flow rate and is designated as the limiting current. The limiting currents at different flow rates are indicated by an arrow in the current-voltage plot of Figure 5.2.

Current efficiency for iron and nickel dissolution determined from anodic weight loss measurements are presented as a function of current density for different flow rates in Figure 5.3. Current efficiency values were computed based on divalent iron and nickel formation. The data of Figure 5.3 show that the current efficiency for iron dissolution depends on current density as well as flow rate. At low current densities,

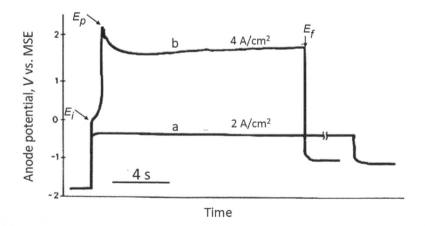

FIGURE 5.1 Potential transients on iron anode at two different current densities corresponding to etching (2 A/cm²) (a) and brightening (4 A/cm²) (b) in 5 M NaCl. Flow velocity: 100 cm/s [14].

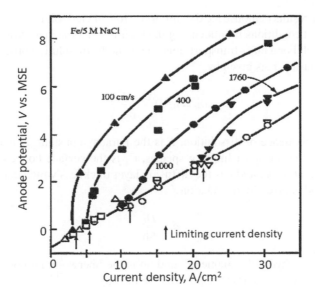

FIGURE 5.2   Galvanostatic anodic polarization curves for iron in 5 M NaCl. Arrow points to limiting current density at different electrolyte flow velocities [14].

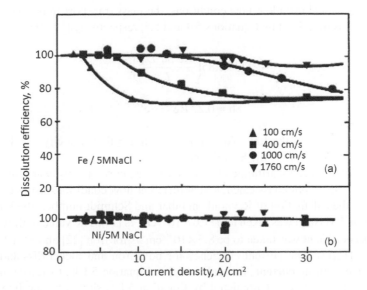

FIGURE 5.3   Current efficiency for (a) iron and (b) nickel dissolution at different current densities and electrolyte flow rates. Electrolyte: 5 M NaCl electrolyte [14].

below the limiting current, the current efficiency is 100%, but beyond the flow-dependent limiting current density, it decreases tending toward a constant value close to 70%. Current efficiency for nickel dissolution, on the other hand, is independent of current density and flow rate and is always close to 100%.

The above results demonstrate the important role of electrolyte flow and hence the mass transport conditions in influencing the high-rate anodic dissolution behavior of iron and nickel. For a mass transport-controlled anodic dissolution process, the limiting current density, i_l, is given by

$$\theta i_l = nFD \frac{C_s - C_b}{\delta} \tag{5.1}$$

where n is the valence of dissolution, F is the Faraday constant, D is the diffusion coefficient of the transport limiting species, C_s is the surface concentration, C_b is the bulk concentration, and δ is the diffusion layer thickness which depends on flow conditions and is given by the relation

$$\delta = \frac{D_h}{Sh} \tag{5.2}$$

where D_h is the "hydraulic diameter" and Sh is the Sherwood number which is the dimensionless average mass transport rate defined by

$$Sh = \frac{i_l D_h}{nFC_s D}. \tag{5.3}$$

Under laminar and turbulent flow conditions, the mass transport in the present flow channel cell is described by Equations 5.4 and 5.5, respectively [19–21]:

$$Sh = 1.85 \left(\frac{Re\ Sc\ D_h}{L} \right)^{1/3} \tag{5.4}$$

$$Sh = 0.22\ Re^{7/8}\ Sc^{1/4} \tag{5.5}$$

where $Re = u\,D_h/v$ is the Reynolds number (with u = flow velocity, v = kinematic viscosity) and $Sc = v/D$ is the Schmidt number.

Figure 5.4 shows a logarithmic plot of the limiting current density as a function of the diffusion layer thickness calculated at different flow conditions using Equations 5.2–5.5. In the calculation of Reynolds number and Schmidt number, the kinematic viscosity of 5 M NaCl was taken to be $1.6 \times 10^{-2}\ cm^2/s$ [14]. The value of the effective diffusion coefficient was taken to be $8.5 \times 10^{-6}\ cm^2/s$ for iron [22] and $9 \times 10^{-6}\ cm^2/s$ for nickel [12,13]. The current efficiency for both iron and nickel dissolution was 100% at the limiting current. The value of θ in Equation 5.1 is, therefore, unity. In Figure 5.5, the slope of -1 predicted by Equation 5.1 is shown by the line drawn. A close correspondence between theoretical prediction and experimental results is observed confirming that the limiting currents for iron and nickel under high-rate dissolution conditions in concentrated NaCl solution is mass transport controlled.

The surface microtexture of the anode after dissolution depended on applied current density and flow rate as shown in Figure 5.5. Surface etching resulted from dissolution at low current densities below the limiting current. Iron surfaces under such conditions exhibit preferred grain boundary dissolution. In the case of nickel,

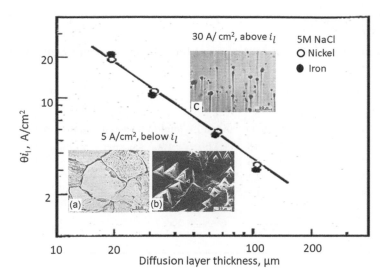

FIGURE 5.4 Limiting current density as a function of the diffusion layer thickness. Inset: Photographs showing typical etched surfaces dissolved below the limiting current density at 5 A/cm² (a) iron surface, (b) nickel surface, and (c) bright iron surface dissolved above the limiting current density at 30 A/cm². Electrolyte flow velocity: 1,000 cm/s [14].

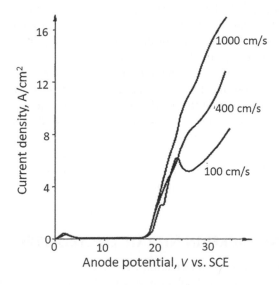

FIGURE 5.5 Potentiodynamic anodic polarization curves for nickel in 5 M $NaNO_3$ + 1 M HNO_3 at different electrolyte flow velocities. Scan rate: 500 mV/s [12].

the dissolved surfaces show crystallographic steps and regularly shaped etch pits indicating that dissolution occurs preferentially from certain crystallographic planes. The onset of surface brightening corresponded to the limiting current. At and above the limiting current density, a change in the surface microtexture occurred; instead

TABLE 5.1

Estimation of Surface Concentration at the Limiting Current Density in 5 M NaCl

Flow Rate (cm/s)	Reynolds Number	Fe				Ni			
		D_{eff} (cm²/s)	δ (μm)	i_l (A/cm²)	C_s (mol/L)	D_{eff} (cm²/s)	δ (μm)	i_l (A/cm²)	C_s (mol/L)
100	536	8.5×10^{-6}	10.5	3.0	1.9	9×10^{-6}	10.7	3.2	2.0
400	2,142		6.6	5.5	2.2		6.7	5.7	2.2
1,000	5,356		3.2	10.5	2.0		3.2	11.0	2.0
1,760	9,427		1.9	20.7	2.4		2.0	19.0	2.2
Avg					2.1	Avg			2.1

Source: Adapted from Ref. [14].

of crystallographic etching or preferred grain boundary attack, uniform dissolution took place and surface brightening was observed. The microtexture of a bright surface exhibited randomly distributed hemispherical pits, few micrometers in diameter, tailing in the flow direction. Such pits may originate from inclusions and other defects in the material which will be discussed later in this chapter.

Estimated surface concentrations at different limiting current densities are given in Table 5.1. As stated above, the limiting current density at each flow rate corresponded to the onset of brightening; the calculated surface concentration at the onset of brightening is almost the same for all flow conditions, with its average value being 2.1 mol/L for both metals. This value corresponded to that reported by Kuo and Landolt [22], who studied the anodic dissolution of iron in 5 M NaCl solution under potentiostatic conditions using a rotating disk electrode and found a value of $C_s = 2.16$ mol/L at the anodic limiting current. The close agreement between the two results confirms that the onset of surface brightening coincides with the formation of a thin salt film on the anode and it demonstrates the usefulness of using flow channel cells for the study of high-rate metal dissolution phenomena under controlled mass transport conditions.

5.3 IRON AND NICKEL IN NITRATE ELECTROLYTES

Figure 5.5 shows the anodic polarization behavior of nickel in 5 M $NaNO_3$ + 1 M HNO_3 electrolyte. An acidified solution was used to reduce the effect of local pH changes at the anode due to the expected oxygen evolution in this system. The potentiodynamic data of Figure 5.5 were obtained at different electrolyte flow velocities by applying a potential scan rate of 500 mV/s. The curves exhibit a typical active-passive-transpassive behavior. The passivation potential is 0.250 V vs. SCE, the critical current density for passivation is approximately 350 mA/cm². This current density is smaller than those applied in subsequent galvanostatic dissolution experiments (0.5–30 A/cm²) which therefore refer exclusively to the transpassive potential region and correspond to potentials higher than 1.7 V vs. SCE. Flow rate has no appreciable effect on the active-passive behavior of nickel in sodium nitrate electrolyte, but it influences the polarization curves in the transpassive potential region. The application of a constant current pulse to the anode resulted in two distinct types of potential transients as observed earlier in chloride solution (Figure 5.4). At low current densities, the steady state was reached almost immediately and the potential stayed constant throughout the run. At high current densities, the potential jumped to an initial value, and then increased to a maximum before decaying to a steady value. Which type of potential transient occurred during an experiment depended not only on the magnitude of the applied current density but also on the flow rate. In Figure 5.6, initial values and peak values of measured potential transients are plotted as a function of current density for different flow rates. As noted earlier, the difference between peak potentials and initial potentials is due to the presence of an anodic layer which increases the effective resistance at the anode surface. This layer is different from the usual passive film present on nickel at lower current density. In Figure 5.6, the current density at which the current-voltage curve deviates from linearity is designated as the limiting current corresponding to the formation of an anodic surface layer. The limiting current densities, i_l, at different electrolyte flow rates are shown by an arrow in Figure 5.6.

The current efficiency for metal dissolution, θ, was determined by weight loss measurements at different flow rates and current densities up to 30 A/cm². Current efficiency values were calculated based on Ni^{2+} formation and are given as a function of current density in Figure 5.7. As seen in the figure, the dissolution efficiency increases with increasing current density until a maximum of around 90%–95% is reached. The initial part of the current efficiency vs. current density curve is independent of flow conditions, but the current density at which maximum efficiency is attained increases with flow rate. After the maximum, the apparent current efficiency for dissolution decreases slightly and levels off at a constant value of around 85%–90%.

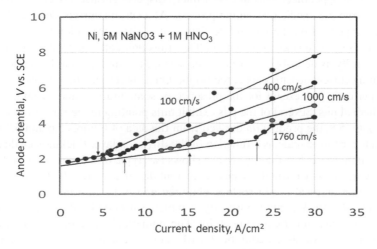

FIGURE 5.6 Galvanostatically measured anode-potential curves for nickel dissolution in 5 M NaNO₃ + 1 M HNO₃ at different flow velocities [12].

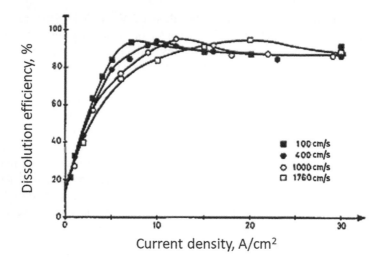

FIGURE 5.7   Current efficiency for nickel dissolution in 5 M $NaNO_3$ + 1 M HNO_3 as a function of current density at different electrolyte flow velocities [12].

Two distinctly different types of surface texture resulted from anodic dissolution. One type of surface, obtained at current densities below the limiting current, appeared grayish-black to the eye and gave the impression of a surface covered by a thick porous film. This impression was not consistent with reality, however. First, no deposits could be removed from the black-appearing surface by mechanical or chemical means. X-ray diffraction analysis of the black surface showed only the diffraction pattern characteristic for metallic nickel, but neither oxides nor salts could be identified on the surface. While this does not exclude the presence of a thin passive film, it excludes the presence of films of greater thickness. It is therefore concluded that the black surface appearance was due to a finely disperse surface texture of the metal resulting from the anodic dissolution process. At a certain current density, the value of which depended on the electrolyte flow rate, a sharp transition took place from surface etching to surface brightening. The transition corresponded to the limiting current density as described above and the surfaces resulting from dissolution above the limiting current densities appeared bright. On a microscale, the surfaces appeared almost perfectly flat, except for the presence of some grain boundary attack and of hemispherical pits. Several of the pits exhibit tails pointing in the flow direction.

In Figure 5.8, the product θi_l derived from the data of Figures 5.6 and 5.7 is plotted on a logarithmic scale as a function of the diffusion layer thickness evaluated by using Equations 5.2–5.5. Figure 5.8 also shows two typical surfaces described above, one showing an etched rough surface at current densities below the limiting current and the other atypical bright surface obtained above the limiting current. For the calculation of δ, the kinematic viscosity, v, which is needed for the calculation of Re and Sc, was determined experimentally. The numerical value of D needed for the calculation of Sc was assumed to be 9×10^{-6} cm^2/s. A slope of -1 is the same value as predicted by Equation 5.1. Therefore, the onset of surface brightening on nickel dissolving in sodium nitrate solution is mass transport controlled.

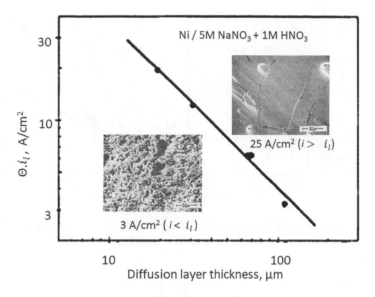

FIGURE 5.8 Limiting current density as a function of diffusion layer thickness for anodic dissolution of Ni in 5 M $NaNO_3$ + 1 M HNO_3. Inset: Photographs showing typical etched surface dissolved below the limiting current density at 3 A/cm² and a bright surface dissolved above the limiting current density 25 A/cm². Electrolyte flow velocity: 400 cm/s [12].

5.4 IRON AND NICKEL IN CHLORATE ELECTROLYTES

Anodic potential transients resulting from the application of a constant current pulse were of similar shape as those described above. At relatively low current density, the measured anode potential reached a steady state within a few milliseconds. Above a critical current density, the value of which depended on the flow rate, the measured anode potentials jumped to an initial value then slowly increased to a maximum before decreasing to a quasi-steady state value lying between initial and maximum values. The observed difference between the initial value and peak value of the anode potential is attributed to a potential drop at the anode surface due to the formation of a film. All experiments reported here refer to transpassive dissolution conditions and in analogy with previous findings [12–15], the potential increase is attributed to the formation of a salt film. The initial and peak potential values of transients are shown as a function of applied current density for different flow rates in Figure 5.9. The deviation from a linear relation in current potential curves is shown by an arrow and is referred here as the limiting current, which increases with the increase in electrolyte flow velocity. For the dissolution of iron at current densities beyond the limiting current, potential fluctuations were observed. The frequency of these oscillations increased with increasing current density varying between 5 and 100 Hz. The amplitude was typical of the order of 3–4 V. Such oscillations were not observed for nickel dissolution.

The current efficiency for iron and nickel dissolution was determined from anodic weight loss measurements and computed based on divalent metal ion formation.

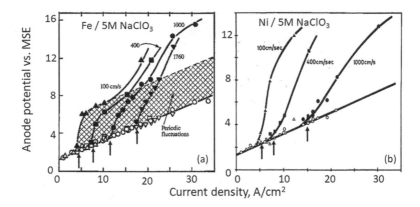

FIGURE 5.9 Galvanostatically measured anode-potential curves for (a) iron and (b) nickel dissolution in 5 M NaClO₃ solution at different flow velocities [15].

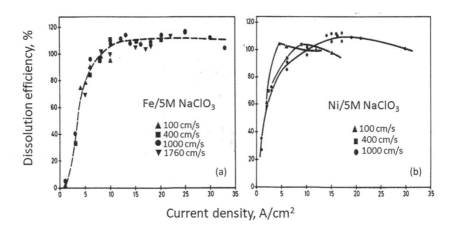

FIGURE 5.10 Current efficiency for (a) iron and (b) nickel dissolution in 5 M NaClO₃ at different flow velocities [15].

Figure 5.10 shows that current efficiency for iron dissolution is independent of flow rate; it increases with current density to a constant value of 110%. For nickel dissolution, the current efficiency increases with current density to a flow rate dependent maximum, the value of which is also more than 100%. Such anomalous behavior may result from an intergranular attack, leading to the removal of metallic particles from the anode, or they may be the result of chemical reactions occurring at the anode surface. X-ray diffraction analysis of dissolution products has not shown any evidence of the presence of metallic particles [23]. The low dissolution valence values have been attributed to chlorate reduction at active sites on the anode by the following corrosion reactions for iron:

$$2Fe + ClO_3^- + 6H^+ \rightarrow 2Fe^{3+} + Cl^- + 3H_2O \qquad (5.6)$$

and

$$6Fe^{2+} + ClO_3^- + 6H^+ \rightarrow 6Fe^{3+} + Cl^- + 3H_2O. \qquad (5.7)$$

For nickel, the reaction is:

$$3Ni + ClO_3^- + 6H^+ \rightarrow 3Ni^{2+} + Cl^- + 3H_2O. \qquad (5.8)$$

Solution analysis, indeed, showed an accumulation of Cl^-, providing conclusive evidence of chlorate reduction. In order to assure that no reduction of ClO_3^- took place at the cathode, a blank experiment was carried out with a platinum anode at a current density of 15 A/cm^2 [15]. The solution analysis showed that no chloride ion was cathodically produced in the blank experiment. This confirmed that the chloride ions were produced due to the decomposition of chlorate ions on the active surface produced during the high-rate anodic dissolution of iron and nickel, and thus also providing an explanation for the experimentally obtained anomalous dissolution efficiency [15].

Microtexture of the surfaces resulting from dissolution depended on the applied current density and the electrolyte flow velocity. As shown in Figure 5.11, two distinctly different surfaces obtained were either etched or bright. Under etching conditions, nickel anodes after dissolution appeared black to the naked eye because anodic dissolution yielded a finely disperse surface microtexture (Figure 5.11a). Iron surfaces resulting from dissolution under etching conditions were covered by a reddish-brown precipitate of iron hydroxide which adhered to the surface and could only be removed by application of a strong water jet. The metal surface underneath the hydroxide layer was of a gray dull appearance corresponding to a microstructure illustrated by Figure 5.11b. It exhibited severe grain boundary attack and flow-independent

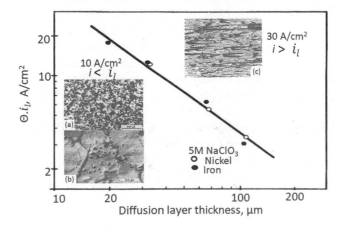

FIGURE 5.11 Limiting current density as a function of the diffusion layer thickness. Inset: Photographs showing etched surfaces at 10 A/cm^2, below the limiting current density (a) iron surface, (b) nickel, and (c) bright iron surface at 30 A/cm^2, above the limiting current density. Flow velocity: 1,000 cm/s [15].

dissolution. Figure 5.11c of a bright iron surface shows that crystallographic features are absent, but many small hemispherical pits are randomly distributed on the surface. The pits exhibit marked tails in the flow direction.

Figure 5.11 includes a plot of the product, θi_l, calculated from experimental data for iron and nickel dissolution in 5 M $NaClO_3$ as a function, δ, the diffusion layer thickness. For the calculation of δ, the kinematic viscosity of 5 M NaCl was taken to be 1.6×10^{-2} cm^2/s [14]. The value of the effective diffusion coefficient was taken to be 8.5×10^{-6} cm^2/s for iron [22] and 9×10^{-6} cm^2/s for nickel [12,13]. The figure also presents the surfaces resulting from dissolution below the limiting current showing etched surfaces and above the limiting current showing a typical bright surface. The data provided in Figure 5.11 thus present a relationship between the surface finish and the limiting current. The plot of θi_l vs. δ presented in Figure 5.11 yields a straight line with a slope of -1 confirming that the surface brightening during high-rate dissolution of iron and nickel in 5 M $NaClO_3$ is mass transport controlled. The average surface concentration of iron and nickel ion at the anode corresponding to the limiting currents was estimated from Equation 5.1 to be 2.2 mol/L for both metals. The solubility of $Fe(ClO_3)_2$ and $Ni(ClO_3)_2$ in 5 M $NaClO_3$ is not known but the solubility of $Ni(ClO_3)_2$ in the water at 18°C is 4.47 mol/L [15]. Solubility in 5 M $NaClO_3$ will be lower than this value. For example, the solubility of $Ni(NO_3)_2$ in H_2O is 4.4 mol/L while that in 5 M HNO_3 is 2.8 mol/L [12]. Because of this, the value of 2.2 mol/L found for the surface concentration of iron and nickel ion at the onset of brightening may be considered to correspond well to the saturation concentration. The transition from etching to brightening, therefore, coincides with the precipitation of a salt film in agreement with behavior found for other metal electrolyte systems [12–15].

5.5   FeCr ALLOYS IN CHLORIDE AND NITRATE ELECTROLYTES

Rosset et al. [24] studied the high-rate dissolution behavior of Fe, Cr, Fe13Cr, Fe24Cr, and AISI type 304 stainless steel in a flow channel cell. Figure 5.12 shows potentiodynamic anodic current-voltage curves for these materials in 5 M NaCl electrolyte measured at a scan rate of 50 mV/s. Both Fe and Fe13Cr exhibit a current plateau. Similar flow-dependent, mass transport current plateau has been reported for Fe [15,22] and for Fe15Cr [25]. Fe24Cr and Cr dissolve in the transpassive potential region but do not exhibit a current plateau. The anodic behavior of SS304 resembles that of Fe13Cr but the plateau is reached at a higher current density and presents a maximum. This behavior is possibly due to the formation of a complex surface film consisting of mixed iron and nickel chlorides.

Galvanostatic experiments were performed to determine the apparent valence of dissolution, n, by weight loss measurements using Equation 5.9:

$$n = \frac{MIt}{\Delta WF} \qquad (5.9)$$

where M is the atomic weight, I is current, t is the dissolution time, ΔW is the weight loss, and F is the Faraday constant. The atomic weight of the alloys was calculated by using $M_{alloy} = \Sigma x_j M_j$, with x_j being the mole fraction of the component j and M_j

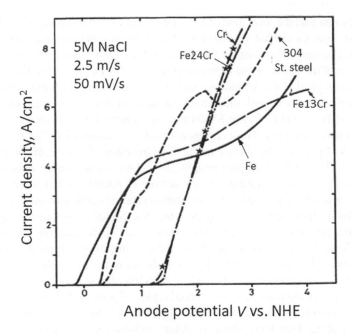

FIGURE 5.12 Potentiodynamic anodic polarization curves for Fe, Cr, and their alloys in 5 M NaCl. Electrolyte flow velocity: 2.5 m/s; scan rate: 50 mV/s [24].

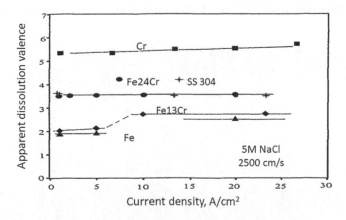

FIGURE 5.13 The apparent valence of dissolution in 5 M NaCl derived from galvanostatic experiments at 2.5 m/s [24].

its atomic weight. Figure 5.13 shows the apparent valence of dissolution as a function of current density for different materials studied. For Fe and Fe13Cr, below the limiting current, n is 2.0 and 2.1, respectively; and above the limiting current, $n = 2.5$ and $n = 2.7$, respectively. Thus, for Fe dissolution, Fe^{2+} forms below the limiting current, whereas both Fe^{2+} and Fe^{3+} form above the limiting current. The measured values for Fe13Cr are explained by postulating the formation of Fe^{2+}, Cr^{2+}, and Cr^{3+} below the limiting current and the formation of Fe^{2+}, Fe^{3+}, and Cr^{6+} (chromate) above the limiting current. Pure Cr yielded a value of n between 5 and 6 indicating mostly chromate formation. The deviation from the theoretical value of $n = 6$ could, in part, be due to the presence of inclusions in the chromium used. For the Fe24Cr alloy, the weight loss measurements yielded $n = 3.6$ independent of current density. This measured value is in reasonable agreement with the theoretical value of $n = 3.76$ corresponding to the formation of Fe^{3+}, and Cr^{6+}. For SS304 the experimental data yielded $n = 3.5$ independent of current density. This corresponds to the formation of Fe^{3+}, Cr^{6+}, and Ni^{2+}.

Figure 5.14 presents the surface morphology resulting from the dissolution of the three alloys. The SEM microphotographs (a–c) show surfaces after dissolution at a current density below the limiting current density (1 A/cm²) for Fe13Cr, Fe24Cr, and SS304, respectively. The Fe13Cr alloy exhibits crystallographic etch pits; Fe13Cr alloy shows a finely dispersed etched surface while the SS304 shows uniformly distributed circular pits. Dissolution at a current density above the limiting current density (20 A/cm²) yielded a similar surface finish for all three alloys. Typical surface finish obtained above the limiting current density is shown in the microphotograph (d).

Anodic dissolution of Fe13Cr, Fe24Cr, and SS304 alloys in sodium nitrate electrolyte takes place in the transpassive potential region. Figure 5.15 shows typical anodic potential transients for SS304 in 6 M $NaNO_3$ at a flow velocity of 2.5 m/s at different current densities. Transients above 10 A/cm² exhibit the characteristic overshoots associated with the salt film formation. Figure 5.15 also presents scanning electron microphotographs of surfaces after dissolution. Dissolution at 1 A/cm² yields a surface morphology which suggests that a pitting-type breakdown of the passive film occurred. Dissolution at 28 A/cm² yields a smooth and bright surface. The observed

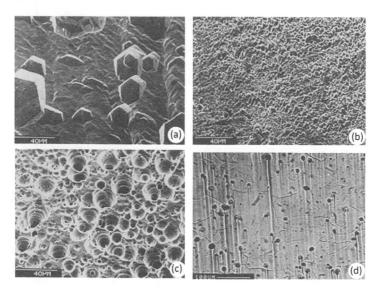

FIGURE 5.14 Microphotographs of surfaces after anodic dissolution in 5 M NaCl. Etched surfaces at 1 A/cm² (a) Fe13Cr, (b) Fe24Cr, (c) SS304. Typical bright surface at 20 A/cm² (d). Electrolyte flow velocity: 2.5 m/s [24].

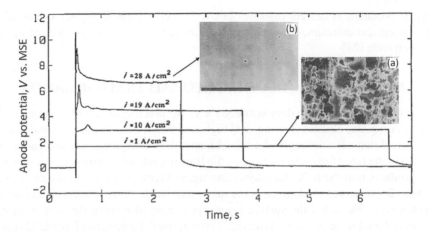

FIGURE 5.15 Potential transients measured upon application of current pulses to an SS304 anode in 6 M NaNO$_3$. Photographs of an etched surface dissolved at 1 A/cm^2 (a) and a bright surface dissolved at 28 A/cm^2 (b). Flow rate: 2.5 m/s [24].

behavior is similar to those for transpassive dissolution of nickel and iron in sodium nitrate where the transition from surface roughening to brightening is mass transport controlled and coincides with salt film precipitation [12,13,15].

Figure 5.16 shows the apparent valence of dissolution for Fe, Cr, and SS304 calculated from weight loss measurements. In 6 M NaNO$_3$, chromium dissolves in the hexavalent state at all current densities. The high values of dissolution valence for iron at low current densities indicate that the dissolution is accompanied by oxygen evolution. The dissolution valence drops down to a value between 2 and 3 at around 15 A/cm^2 indicating the formation of Fe^{2+} and Fe^{3+}. At still higher current densities, the higher values of dissolution valence are indicative of additional reactions including renewed oxygen evolution. The apparent valence of dissolution for SS304 in 6 M NaNO$_3$ is $n = 3.4$, independent of current density. This corresponds to the formation of Fe^{3+}, Ni^{2+}, and Cr^{6+}; the theoretical value of n is 3.48. In 2 M NaNO$_3$, dominant

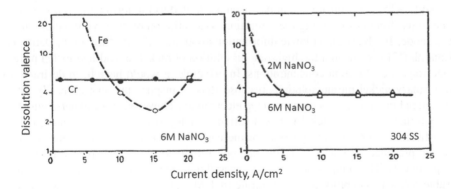

FIGURE 5.16 The apparent valence of dissolution for anodic dissolution of Fe, Cr, and 304 stainless steel in NaNO$_3$ electrolytes derived from galvanostatic experiments at 2.5 m/s [24].

oxygen evolution at current densities below 5 A/cm^2 leads to higher values of n. At high current densities, however, the dissolution valence is independent of nitrate concentration [24].

5.6   MASS TRANSPORT-CONTROLLED SALT FILM FORMATION

Several investigations have demonstrated the important role of salt film formation in pitting corrosion, electropolishing, and high-rate metal dissolution processes under ECM conditions. Precipitation of a salt film on a dissolving metal surface can occur whenever the rate of dissolution reaches the limiting rate of transport of the dissolution products into the bulk electrolyte. During an investigation of pitting corrosion, Franck [26] observed the bottom of the hemispherical to be covered with a resistive surface layer. He called the surface layer a polishing film since the bottom of the pits were found to be polished. Issacs [27] investigated the growth of artificial hemispherical pit in stainless steel and found the presence of a salt film on the pit surface that acted as a resistive layer and limited the dissolution rate of the pit. The important role of a salt film for pit growth was further emphasized by several authors [28–34]. According to Beck and Alkire [29,30], a saturated solution forms in the pit within an extremely short period after pit nucleation thus leading to salt film precipitation which plays an important role in pit growth and stability. Frankel et al. presented a mathematical framework for the role of salt film in pit growth [34].

In electropolishing systems, the value of the limiting current is governed by the rate of transport of dissolution products from the anode into the bulk [35–40]. In concentrated acid electrolytes, used in electropolishing systems, the estimated surface concentration exceeds the value of the equilibrium saturation concentration, indicating that metastable species are formed at the surface. There is considerable evidence that a salt film is present on the anode at the limiting current. According to Hoar [35], the presence of such an ionically conducting film could accentuate the microsmoothing process by presenting a barrier that suppresses crystallographic effects.

High-rate anodic dissolution studies of iron, nickel, and their alloys in ECM electrolytes have demonstrated that the transition from surface etching to surface brightening corresponds to mass transport-controlled limiting current in the anodic polarization curve [12–15]. This phenomenon has been attributed to the formation of a salt layer due to reaching or exceeding the saturation concentration of dissolving metal ions at the anode. To obtain quantitative information about the salt film formation, Datta and Landolt [12] investigated the high-rate dissolution of nickel in acidified nickel nitrate electrolytes of different concentrations. In principle, it would have been desirable to use binary nickel nitrate solutions because mass transport data can be easily interpreted in binary solutions than in mixed electrolytes. However, some experiments had shown that unless the nickel nitrate solutions employed were strongly acidic, metal, and hydroxide deposition processes at the cathode rendered results unreproducible. In the presence of 1 M HNO$_3$, the cathode remained clean of deposits during the experiments. The composition of solutions and their measured properties are given in Table 5.2. The solubility of nickel nitrate in 1 M HNO$_3$ was determined experimentally to be 4.14 mol/L. The concentration of the nickel nitrate solutions of Table 5.2 corresponds therefore to 12%, 60%, and 85% of saturation, respectively.

TABLE 5.2

Physical Properties of Nitrate Electrolytes at 25°C

Properties	0.5 M Ni(NO₃)₂ + 1 M HNO₃	2.5 M Ni(NO₃)₂ + 1 M HNO₃	3.5 M Ni(NO₃)₂ + 1 M HNO₃
Density (g/cm³)	1.103	1.387	1.517
Viscosity (c.p.)	1.181	2.758	2.169
Kinematic viscosity (cm²/s)	1.07×10^{-2}	1.99×10^{-2}	3.40×10^{-2}

Source: Adapted from Ref. [12].

Figure 5.17 shows current-voltage curves measured at a constant flow rate of 1,000 cm/s in nickel nitrate solutions of different concentrations. The data were derived from galvanostatic transients in the same way as described above for sodium nitrate solution. As before, the change in slope in the current-voltage curve corresponded to the transition from surface etching to surface brightening. The transition depended on the nickel nitrate concentration, it occurred at a lower current density in concentrated nickel nitrate concentration. This is qualitatively consistent with a salt film precipitation mechanism of surface brightening. According to this model, the anode surface concentration of nickel nitrate at the onset of brightening corresponds to the saturation concentration and the driving force for diffusion (Equation 5.10) is the concentration difference of nickel ion between anode surface and bulk:

$$\Delta C = C_s - C_b. \tag{5.10}$$

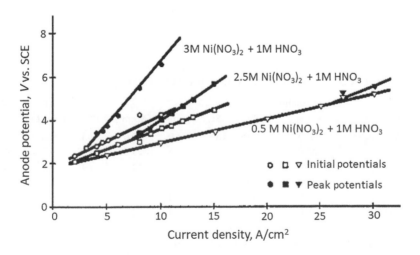

FIGURE 5.17 Anode potential as a function of current density for nickel dissolution in different concentrations of nickel nitrate + nitric acid solutions. Flow rate: 1,000 cm/s [12].

The saturation concentration, C_s, is the same for the three nickel nitrate solutions employed. Therefore, by increasing the bulk concentration, C_b, the current density for the onset of brightening becomes smaller according to Equation 5.1.

A more quantitative test of the proposed model is provided by the data of Table 5.3, where numerical values of ΔC and C_s for the experiments performed in a nickel nitrate solution were evaluated by an iteration procedure. Different values of D were assumed, and corresponding C_s values were calculated using the equations given above. The best numerical value for D was presumed to be the one giving identical C_s values for all different nickel nitrate solutions employed. The best value for D was thus found to be 9×10^{-6} cm²/s. The ionic diffusion coefficient of Ni^{2+} at infinite dilution evaluated from mobility data is 6×10^{-6} cm²/s. The diffusion coefficient employed here is an effective diffusion coefficient which includes the contribution of migration to mass transport and, therefore, its value should be higher than that of the ionic diffusion coefficient. The numerical value of D cited above, therefore, is very reasonable. It also compares favorably with effective diffusion coefficients reported for iron dissolution in chloride media [22]. According to the data of Table 5.3, the surface concentration is the same for the three nickel nitrate solutions and its value lies close to the experimentally determined saturation concentration of nickel nitrate in 1 M HNO_3. All these facts provide quantitative support for the proposed salt film precipitation model for surface brightening of nickel in nitrate electrolytes.

Kuo and Landolt [22] studied the anodic dissolution of iron in NaCl solutions using a rotating disk electrode system. In this study, they observed rotation-dependent limiting current densities, based on which they concluded that the diffusion-limiting species is the divalent iron ion. Upon anodic dissolution, a concentration gradient builds up near the anode and the local concentration of Fe^{2+} near the electrode surface increases with increasing current density. However, the Fe^{2+} concentration at the surface cannot increase indefinitely because precipitation of $FeCl_2$ occurs. The salt film thus formed on the anode surface corresponds to the limiting current density. To provide evidence to this claim, they performed anodic dissolution experiments

TABLE 5.3
Estimation of Surface Concentration at the Limiting Current

Electrolyte	Flow Velocity (cm/s)	Re	Sc	δ (μm)	i_l (A/cm²)	θ (%)	ΔC (mol/L)	C_s (mol/L)
0.5 M Ni(NO₃)₂ +	100	855	1,189	11.0	9.0	63	3.6	4.1
1 M HNO₃	1,000	8,550		2.5	29.0	88	3.7	4.2
2.5 M Ni(NO₃)₂ +	100	460	2,209	11.0	2.5	89	1.4	3.9
1 M HNO₃	1,000	4,600		3.7	9.0	97	1.9	4.4
3.5 M Ni(NO₃)₂ +	100	269	3,378	11.4	1.0	93	0.6	4.1
1 M HNO₃	1,000	2,690		5.4	3.0	100	0.9	4.4
Average C_s								4.2
Experimentally determined solubility of nickel nitrate in 1 M HNO₃								4.14

Source: Adapted from Ref. [12].

in binary $FeCl_2$ electrolytes. They also determined the saturation concentration of $FeCl_2$ in water, which they reported being 4.25 ± 0.05 mol/L. The anodic polarization curves for iron were measured in different concentrations of the $FeCl_2$ electrolyte. Figure 5.18 shows the rotation speed-dependent plateau currents observed for iron dissolution in a 3 M $FeCl_2$ solution. The measured plateau currents for different concentrations of $FeCl_2$ solutions measured at different rotations speeds are plotted in Figure 5.19. In principle, extrapolation of the plotted data to the zero current $(C_s = C_b)$ provides the surface concentration. The intercept of the data with the x-axis provides C_s value to be 4.2 mol/L. This value is the same as the experimentally determined saturation concentration of $FeCl_2$ in water. These data provide further evidence of the salt film formation by reaching saturation concentration of the dissolving species during high-rate anodic dissolution processes.

AC impedance study of salt films: As demonstrated above, the precipitation of a salt film occurs during anodic metal dissolution when the surface concentration of the salt exceeds its solubility. In ECM and polishing, when the dissolution process is mass-transport controlled, the formation of such films is responsible for surface leveling and brightening. In corrosion, the salt films are a precursor to passivation and the formation of such films at the bottom of pits control the pit growth rate. Several investigators have studied the properties and structure of salt films formed under corrosion conditions. Vetter and Strehblow [41] studying pitting corrosion considered the films to be nonporous. In a later paper, Strehblow and Wenners [42] attributed deviations from ohmic behavior of iron and nickel in chloride solutions to high-field conduction through the film. Based on transient experiments in artificial pits, Isaacs [27,43] and Hunkeler et al. [44] postulated a low-field conduction relationship for the films. For anodic salt films formed on iron in chloride solution, Beck [45,46] proposed the film to consist of two parts, a compact film through which

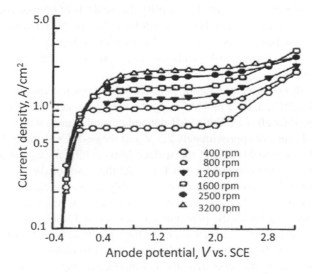

FIGURE 5.18 Anodic polarization curves for iron dissolution in 3 M $FeCl_2$ solution at different rotation speeds [22].

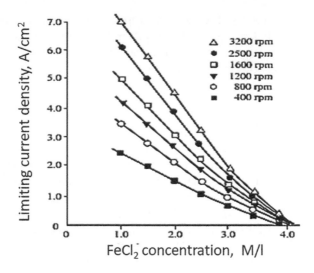

FIGURE 5.19 Limiting current density for anodic dissolution of iron as a function of $FeCl_2$ concentration at different rotation speeds [22].

ionic transport occurs by high-field conduction and a porous layer which contributes to additional ohmic resistance. Salt films formed on nickel in concentrated chloride solutions were studied by Clerc and Landolt [47] who proposed a duplex structure for the film, consisting of an outer, porous layer exhibiting low-field conduction and an inner, compact layer having semiconductor properties.

Grimm et al. [25,48] investigated the salt films formed on iron and iron-chromium alloys using the AC impedance technique which facilitated the separation of the ohmic and nonohmic resistances. To simplify the analysis of convective diffusion in the solution, they used a binary $FeCl_2$ electrolyte. Anodic polarization experiments and impedance measurements were carried out under well-controlled hydrodynamic conditions using a rotating disk electrode. High-purity Fe, Fe-15Cr, and Fe-25Cr alloys were employed.

Figure 5.20 shows potentiodynamic polarization curves at a scan rate of 5 mV/s for Fe, Fe-15Cr, Fe-25Cr, and Cr in 4 M $FeCl_2$ at a rotation speed of 200 rpm [25]. Pure iron and the Fe-15Cr alloy exhibit a well-defined current plateau which for Fe extends over a potential range of approximately 2.5 V and for Fe-15Cr of more than 5 V. Such high potentials are possible only if the surface films exhibit no significant electronic conduction as may be expected for salt films. At the onset of the current plateau, a pronounced current maximum is observed, probably due to oversaturation at the onset of salt film precipitation. The current plateau for Fe-25Cr is less well defined. At low potentials passivation limits the peak current density and hinders the establishment of a mass transport controlled current and at high potentials transpassive dissolution of chromium limits the extent of the plateau. Pure chromium does not exhibit any current plateau, and it undergoes activation-controlled transpassive dissolution.

For AC impedance studies, a duplex salt film model shown in Figure 5.21 was assumed which consists of a compact film through which ions migrate by a high-field

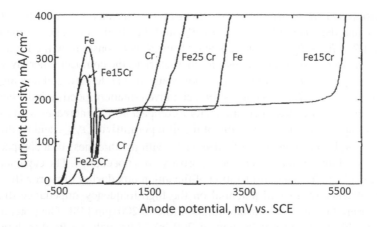

FIGURE 5.20 Potentiodynamic current-voltage curves of Fe, Fe-15Cr, Fe-25Cr, and Cr in 4 M FeCl₂. Rotation rate: 200 rpm, sweep rate: 5 mV/s [25].

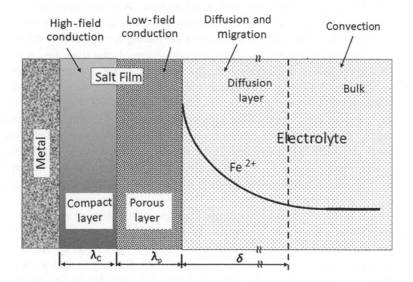

FIGURE 5.21 Schematic presentation of Duplex Layer for salt film consisting of a compact layer (λ_c) and a porous layer (λ_p); δ is the diffusion layer thickness [25].

conduction mechanism and a porous film of constant porosity through which all the current is due to migration. In Figure 5.21, λ_c and λ_p are the thickness of the compact and the porous layers, respectively, and δ is the diffusion layer thickness. The porous film behaves like an ohmic resistor with a fixed resistance per unit length R_p. For both types of film, the high-frequency part of the AC impedance, Z, can be presented in the form of Equation 5.11:

$$Z = \frac{\Delta V}{\Delta i} = R_1 + \frac{1}{jwC_2} + \frac{R_2}{1 + jwR_2 C_1} \tag{5.11}$$

In Equation 5.11, R_1 is the sum of the electrolyte resistance between the reference electrode and the outer salt film surface R_{ele} and the ohmic resistance of the porous film R_p, $\left(R_1 = R_{\text{ele}} + R_p\lambda_p\right)$. For a compact film, R_1 is a constant, while for a porous film, it increases with film thickness. R_2 characterizes the compact inner layer and is defined as $R_2 = \lambda_c/i\alpha$, where i is the steady-state current density, and α is high-field conduction constant. R_2 for a porous film, at a given rotation speed, is independent of porosity and film thickness. In contrast, for a compact film R_2 increases with its thickness. The capacitance, C_1, is the ratio of the film permittivity, $\varepsilon_{\text{perm}}$, and its thickness $\left(C_1 = \varepsilon_{\text{perm}}/\lambda_c\right)$. The capacitance C_1 decreases with the thickness of a compact layer but is constant for a porous layer whose porosity does not change. The capacitance C_2 can be calculated from a consideration of the anion flux during film growth.

The influence of applied potential on the high-frequency impedance diagrams were measured for Fe at a constant rotation rate of 200 rpm [48]. The potential varied from −400 to 2,600 mV in steps of 200 mV. Although the Fe dissolution was mass-transfer controlled, no Warburg or Nernst impedance was observed. This was attributed to the fact that the iron chloride concentration at the electrode surface is constant and is equal to the saturation concentration. Warburg impedance, on the other hand, requires modulation of the concentration. The influence of rotation rate was investigated at a constant potential (after correcting for the ohmic drop in the electrolyte). Increasing the rotation rate shifted the impedance diagram on the real axis toward lower values. Variations with applied potential and disk rotation speed of AC impedance spectra at high frequencies validated the duplex structure for the salt film formed on iron dissolving in ferrous chloride solutions. AC impedance studies led to the following conclusions: The duplex film consists of a compact, inner layer exhibiting high-field conduction, and a porous outer layer involving low-field conduction through the pores. The thickness of the porous layer decreases with increasing rotation rate. The thickness of the inner layer increases linearly with the applied potential. The ionic-charge-carrier density in the film increases with increasing limiting-current density.

For Fe-15Cr, the impedance diagrams [25] resembled those for iron described above [48]. The resistance, R_1, measured for Fe-15Cr, was found to be higher than that for Fe and it increased more steeply with potential. This suggests that the outer layer of the film formed on Fe-15Cr either has a lower porosity than that on iron or the electrolyte conductivity in the pores is lower, possibly due to the presence of complexed chromium ions. An increase in the rotation rate decreased the value of R_1, suggesting that an increasing rotation rate leads to thinning of the porous layer and eventually to film removal due to increased mechanical shear.

The AC impedance studies showed that the dissolution behavior of Fe and Fe-15Cr under limiting current conditions is governed by the same mechanism [25,48]. However, the most striking difference between the two anode materials concerns the width of the current plateau. For Fe-15Cr, it extends over twice the potential range of iron. This behavior could be related to the fact that the ionic carrier density in the compact layer formed on Fe-15Cr is higher than on Fe. The electric field strength in the compact film formed on Fe-15Cr under otherwise comparable conditions, therefore, is lower than on Fe. This view is consistent with the observation that a relatively larger part of the total potential drop at the interface occurs in the outer porous layer.

Since stresses during film growth and thermal stresses due to Joule heating in the compact film are expected to increase with the potential drop across the compact layer, the above argument qualitatively explains the increased stability at high potentials of the salt films formed on Fe-15Cr compared to those formed on Fe.

5.7 PULSED DISSOLUTION

The use of pulsating current allows one to apply a high instantaneous current density without the necessity of a high electrolyte flow rate since each current pulse is followed by an off-time such that the system relaxes between two consecutive pulses. One of the attractive features of pulsed dissolution is that three parameters, namely peak current density (i_p), pulse-on time (t_p), and pulse-off time (t_{pp}), can be varied independently while in conventional direct current dissolution only average current (i_a) can be chosen. In the following results and benefits of pulse dissolution studies are discussed for anodic nickel dissolution in chloride and nitrate electrolytes [49,50].

Anode potential: The shape of the anode potential transients, resulting from the application of repetitive current pulses for a given hydrodynamic condition depends on applied current density, pulse time, and duty cycle. Typical anode potential transients during the pulsed current dissolution of nickel in 5 M NaCl and 6 M NaNO$_3$ solutions are shown in Figure 5.22. At low current densities, the anode potential remains almost constant throughout the pulse time (Figure 5.22 Aa and Ba).

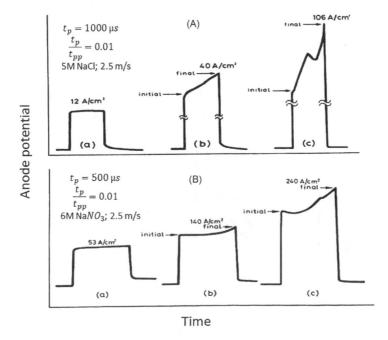

FIGURE 5.22 Anode potential transients resulting from high-rate anodic dissolution of nickel at different peak current densities. (A) 5 M NaCl, (B) 6 M NaNO₃; electrolyte flow velocity: 2.5 m/s [49,50].

Beyond a certain current density, the value of which depends on the pulse time and duty cycle, the anode potential after an initiation period increases with time (Figure 5.22 Ab and Bb) because of the formation of a salt film on the anode as described above. At still higher current densities the anode potential jumped to very high values (Figure 5.22 Ac and Bc) and eventually led to sparking. Figure 5.23 shows initial and final values of anode potential as a function of applied current density for different pulse times of 100, 500, and 1,000 μs and a constant duty cycle of 0.01 for nickel dissolution in 6 M $NaNO_3$. Similar data were obtained for nickel dissolution in 5 M NaCl. In Figure 5.23 the initial potential values fall on a straight line and are largely dominated by the ohmic drop in the solution between the capillary and the anode. The final values which deviate from the straight-line behavior include the additional resistance due to the formation of an anodic film. The current density beyond which the potential deviates from linearity is the limiting pulse current density, i_{pl}, which corresponds to the onset of salt film formation at the anode caused by exceeding the limiting rate of mass transport of dissolution products. Experimental data indicated that i_{pl} is higher for small pulse times and small duty cycles [49,50].

Surface microtexture: Based on earlier results from direct current dissolution studies [14,15], one would expect surface etching to occur for $i_p < i_{pl}$ and surface brightening for $i_p > i_{pl}$. This expectation is qualitatively born out as evidenced by the micrographs of Figure 5.24 for nickel dissolution in 5 M NaCl. At a duty cycle of 0.01 and a pulse time of 500 μs etching is observed at 20 A/cm² (Figure 5.24a) and brightening with pitting is observed at 120 A/cm² (Figure 24b). In Figure 5.25 the

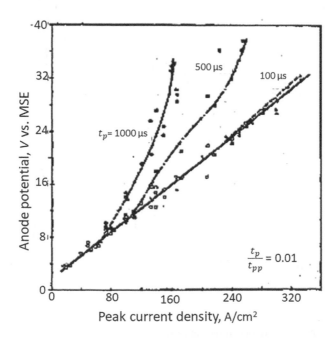

FIGURE 5.23 Anode potentials as a function of peak current density at different pulse times for high-rate dissolution of nickel in 6 M $NaNO_3$ [50].

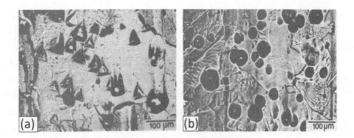

FIGURE 5.24 Microphotographs of surfaces resulting from dissolution in 5 M NaCl at different peak current densities: (a) 20 A/cm^2 (b) 120 A/cm^2. t_p = 500 μs, t_p/t_{pp} = 0.01, u = 2.5 m/s [49].

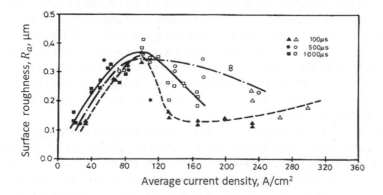

FIGURE 5.25 Average surface roughness, R_a, as a function of peak current density for nickel surfaces dissolved in 6 M NaNO$_3$; duty cycle: 0.01, closed symbols: etched surfaces, open symbols: bright surfaces [50].

measured average surface roughness, R_a, is plotted as a function of applied peak current density for different pulse times obtained for nickel dissolution in 6 M NaNO$_3$ solution. All R_a values lie between 0.1 and 0.4 μm. At low current densities, the value of R_a increases with increasing peak current density. The R_a value goes through a maximum and finally drops down at $i_p \gg i_{pl}$. An important aspect of the microstudies is the observation that bright surfaces resulting from pulsed dissolution do not show flow streaks, which is commonly observed in direct current processing [14,15]. As mentioned earlier, during direct current dissolution, pits formed on the surface lead to flow separation creating small vortices and eddies that increased the local mass transport rate at the down-stream end. Under pulsed dissolution conditions, the effective dissolution layer thickness is much smaller than under DC conditions and is essentially determined by the pulse parameters. Therefore, local variations of hydrodynamic conditions have little influence on the dissolution rate, hence pit tails and flow streaks are largely absent in pulsed dissolution.

Dissolution efficiency: Current efficiency for nickel dissolution determined from anodic weight loss measurements using Faraday's law based on divalent nickel ion formation indicated that in 5 M NaCl solution the dissolution efficiency is 100% independent of pulse parameters [49]. In nitrate electrolyte, the current efficiency for nickel dissolution increases with increasing peak current density up to a maximum value close to 100% then decreases slowly. This behavior is like that obtained during high-rate dissolution of nickel in nitrate electrolytes using direct current [15]. Figure 5.26 shows the obtained current efficiency data in 6 M sodium nitrate electrolyte as a function of applied peak current density at different pulse times of 100, 500, and 1,000 μs and a constant duty cycle of 0.01. In the current density region where current efficiency is an increasing function of current density, the dissolution efficiency is higher for longer pulses. Other experimental data (not shown here) indicated that a higher duty cycle leads to higher dissolution efficiency, whereas a linear relationship exists between current efficiency and pulse time, the slope being a function of current density. These results indicate that compared to direct current ECM, the use of pulsating current provides additional variables which influence the current efficiency for metal dissolution in passivating electrolytes. In Figure 5.27, the pulsed dissolution current efficiency data are replotted as a function of average current density and compared with direct current in the same flow channel cell and at the same electrolyte flow rate. It follows that the current efficiency in pulsed dissolution is significantly high even at average current densities where very little dissolution takes place with direct current.

From the above discussion, it is evident that mass transport processes influence the anodic polarization behavior and surface finish during the pulsed dissolution of nickel in passivating and nonpassivating solutions. Besides, the dissolution current efficiency for nickel dissolution in nitrate electrolytes is also governed by mass transport processes at the anode. In pulsed dissolution, the mass transport processes are determined by the pulsating diffusion layer thickness, δ_p, which for small duty cycles, $t_p/t_{pp} \ll 1$, is given by

$$t_p = \sqrt{2Dt_p}. \tag{5.12}$$

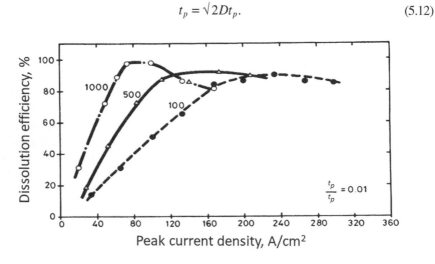

FIGURE 5.26 Current efficiency for nickel dissolution in 6 M NaNO₃ as a function of peak current density at different pulse times [50].

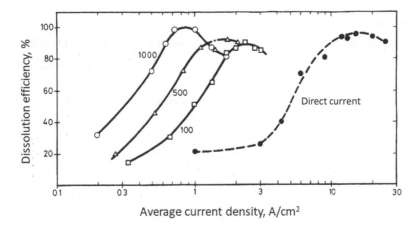

FIGURE 5.27 Current efficiency for nickel dissolution as a function of average current density. Direct current data are compared with pulsed current data at different pulse times [50].

Since in pulsed dissolution, the pulse time is kept very small, the peak currents are much higher than the steady-state limiting current for a given hydrodynamic condition. Under these conditions in a nitrate electrolyte, the metal dissolution efficiency increases with time within each repetitive pulse. Eventually, the limiting rate of mass transport of dissolved metal ions away from the anode is reached leading to the formation of salt film at the surface. The current density at which the surface concentration reaches the saturation concentration of dissolved species at the end of a pulse is the pulse limiting current density, i_{pl}, which for metal dissolution at a 100% current efficiency is given as

$$i_{pl} = i_l \left[\frac{\delta_p}{\delta} \left(1 - \frac{t_p}{t_{pp}} \right) + \frac{t_p}{t_{pp}} \right]^{-1} \tag{5.13}$$

where i_l is the steady-state limiting current density defined by $i_l = nFD\left(C_{sat}/\delta\right)$, and δ is the steady-state diffusion layer thickness defined by the hydrodynamic conditions at the anode.

Calculated and experimentally determined i_{pl} values determined from current-potential curves are given in Table 5.4. The solubility of nickel chloride in 5 M NaCl and nickel nitrate in 6 M NaNO$_3$ is approximately the same around 2.1 M/L. Therefore, one would expect the i_{pl} values to be of similar magnitude. The data of Table 5.4 show the measured value of i_{pl} in nitrate solution to be higher than those calculated and measured in chloride solution. This is attributed to the fact that in nitrate solution the current efficiency for dissolution is an increasing function of time within a pulse until it reaches its maximum value at the end of the pulse corresponding to i_{pl}. For this reason, the nickel ion build-up at the anode in nitrate solution is slower than in chloride solution for which the calculated values correspond well to the measured values.

TABLE 5.4

Theoretical and Experimental Values of Pulse-Limiting Current Density

| | | | | | | | Experimental i_{pl} (A/cm²) | |
t_p (µs)	$\dfrac{t_p}{t_{pp}}$	u (m/s)	δ (µm)	δ_p (µm)	i_l (A/cm²)	i_{pl} (A/cm²) (Equation 5.13)	6 M NaNO₃ [50]	5 M NaCl [49]
100	0.1	2.5	5.8	0.40	6.3	39	85 ± 15	50 ± 10
100	0.01			0.42		77	230 ± 30	60 ± 20
100	0.001			0.42		85	260 ± 30	90 ± 20
500	0.01			1.0		37	110 ± 10	35 ± 10
1,000	0.1			1.3		21	45 ± 5	20 ± 5
1,000	0.01			1.3		27	65 ± 5	20 ± 5
1,000	0.01	10	3.2	1.3	11	27	Not available	25 ± 5

Source: Adapted from Refs. [49,50].

Sparking: Attainable dissolution rates in pulsed dissolution are limited by the onset of sparking. Sparking observed during the experiments was always preceded by a steep rise in anode potential as shown in the traces of Figure 5.22 (Ac and Bc) and was followed by potential fluctuations (not shown) and a drop in current density due to overload of the pulse generator. For a constant duty cycle and electrolyte flow rate, the current density at which sparking set in depended on pulse time and electrolyte conductivity. This is shown in Figure 5.28 where sparking current density has been plotted as a function of pulse time for different NaCl solutions having different conductivities: 5 M NaCl (0.22/Ω/cm), 2 M NaCl (0.14/Ω/cm), and 1 M NaCl (0.08/Ω/cm). These data were obtained at a constant duty cycle and electrolyte

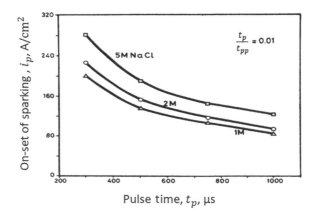

FIGURE 5.28 Influence of pulse time on the peak current density corresponding to the onset of sparking in NaCl solution of different concentrations, $u = 2.5$ m/s [49].

TABLE 5.5

Influence of Pulse Parameters on the Onset of Sparking

			Onset of Sparking			
			6 M NaNO₃ [50]		5 M NaCl [49]	
t_p (μs)	$\dfrac{t_p}{t_{pp}}$	u (m/s)	i_p (A/cm²)	i_a (A/cm²)	i_p (A/cm²)	i_a (A/cm²)
100	0.1	2.5	300	30.0	225	22.5
100	0.01		*	*	*	*
100	0.001		*	*	*	*
500	0.01		267	2.67	185	1.85
1,000	0.1		153	15.3	122	12.2
1,000	0.01		167	1.67	124	1.24
1,000	0.01	10.0	–	–	180	1.80

Source: Adapted from Refs. [49,50].
* No sparking

flow rate of 0.01 and 2.5 m/s respectively. Table 5.5 summarizes the influence of pulse parameters on the onset of sparking. The data show that the onset of sparking in nitrate solution occurs at a higher current density than in chloride solution. The strong dependence of sparking current on the electrolyte conductivity and pulse time suggests that sparking is related to instantaneous electrolyte heating in the interelectrode gap.

To estimate instantaneous heating, the electrolyte within the gap can be considered as stationary since pulse time will normally be smaller than the average residence time of the electrolyte in the gap of the flow cell. The instantaneous temperature rise, ΔT_p, due to ohmic heating during a single pulse is given by:

$$\Delta T_p = \frac{i_p^2 t_p}{\kappa \rho C_p} \quad \left(t_p << \frac{L}{u} \right) \tag{5.14}$$

where i_p, κ, ρ, and C_p are the peak current density, specific conductivity, density, and specific heat of the electrolyte, respectively, L is the electrode length, u is the electrolyte flow velocity.

The instantaneous temperature rise calculated by ohmic heating using Equation 5.14 is less than 40°C in all the experiments [49,50]. It is evident that simple ohmic heating severely underestimates the local heating responsible for sparking. Under steady-state dissolution conditions, optical observation of gases evolving at the cathode revealed that the size of gas bubbles increased with increasing current density [51]. The gas bubbles tend to stick to the cathode surface and in extreme cases, large gas patches may be formed which shield the cathode and lead to an electric breakdown. Such phenomena may contribute to the onset of sparking but the dynamics of gas evolution under pulse dissolution conditions are not known.

It is interesting to note that the current efficiency for metal dissolution at a given average current density is higher in PECM than in DC electrolysis (Figure 5.27). From the ECM point of view, this behavior allows one to dissolve passivating metals under transpassive conditions with high efficiency without the need for high average current densities and correspondingly high electrolyte flow rates. Good surface finish without pit tails and flow streaks may thereby be obtained. These concepts have been used to develop an electrochemical saw for metal cutting [52]. The benefits of employing pulsating current in electrochemical micromachining processes for the fabrication of microelectronic, microfluidic, and MEMS components will be discussed in later chapters. Pulsed dissolution is particularly suitable for the through-mask micromachining of thin films, an application in which fine surface finish and low dissolution rates are desirable to achieve control over the machining process [53].

5.8   INFLUENCE OF METALLURGICAL FACTORS

High-rate anodic dissolution behavior of metals and alloys is controlled by tangled interactions among several parameters such as the nature of electrolyte (including its composition and temperature), applied current density (voltage), hydrodynamics, and metallurgical factors. In the previous sections, the interaction of the first three items has been discussed to a great extent; in this section, the role of metallurgical factors will be discussed.

All metals and alloys in practical use are polycrystalline; they are made up of many individual crystals (grains) that are joined together at the atomic level. The orientations of different grains are random with respect to each other. The grain boundaries are energetically sites of impurities and other imperfections. The grain boundaries are preferred sites for accumulation (segregation) of solute atoms since they have a higher solubility in the grain boundary regions. Solute atoms and impurities tend to congregate at defects within the grains as well. Within individual crystals, imperfections arise from several sources [54]. Point defects are atom-sized imperfections in crystals. These include interstitial atoms, vacancies, and foreign atoms in lattice positions. Point defects may be present in the metal from the time of its solidification or introduced later by heating, plastic deformation, or bombardment with high-energy radiation. Defects such as dislocations can be created by plastic deformation of a crystal, and their movements are important in determining the deformation properties of a material [54]. Stacking faults can arise during crystal growth or because of plastic deformation but can only occur along the close-packed planes within a crystal. Inclusions are three-dimensional defects consisting of insoluble particles of foreign material in the metal and may be either intentionally or accidentally incorporated. Inclusions may include nonmetallic phases in the form of particles of insoluble foreign material in the matrix. Particles composed of various oxides, sulfides, and silicates are common inclusions in metals and alloys. A good example is that of manganese sulfide inclusion in stainless steels. Other forms of inclusion include the presence of intermetallic compounds and metalloids such as carbides. Other defects include voids that are empty or gas-filled spaces within the metal. They may be formed by trapped bubbles of gas in the metal during solidification or by coalescence of several vacancies in the metal. Inclusions and voids are normally orders of magnitude larger than single-impurity atoms or vacancies. These metallurgical factors play a significant role in the surface finish resulting from high-rate anodic dissolution processes.

5.8.1 PITTING AND GRAIN BOUNDARY ATTACK

The influence of the presence of small amounts of impurities on high-rate anodic dissolution of nickel is presented in Figure 5.29, which shows SEM photographs of surfaces resulting from active (below the limiting current) dissolution of nickel 200 (99.5%) and high-purity nickel (99.99%) anodes after passing a constant charge of 10C [55]. The chemical composition of two nickel materials in weight percent are given as [55]: Ni 200: 0.096 Fe, 0.047 Cu, 0.102 Mg, 0.081 Si, 0.159 Mn, 0.110 Co, 0.008 S, rest Ni; Ni pure: 0.020 Fe, 0.083 Si, 0.021 C, rest Ni. On nickel 200, pronounced pitting is observed (Figure 5.29a) but not on pure nickel (Figure 5.29b). This indicates that impurities present in nickel 200 are responsible for the localized attack. Indeed, many pits in Figure 5.29a reveals the presence of inclusions in the middle of the pits. Similar results were obtained for transpassive dissolution (above the limiting current) which showed randomly distributed pits on an otherwise fiat surface on nickel 200 but not on pure nickel. Pits on bright nickel 200 surfaces were generally hemispherical while those on etched surfaces usually exhibited crystallographic features. Similar observations have been made with copper where the influence of crystallographic factors on active dissolution has been systematically investigated using monocrystals [10]. During high-rate dissolution of copper in the active mode, submicroscopic facets lead to the development of ridges, which give the surface an etched appearance. This appearance depends on crystallographic orientation and results in differentiation between grains in a polycrystalline material. Active dissolution of [1 0 0] and [1 1 1] faces does not result in the formation of facets. These observations agree with a dissolution mechanism based on the motion of atomic ledges on tightly packed lattice planes. Transpassive dissolution proceeds in the presence of anodic layers, which diminish the effect of crystal orientation and lead to brightening. However, flow streaks are associated with transpassive dissolution, while the surface resulting from active dissolution is independent of flow direction.

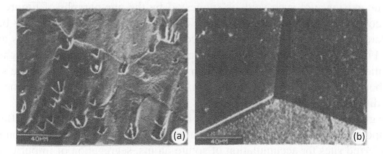

FIGURE 5.29 SEM photographs of nickel 200 (a) and high-purity nickel (b) surfaces resulting from active dissolution at a constant charge of 10C [55].

FIGURE 5.30   (a–c) SEM photographs of typical pits formed on nickel 200 under active dissolution conditions [55].

Typical pits formed on nickel 200 under active dissolution conditions are presented at higher magnification in Figure 5.30. Figure 5.30a shows an inclusion that led to the growth of a hemispherical pit with a void at the bottom between the inclusion and the metal matrix. In some cases, the inclusions are broken or washed away (Figure 5.30b). The triangular pit of Figure 5.30c shows crystallographic steps on its sidewalls and a perfectly flat bottom. This type of pit may have been initiated at inclusions of smaller dimensions that are washed away as dissolution proceeds. The energy-dispersive X-ray microprobe analysis of the different inclusions showed large peaks of sulfur, magnesium, and nickel [55]. These data demonstrate the role of impurities in the formation of pits during high-rate dissolution of metals. It is well known that to eliminate brittleness caused by the presence of sulfur, magnesium is added to molten nickel just before casting since magnesium preferentially combines with sulfur. Magnesium sulfide, which is insoluble in nickel, forms inclusions that are trapped within the grains during solidification. These inclusions act as nucleation sites of pits during high-rate dissolution. The behavior is similar to that observed during corrosion of steels and stainless steels where pitting is frequently initiated at sulfide inclusions [56–59]. Different hypotheses can be advanced for explaining pit formation at inclusions. Firstly, assuming the inclusion to be a poor conductor, the current density at the boundary between inclusion and metal is increased due to the nonuniform local current distribution. This could lead to preferential attacks around the inclusion and thus pit nucleation. A second possibility is that voids existing between the inclusions and the metal may form the nuclei of pits. The existence of such voids is suggested by the photographs of Figure 5.30a. Voids could have been formed during the cooling step of metal processing because of differences in the thermal expansion coefficient between the inclusions and the metal. Finally, the inclusions themselves may undergo chemical or electrochemical dissolution [58,59]. Such a process, which may preferably proceed near the sulfide-metal boundary, would liberate sulfide ions that are known to be catalysts for anodic nickel dissolution [60].

The grain boundary attack as shown in Figure 5.29b results from local differences in composition between the grain and its adjacent boundary. Since grain boundaries are preferred sites for accumulation of solute atoms, segregation of impurities leads to precipitation of a secondary phase at the grain boundaries. The grain boundary region is, therefore, electrochemically different from the bulk grain microstructure.

In corrosion literature, it is well known that intergranular corrosion occurs when the metal adjacent to grain boundaries dissolve preferentially to grain interiors. The grain boundary region acts as an anode while the grain interior acting as a cathode, thus forming a galvanic cell with a large cathode and a small anode in which rapid corrosion occurs at the anode. By the same analogy, in this case, during anodic dissolution, the grain boundary region is more active compared to the grain interior such that the preferential dissolution of the active region leads to grain boundary attack. Under severe conditions, the preferential concentration of anodic currents in the area adjacent to the grain boundary may lead to excessive dissolution until eventually the grain may be undercut and fall out. Such phenomena lead to anomalous current efficiency for metal dissolution.

Figure 5.31 shows the bright surfaces resulting from the transpassive dissolution of nickel 200 anodes using direct current (compare this to the bright surface resulting from transpassive dissolution using pulsating current, Figure 5.24b). For dissolution with direct current, pits formed on nickel exhibit characteristic tails in the flow direction. The phenomenon has also been observed with other metals [10,14] and it has been shown that these tails may give rise to so-called flow streaks mentioned in the ECM literature [61]. On the other hand, surfaces dissolved in the active mode or during transpassive dissolution with pulsating current did not show pit tails (Figures 5.24b and 5.29).

5.8.2 FLOW STREAK FORMATION

An insight on the role of local hydrodynamic conditions for pit tailing can be obtained from an experiment performed with an artificial pit (0.35 mm deep and 0.38 mm in diameter) using direct current [55]. A charge of 40C was passed at a constant current density of 15 A/cm^2 and a flow rate of 10 m/s. The experimental conditions

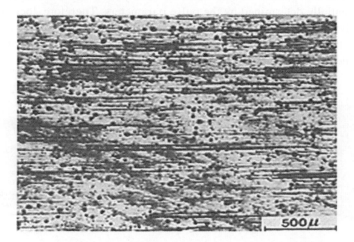

FIGURE 5.31 Optical microphotograph of nickel 200 surface resulting from transpassive dissolution exhibiting pits with characteristic tails in flow direction, $i = 20$ A/cm^2, $u = 10$ m/s, $Q = 18C$. Flow direction: left to right [55].

were chosen such that the applied current density was just slightly above the average limiting current for salt precipitation at which surface brightening sets in [55]. Thus, by observing the surface microscopically, one could distinguish surface regions that dissolve above or below the limiting current. Figure 5.32 shows that a marked tail is indeed formed downstream of the artificial pit. The surface within the tail exhibits etching in the immediate vicinity of the pit while further downstream brightening is observed. The pit bottom also exhibits etching (Figure 5.32b).

Pit tailing and flow streak formation are observed only during transpassive dissolution with the direct current when anodic dissolution is entirely, or partially mass transport controlled. The formation of pit tails as an initial step to flow streak formation is therefore due to a local disturbance of hydrodynamic conditions introduced by the presence of a pit. The experiment with an artificial pit presented in Figure 5.32 provides some qualitative information on this effect. By working just above or below the average limiting current, differences in local mass transport conditions can be visualized by local differences in resulting surface texture. In Figure 5.32, the pit bottom and the initial part of the formed pit tail are etched (dissolution below limiting current) while the remainder of the surface is bright (dissolution at or above the limiting current). It may be concluded that due to larger ohmic drops, the limiting current was not reached at the pit bottom. The situation is different for the initial part of the pit tail where active dissolution is also observed, indicating that the local limiting current density must be significantly higher than on the rest of the surface. Such is possible if flow separation [62] caused by the presence of pits creates small vortices and eddies that increase the local mass transport rate at the downstream end. The experiment described here provides strong evidence that such is indeed the case. The fact that under pulsating current no tails are observed even under transpassive dissolution conditions is also consistent with this interpretation. The effective diffusion layer thickness during dissolution with the pulsating current is much smaller than the steady-state diffusion layer thickness and is essentially determined by the pulse parameters. Therefore, local variations of hydrodynamic conditions have little influence on the dissolution rate. These results further indicate the advantages of employing pulsating current in metal shaping and finishing operations.

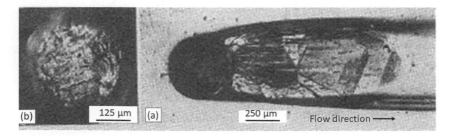

FIGURE 5.32   (a) Tail formed at the downstream end of an artificial pit. (b) Surface microtexture of the pit bottom. Transpassive dissolution conditions: $i = 15$ A/cm^2, $u = 10$ m/s, $Q = 40$C [55].

REFERENCES

1. M. L. McMillan, M. A. LaBoda, *J. Electrochem. Technol.*, 5, 340 (1967).
2. M. L. McMillan, M. A. LaBoda, *J. Electrochem. Technol.*, 5, 346 (1967).
3. J. P. Hoare, M. A. LaBoda, M. L. McMillan, A. J. Wallace, Jr., *J. Electrochem. Soc.*, 116, 199 (1969).
4. K. W. Mao, *J. Electrochem. Soc.*, 118, 1870 (1971).
5. K. W. Mao, *J. Electrochem. Soc.*, 118, 1876 (1971).
6. D. T. Chin, A. J. Wallace, *J. Electrochem. Soc.*, 120, 1487 (1973).
7. K. W. Mao, D. T. Chin, *J. Electrochem. Soc.*, 121, 191 (1974).
8. D. Landolt, R. H. Muller, C. W. Tobais, *J. Electrochem. Soc.*, 116, 1384 (1969).
9. K. Kinoshita, D. Landolt, R. H. Muller, C. W. Tobais, *J. Electrochem. Soc.*, 117, 1246 (1970).
10. D. Landolt, R. H. Muller, C. W. Tobais, *J. Electrochem. Soc.*, 118, 36 (1971).
11. D. Landolt, R. H. Muller, C. W. Tobais, *J. Electrochem. Soc.*, 118, 40 (1971).
12. M. Datta, D. Landolt, *J. Electrochem. Soc.*, 122, 1466 (1975).
13. M. Datta, D. Landolt, *J. Appl. Electrochem.*, 7, 247 (1977).
14. M. Datta, D. Landolt, *Electrochim. Acta*, 25, 1255 (1980).
15. M. Datta, D. Landolt, *Electrochim. Acta*, 25, 1263 (1980).
16. M. Datta, H. J. Mathieu, D. Landolt, *Electrochim. Acta*, 24, 843 (1979).
17. M. Datta, H. J. Mathieu, D. Landolt, *J. Electrochem. Soc.*, 131, 2484 (1984).
18. M. Datta, H. J. Mathieu, D. Landolt, *Appl. Surf. Sci.*, 18, 299 (1984).
19. D. Landolt, *Rev. Sci. Instrum.*, 43, 592 (1972).
20. J. Newman, *Ind. Eng. Chem.*, 60, 12 (1968).
21. P. Van Shaw, L. P. Reiss, T. J. Hanratty, *Am. Inst. Chem. Eng.*, 9, 362 (1953).
22. H. C. Kuo, D. Landolt, *Electrochim. Acta*, 20, 393 (1975).
23. M. Datta, D. Landolt, *Corros. Sci.*, 13, 187 (1973).
24. E. Rosset, M. Datta, D. Landolt, *J. Appl. Electrochem.*, 20, 69 (1990).
25. R. D. Grimm, D. Landolt, *Corros. Sci.*, 36, 1847 (1994).
26. U. Franck, *Werkst. Korros.*, 11, 401 (1960).
27. H. S. Isaacs, *J. Electrochem. Soc.*, 120, 1456 (1973).
28. K. J. Vetter, H.-H. Strehblow, *Proc. of U.R. Evans International Conf. on Localized Corrosion*, R. W. Staehle ed., p. 240, NACE, Houston, (1974).
29. R. Alkire, D. Ernsberger, T. R. Beck, *J. Electrochem. Soc.*, 125, 1382 (1978).
30. T. R. Beck, R. C. Alkire, *J. Electrochem. Soc.*, 126, 1662 (1979).
31. J. Galvele, *Corros. Sci.*, 21, 551 (1981).
32. T. Li, J. R. Scully, G. S. Frankel, *J. Electrochem. Soc.*, 165, C762 (2018).
33. N. J. Laycock, M. H. Moayed, R. C. Newman, *J. Electrochem. Soc.*, 145, 2622 (1998).
34. T. Li, J. R. Scully, G. S. Frankel, *J. Electrochem. Soc.*, 166 (6), C115 (2019).
35. T. P. Hoar, *Corros. Sci.*, 7, 341 (1967).
36. K. Kojima, C. W. Tobias, *J. Electrochem. Soc.*, 120, 1026 (1973).
37. D. Landolt, *Electrochim. Acta*, 32, 1 (1987).
38. L. Ponto, M. Datta, D. Landolt, *Surf. Coat. Tech.*, 30, 265 (1987).
39. M. Datta, V. Vercruyesse, *J. Electrochem. Soc.*, 137, 3016 (1990).
40. M. Datta, L. F. Vega, L. T. Romankiw, P. Duby, *Electrochim. Acta*, 37, 2469 (1992).
41. K. J. Vetter, H.-H. Strehblow, *Ber. Bunsenges. Phys. Chem.*, 74, 1024 (1970).
42. H.-H. Strehblow, J. Wenners, *Electrochim. Acta*, 22, 421 (1977).
43. J. W. Tester, H. S. Isaacs, *J. Electrochem. Soc.*, 122, 1438 (1975).
44. F. Hunkeler, A. Krolikowski, H. Bohni, *Electrochim. Acta*, 32, 615 (1987).
45. T. R. Beck, *J. Electrochim. Acta.*, 30, 725 (1985).
46. T. R. Beck, *Chem. Eng. Commun.*, 38, 393 (1985).
47. C. Clerc, D. Landolt, *Electrochim. Acta*, 33, 859 (1988).

48. R. D. Grimm, A. C. West, D. Landolt, *J. Electrochem. Soc.*, 139, 1622 (1992).
49. M. Datta, D. Landolt, *Electrochim. Acta*, 26, 899 (1981).
50. M. Datta, D. Landolt, *Electrochim. Acta*, 27, 385 (1982).
51. D. Landolt, R. Acosta, R. H. Muller, C. W. Tobias, *J. Electrochem. Soc.*, 117, 839 (1970).
52. M. Datta, D. Landolt, *J. Appl. Electrochem.*, 13, 795 (1983).
53. M. Datta, *IBM J. Res. Dev.*, 42 (5), 655 (1998).
54. L. E. Samuels, *Metals Engineering: A Technical Guide*, pp. 5, 6, 238, ASM International, Materials Park, Ohio, (1988).
55. M. Datta, D. Landolt, *J. Electrochem. Soc.*, 129, 1889 (1982).
56. Z. Szklarska-Smialowska, *Corrosion*, 27, 223 (1971).
57. G. Wranglen, *Corros. Sci.*, 9, 585 (1969).
58. P. Sury, *Corros. Sci.*, 16, 879 (1976).
59. G. S. Eklund, *J. Electrochem. Soc.*, 121, 467 (1974).
60. P. Marcus, J. Oudar, I. Olefjord, *Mat. Sci. Eng.*, 42, 191 (1980).
61. J. A. McGeough, *Principles of Electrochemical Machining*, Chapman and Hall, New York, (1974).
62. P. K. Chang, *Separation of Flow*, Pergamon Press, Elmsford, NY, (1970).

6 High-Rate Anodic Dissolution of Ti, W, and Their Carbides

6.1 INTRODUCTION

Titanium, tungsten, and their carbides possess some unique properties because of which these materials have found many industrial applications. Titanium and titanium-based alloys find special applications in the aerospace and biomedical industries. These materials are used as prostheses and implants due to their excellent corrosion resistance and biocompatibility. Also, titanium aluminide (TiAl), an intermetallic compound of light metals titanium and aluminum, is used in jet engines due to its lightweight, high strength-to-density ratio, and its excellent high-temperature performance. Tungsten is the most important metal for thermo-emission applications because of its high thermal and chemical stability. Furthermore, tungsten is a very stiff material with very high Young's modulus (411 GPa), thus allowing the fabrication of very small size products. Tungsten is thus used in the fabrication of precision probes for surface analytical tools. Metal carbides and borides are used in the aerospace industry because of their very high melting points and extreme hardness. Uses of TiC include tools for metal cutting and dies for drawing wire. Cemented carbides account for the majority of cutting tool material. Sintered carbides with high elastic coefficient and high elastic limit are suitable material for long-lasting, high precision forging dies. However, it is difficult to machine these materials using conventional machining methods. Electrical discharge machining (EDM) is also not applicable since the machining speed is slow and it makes microcracks on the machined surface. Electrochemical machining (ECM) is perhaps the perfect technique for machining these hard materials since the process is independent of the hardness of the workpiece. This chapter focuses on the discussion of the anodic dissolution behavior of these materials under ECM conditions.

6.2 HIGH-RATE ANODIC DISSOLUTION OF TITANIUM

Titanium is one of the most abundant metals in the earth's crust. It is lightweight, strong, extremely resistant to corrosion, and is nontoxic in the human body. Titanium is about 45% lighter than steel and while it is 60% heavier than aluminum, it is twice as strong. The melting point of titanium is 1,670°C (melting point of steel is 1,450°C–1,520°C and that of Al is 660°C). The coefficient of thermal expansion of titanium is relatively low at about one third that of aluminum. The metal has a very low electrical and thermal conductivity and is paramagnetic. It exists in two crystal structures: below 883°C (1,621°F), hexagonal close-packed (alpha); above 883°C, body-centered cubic (beta). Titanium can be easily alloyed with aluminum, molybdenum, iron, manganese, and many other metals.

Due to the above-described properties, titanium and titanium-based alloys are the materials of choice in many areas of application including aerospace, biomedical, electronics, and photovoltaics. Because of their high strength-to-density ratio and their excellent high-temperature performance, titanium alloys are used in the aerospace industry where they provide weight savings and improved operational efficiency. Another key application area is in the biomedical industry where due to their outstanding biocompatibility and corrosion resistance properties, titanium and titanium alloys are used for various applications such as artificial hip and knee joints, dental prosthetics, vascular stents, and heart valves [1]. Also, enhancement of bone formation is a desired feature of a metallic implant developed through adequate surface treatments to obtain proper osseointegration [1]. The important criteria in most of these applications are that the fabrication processes involving machining, shaping, and finishing operations should produce components that are free from machining process-induced problems such as surface defects, thermally or mechanically stressed surfaces, or hardened surface layers. ECM meets these criteria since in this process there is no physical contact between the tool and the workpiece. However, ECM of titanium is challenging due to its great affinity for oxygen which leads to the formation of an adhering, dense oxide layer on the surface when it is exposed to oxidizing media. Therefore, the anodic behavior of titanium in different electrolytes has been investigated to understand the film breakdown phenomena leading to pitting and to high-rate dissolution under ECM conditions. The results of some of these investigations are discussed in the following paragraphs.

6.2.1 ANODIC BEHAVIOR OF TI

The corrosion resistance of titanium is known to result from the existence of a natural and cohesive passive film, with a thickness of typically 5 nm, which is formed spontaneously when it is exposed to air or water. The reasons why titanium becomes irreversibly corroded depend on the ability of the passive film to reform (self-healing) sufficiently after its eventual breakdown. Information about the suitability of a specific electrolyte for ECM of titanium can be obtained from its anodic behavior data.

Trompette et al. [2] studied the anodic polarization behavior of titanium in different electrolytes. The electrolytes used included 0.1 m/L solutions of sodium fluoride (NaF), sodium chloride (NaCl), sodium bromide (NaBr), sodium iodide (NaI), sodium nitrate ($NaNO_3$), sodium thiocyanate (NaSCN), and sodium perchlorate ($NaClO_4$). Figure 6.1 shows that in all these electrolytes, the anodic polarization curves for titanium exhibit a sharp increase of the current density, except for the thiocyanate ion (SCN), for which the increased current is not observed even for potentials above 110 V. An increase in anodic polarization of titanium leads to several anodic reactions which may take place simultaneously: the formation and repair of an anodic film of TiO_2, water oxidation leading to oxygen evolution, and the onset of the breakdown at a critical potential. In some cases, a slight increase in current density is followed by a decrease of the current density due to the self-healing of the

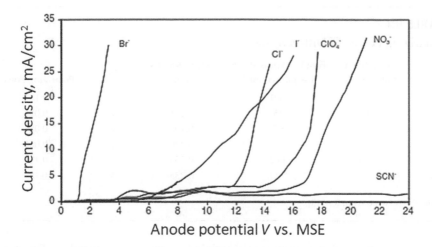

FIGURE 6.1 Anodic polarization curves for titanium showing the variation of current density as a function of potential in different electrolytes [2].

breakdown sites to form the passive film on the underlying metal. In Figure 6.1, the irreversible breakdown of the passive film is evidenced by the abrupt increase of the current density. Bromide and iodide ions were found to be more efficient in the film breakdown process than the chloride ions which requires a greater potential for the breakdown of the passive film. Fluoride solution, on the other hand, required a much higher potential to be applied, around 90 V, to observe an apparent breakdown leading to sparking. Results of Figure 6.1 shows that oxidizing anions such as nitrate and perchlorate ions are also able to cause the rupture of the passive film on titanium at slightly higher potentials.

The above results agree well with the results of Dugdale and Cotton [3] who reported lower breakdown voltages in bromide, chloride, and iodide solutions and higher breakdown voltage values for fluoride solutions. On the other hand, sulfate and phosphate solutions could withstand very high voltages until dielectric breakdown at high voltages. Addition of sulfte ions to the halide ions inhibits the breakdown of the oxide film and the titanium behaves as if it were in pure sulfate solution, where a dielectric breakdown is observed at between 80 and 100 V. Based on their data, Dugdale and Cotton [3] presented a model according to which the oxide film formation is encouraged by an electrostatic effect caused by the adsorbed ions at the oxide/solution interface. The extent of this electrostatic effect is governed by the charge/area ratio of adsorbed anion termed as "polarizing power." Table 6.1 gives the polarizing power of different anions studied. According to these authors, if the polarizing power is high enough the passive oxide film will remain intact, but if it falls below a critical value pitting corrosion will set in. Thus bromide, iodide, and chloride ions with low polarizing powers show breakdown at low potentials, whereas fluoride, sulfate, and phosphate do not.

Beck [4] investigated the occurrence of pitting at the edges and on the circumference of holes in titanium foil in bromide solutions under potentiostatic conditions.

TABLE 6.1
Polarizing Power of Different Ions

Ion	Breakdown Voltage (V)	Film Thickness at the Breakdown (nm)	Anion Radius (nm)	Charge	Polarizing Power
F^-	90	160	0.133	1	0.181
Cl^-	14	25	0.181	1	0.097
Br^-	3.5	5.3	0.197	1	0.084
I^-	4.2	7.8	0.220	1	0.066
SO_4^{2-}	>100	110	0.235	2	0.115
PO_4^{3-}	>120	200	0.235	3	0.180

Source: Adapted from Ref. [3].

The valence of dissolution, measured by weight loss, was approximately 4. The effect of potential, temperature, concentration, pH, and solution flow on current density and pitting potential was determined. Some of the results are summarized in Figure 6.2. Figure 6.2a shows steady-state pitting potential as a function of the concentration of HBr and KBr solutions. The data show a decreasing trend in pitting potential with increased anion concentration. The average slope is −110 mV per decade of concentration on the semi-logarithmic scale and independent of bulk pH in neutral to acid solutions. The effect of temperature on the steady-state pitting potential in 0.6 M bromide solutions is shown in Figure 6.2b. At room temperature and above, the pitting potential is constant at about 0.9 V. At below room temperatures, the

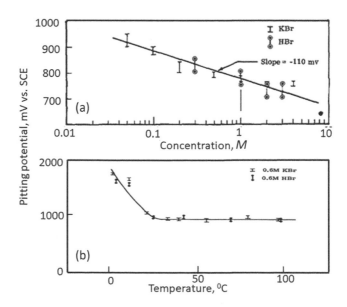

FIGURE 6.2   (a) Effect of bromide concentration on steady-state pitting potential of titanium and (b) effect of temperature on steady-state pitting potential in KBr and HBr solutions [4].

pitting potential rises sharply. This rise in the pitting potential is consistent with the observation that pitting corrosion was not observed in the bromide solution below 15°C. The pitting current decreased with increased flow and the effect was greater with the more dilute bromide solution. A qualitative model combining a salt film on the metal surface with events in the electrolyte diffusion layer was presented to interpret the observed results. The decreased current density with the increased flow was attributed to a decrease in the concentration of hydrolysis products which may be necessary to maintain the salt film and prevent passivation. In another study, Beck [5] conducted experiments with one-dimensional pit in titanium to determine the role of mass transport and salt films on pitting phenomena. The results showed that after a small depth of corrosion, a uniform current density was obtained characteristic of one-dimensional mass transport. The pitting current density was found to be mass-transport limited thus suggesting that salt film forms on the metal surface to absorb most of the potential applied.

Burstein et al. [6] investigated the localized instability of passive titanium in chloride solution. Measurement of the passivating current on titanium microelectrodes in dilute HCl solution showed the evidence of microscopic instability corresponding to microscopic breakdown at potentials well below the pitting potential. The passive surface produced a sequence of sharp anodic current spikes. These events were attributed to microscopic depassivation processes and formed the nuclei of corrosion pits. Most of these nucleated events did not propagate as pits but repassivated immediately through regrowth of the oxide film. A few underwent incomplete repassivation and propagated briefly as metastable pits before they too, repassivated. They proposed that chloride ions migrated through the oxide and reacted at the metal-oxide interface, where volume expansion caused a microscopic breakdown of the oxide film.

6.2.2 HIGH-RATE ANODIC DISSOLUTION OF TITANIUM AND TITANIUM ALLOYS UNDER ECM CONDITIONS

It is clear from the above discussions that under ECM conditions, high-rate anodic dissolution of titanium takes place in the transpassive state. Mathieu and Landolt [7–9] investigated the high-rate anodic dissolution behavior of titanium in a flow channel cell using different electrolytes that are commonly used in ECM. Figure 6.3a shows the current density vs. applied cell voltage curves for titanium dissolution in 5 M NaCl, 5 M NaBr, 5 M $NaClO_3$, 5 M $NaClO_4$, 5 M $NaNO_3$, and 5 M NaOH electrolytes at a flow velocity of 10 m/s. Depending on the nature of the electrolyte, the passive films formed on titanium break down at different voltages leading to an increase in current density. Except for $NaClO_3$ electrolyte, the current density shows a linear relationship with the cell voltage, the slope being dependent on the nature of the electrolyte anion. It is not possible to draw any kinetic conclusions from the presented data since the IR drop dominates the current-voltage relationship. However, high anodic current densities can be achieved for titanium in most of these electrolytes except for chlorates. Figure 6.3b shows the apparent valence of dissolution determined by weight loss measurements and by applying Faraday's law. In both NaBr and $NaNO_3$, titanium dissolves with a constant valence of four, independent of cell voltage, while a lower apparent valence is observed in NaCl, $NaClO_4$,

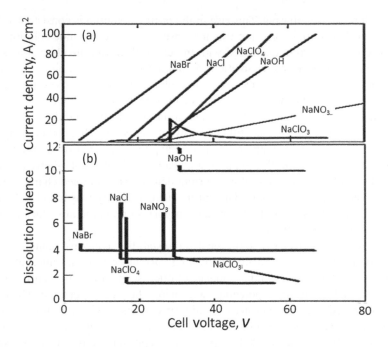

FIGURE 6.3 (a) Current density as a function of cell voltage for dissolution of titanium dissolution in different ECM electrolytes measured in a flow channel cell at a flow rate of 10 m/s. (b) The apparent valence of dissolution vs. cell voltage. Electrolytes: 5 M NaCl, 5 M NaBr, 5 M NaClO$_3$, 5 M NaClO$_4$, 5 M NaNO$_3$, and 5 M NaOH [9].

and NaClO$_3$. The reason for a lower valence, close to 3.5, for NaCl solution is not known. In the case of NaClO$_3$ and NaClO$_4$, the involvement of chemical reactions of the electrolyte could explain the low values of apparent valence. The high values of apparent valence observed in NaOH indicate that excessive oxygen evolution took place concurrently with dissolution. These data indicate that commonly used ECM electrolytes are applicable for high-rate anodic dissolution of titanium.

Madore and Landolt [10] investigated the high-rate dissolution of titanium in 5 M NaBr electrolyte using a rotating disk electrode. In this electrolyte, titanium is known to dissolve with a valence of 4 [7–9]. Figure 6.4a shows the anodic polarization curve measured at 900 rpm. A current plateau is observed, the value of which increased with the increasing rotation rate of the electrode. In Figure 6.4b, the reciprocal of the measured plateau current density is plotted against the reciprocal of the square root of the electrode rotation rate. For mass-transport-limited reaction at a rotating disk electrode, such a plot should yield a straight line going through the origin. The result of Figure 6.4b suggests that the dissolution reaction is partially kinetically, and partially mass transport controlled. The resulting surface morphology after dissolution depended on the applied potential. Dissolution at 2 V yielded a rough surface with randomly distributed pits. A smooth surface was obtained at potentials above 5 V.

Bannard [11] investigated the anodic dissolution behavior of Ti and Ti alloys in potassium bromide-based electrolytes. Two types of Ti materials were used:

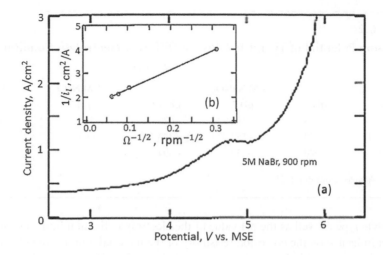

FIGURE 6.4 (a) Potentiodynamic polarization curve for titanium in 5 M NaBr, and (b) reciprocal of the plateau current density plotted against the reciprocal of the square root of the electrode rotation rate [10].

single-phase, commercially pure Ti (99.5%) and Ti2.5Cu alloys, and two-phase Ti6Al4V and Ti4Al4Mo2Sn0.5Si alloys. The different electrolytes used consisted of KBr solutions of different concentrations (1–4 M) and mixtures of KBr with NaCl, KF, and Na_2SO_4. The experiments were conducted to determine the dissolution efficiency and stray current attack, which is defined as that portion of the dissolution taking place on the surface which is not directly opposite to the tool and is unwanted. All the alloys dissolved in the transpassive region of the polarization curve but produced etched surfaces. Evidence was presented to show that the dissolution kinetics is modified by acid generation at the anode. It was found that these materials in general dissolved at efficiencies somewhat greater than 100% based on the four-valent dissolution reaction, although at lower current densities the efficiency was a little lower than 100%. In the case of the multiphase alloys, a little lower-valent dissolution occurred, and some differential dissolution and undermining were also evident. Using sodium chloride as an additive to the potassium bromide solution, slight improvements in the stray-current attack properties of the electrolyte was observed. Sodium chloride-containing bromide solution also yielded a better surface finish.

Baehre et al. [12] investigated the electrochemical dissolution behavior of titanium alloys in different types of electrolytes with varying pH values. The electrolytes consisted of 1 M $NaNO_3$, 1 M NaCl, and 1 M KBr. The materials investigated consisted of commercially available Ti (Ti grade 2), and two types of titanium–aluminum alloys: Ti90Al6V4 (Ti grade 5) and Ti60Al40 (γ-TiAl). The experiments were performed in a microflow cell [13] using different electrochemical methods like cyclic voltammetry, linear sweep voltammetry, and chronoamperometry. The electrolyte flow in the microcell was adjusted to 1.5 mL/min. Corrosion potentials for each material were determined from cyclo-voltammograms designated by the potential that exists at the minimum of the curves. Table 6.2 shows that in addition to the

TABLE 6.2

Corrosion Potential of Ti and Ti Alloys in Different Electrolyte Conditions

| | E_{corr} (mV) | | | | |
| | 1 M NaNO$_3$ | | | 1 M NaCl | 1 M KBr |
Material	pH 1	pH 7	pH 12	pH 7	pH 7
Ti	−428	−491	−672	−546	−638
Ti90Al6V4	−463	−557	−743	−669	−727
Ti60Al40	−546	−650	−894	−799	−816

Source: Adapted from Ref. [12].

electrolyte type as well as the pH value of the electrolyte, the titanium content has a distinct influence on the corrosion resistivity of the material. For a given electrolyte, the E_{corr} increases with the titanium content, indicating that the electrochemical dissolution should be relatively easy for alloys with lower titanium content. Comparison of different electrolytes at pH 7 shows that the order of E_{corr} is KBr < NaCl < NaNO$_3$. Based on this data, it is evident that halide containing electrolytes are suitable for the ECM of the investigated titanium alloys.

TB6 titanium alloy (2.7% Al, 8.76% V, 1.79% Fe) is widely used in aerospace applications due to its high tensile strength and toughness. Liu and Qu [14] investigated the electrochemical machinability of TB6 titanium alloy in NaNO$_3$ electrolytes. Linear sweep voltammetry and cyclic voltammetry measurements were conducted in NaNO$_3$ solutions. The polarization curves for the TB6 titanium alloy in 20% NaNO$_3$ solutions at different temperatures indicated that the breakdown voltage of the passive film decreases as the temperature increases. With an increase in electrolyte temperature from 20°C to 50°C, the breakdown voltage decreased from 10.7 to 8 V in 20% NaNO$_3$ solution. The polarization curves in NaNO$_3$ solutions of different concentrations at 40°C showed that the growth rate of the current density increased with an increase of the electrolyte concentration. An increase in electrolyte concentration led to a decrease in the breakdown voltage; it decreased from 9.6 V in 15% NaNO$_3$ to 8.4 V in the 25% NaNO$_3$ electrolyte. The authors also conducted electrochemical milling experiments in ECM setup in which the tool cathode could be moved freely in XYZ 3D space. The electrolyte used was a 20% NaNO$_3$ solution at 40°C, a machining gap of 0.5 mm, and the constant pressure of 0.4 MPa. The surface morphology resulting from dissolution depended on the applied current density and the dissolution time. Figure 6.5 shows the morphology of TB6 Ti alloy surfaces dissolved for 20 seconds at two different current densities of 80 and 140 A/cm^2 producing rough and relatively flat bright surfaces, respectively. XPS analysis of dissolved surfaces showed TiO$_2$ to be main the component with the additional presence of TiOH on rough surfaces at low current densities. TiOH was not detected on a flat bright surface dissolved at 140 A/cm^2. Grooves and flat surfaces were successfully produced by electrochemical milling of the surface of TB6 titanium alloy samples using NaNO$_3$ solution. Moreover, at high current density, the machined surfaces of the grooves and flat surfaces were uniform and there was less stray corrosion at their edges.

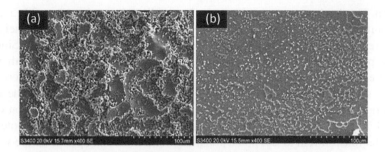

FIGURE 6.5 Surfaces resulting from high-rate dissolution of TB6 Ti alloy for 20 seconds at (a) 80 A/cm^2 and (b) 140 A/cm^2 in 20% NaNO$_3$ at 40°C [14].

6.3 HIGH-RATE ANODIC DISSOLUTION OF TUNGSTEN

Tungsten has the highest melting point of all metals (3,422°C ± 15°C). It is among the heaviest of metals, with its density being 19.25 g/cm^3 while its coefficient of thermal expansion is the lowest of all metals. It has good thermal and electrical conductivity, and high tensile strength and hardness. These outstanding physical properties have led to the widespread use of tungsten in different applications such as cutting tool manufacturing and filament material. In electronics applications, tungsten is used gate and interconnection materials. Tungsten usually contains small concentrations of carbon and oxygen, which give tungsten metal its considerable hardness and brittleness. While these properties may be beneficial for some applications, its hardness makes it very difficult to mechanically machine. ECM is a suitable alternative for tungsten shaping. In the following, some of the investigations on the anodic dissolution behavior of tungsten are discussed.

6.3.1 ANODIC BEHAVIOR OF TUNGSTEN

Investigations of the anodic behavior of tungsten stem from two different technological interests: anodic film formation in acid media and anodic dissolution of tungsten in alkaline solutions. Most of the studies on the electrochemical behavior of tungsten in acid solutions are related to the phenomenon of electrochromism in WO$_3$ films because of their potential use in flat panel display devices and in photoelectrolysis of water for solar energy. Electrochemistry of tungsten in acid solutions also finds application in chemical mechanical polishing for semiconductor device manufacturing. The electrochemical dissolution of tungsten in alkaline solutions has been investigated in connection with ECM, electrochemical polishing, etching, and recycling of tungsten.

Heumann and Stolica [15] studied the anodic dissolution of tungsten in sulfuric acid, citric acid, sodium phosphate, sodium carbonate, and sodium borate solutions in the pH range between 0.3 and 12. In these electrolytes, the electrochemical oxidation of tungsten to hexavalent state can take place at relatively low potentials depending on the pH, but an appreciable dissolution of W takes place only when the anodically formed WO$_3$ (or its hydrate) can dissolve readily in the electrolyte solution. This leads to a current plateau in the anodic polarization curve. The increase of solubility of the WO_3 with increasing pH shows up as an increase in the heights of the current plateau. The dissolution behavior of WO_2 was also examined. Since the current-potential curves of WO_2 are very similar to those of W, it follows that WO_3 is an intermediate in the oxidation of tungsten to tungstate, and the rate-determining step lies somewhere in between. Heumann and Stolica [16] further studied the anodic dissolution of W in NaOH solutions of pH 13–14. The anodic polarization curves showed Tafel behavior at low potentials and a current plateau at higher potentials. The Tafel lines shifted to less noble potentials with increasing pH while the value of the plateau current increased with increasing pH. The plateau current is controlled by the WO_3 film dissolution rate, which is higher in high pH solution. The valence of dissolution was found to be 6.0 as determined by weight loss measurements. Similar results have been reported by Armstrong et al. [17] who investigated the anodic dissolution of tungsten in sodium hydroxide, carbonate, and phosphate solutions over the pH range of 9–14. The current-potential curves showed a Tafel region followed by a limiting current which is dependent on electrolyte pH and stirring. The results have been interpreted by the formation of a WO_3 or a hydrated WO_3 layer on the electrode surface.

Anik and Osseo-Asare [18] examined the anodic behavior of tungsten in a pH range from 0.5 to 13.5, using a rotating disk electrode in H_3PO_4/KOH-buffered solutions. The authors identified five distinct pH regimes (A, B, C, D, E) and the corresponding reaction mechanisms. In the regime A, below pH 1, the main dissolution pathway for the oxide was found to be H^+ assisted dissolution with no diffusion effect. As the pH increased between 1 and 4, in the regime B, H^+ assisted dissolution diminished and the steady-state anodic current vs. pH curve went through a minimum at pH 2.6. At around pH 2.6, dissolution was mainly H_2O assisted. No diffusion effect was observed. In the regime C, typical OH^- assisted dissolution was observed between pH 4.5 and 6.5. The anodic current was under mixed control of reaction kinetics and WO_4^{2-} diffusion. In the regime D between pH 6.5 and 8, the dissolution of the hydrated oxide phase occurred, and at around pH 10.5, OH^- dependence of the anodic current commenced. In the regime E above pH 12.5, the reaction order for OH^- became one. In this highly alkaline regime, the dissolution was controlled by the slow diffusion of OH^- ions to the W surface.

6.3.2 High-Rate Anodic Dissolution of Tungsten under ECM Conditions

Investigating the high-rate anodic dissolution of tungsten, Jaw et al. [19,20] examined several commonly used ECM electrolytes including neutral salt solutions of sodium nitrate and sodium chloride and also sodium hydroxide solution. The anodic polarization curves for tungsten in these electrolytes showed that the dissolution currents in sodium nitrate and chloride are negligibly small with the tungsten surface remaining unchanged (unattacked) even after the passage of 10 V. A layer of tungsten oxide forms on the surface during anodic polarization in these solutions which inhibits the dissolution process. On the other hand, dissolution current for tungsten was significantly higher in sodium hydroxide solution and the surface became smoother and brighter after the full potential scan up to about 10 V.

Figure 6.6 shows the potentiostatic polarization curves for anodic dissolution of tungsten in 1 M NaOH at 30°C using a scanning rate of 1 V/s and at different rotation speeds of the rotating disk electrode. After a steady rise of current, a limiting current is observed which extends over a wide range of potential. The limiting current density increased with increasing rotation speed. Figure 6.7 shows the limiting current density as a function of the square root of the rotation speed. The straight-line curve passing through the origin follows the Levich equation indicating the dissolution process to be mass transport controlled. Weight loss data were obtained to

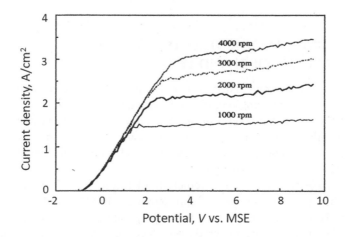

FIGURE 6.6 Anodic polarization curves for tungsten in 1 M NaOH at different rotation speeds [19,20].

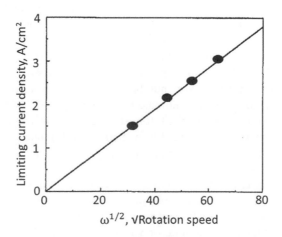

FIGURE 6.7 Limiting current density for anodic dissolution of tungsten in 1 M NaOH as a function of root square of the rotation speed [19,20].

determine the valence of tungsten dissolution at 1,000 rpm in two different regions: below the limiting current at 1 V and at the limiting current at 4 V. In both cases, the current efficiency was found to be 100% based on the value of dissolution valence to be 6. The SEM photographs of surfaces showed that the surface dissolved at the limiting current was shiny while the surface dissolved below the limiting current was rough. These results confirm the suitability of NaOH as the electrolyte for the high-rate dissolution of tungsten. Indeed, sodium hydroxide is the choice solution in ECM and micromachining of tungsten and has been successfully employed in the fabrication of tungsten tip probes with controllable tip profiles for scanning tunneling microscopy [21–25] and in the fabrication tungsten tools and microtools for electro-discharge machining [26,27].

6.4 HIGH-RATE ANODIC DISSOLUTION OF CARBIDES OF Ti AND W

Carbides are carbon compounded with a nonmetal (such as calcium, silicon, or boron) or metal (such as tungsten, titanium, tantalum, cobalt, or vanadium). Metal carbides are extremely hard and resistant to wear, corrosion and heat, making them excellent candidates for coatings for drills and other tools. They possess other valuable properties in combination with toughness, such as electrical conductivity, low thermal expansion, and abrasiveness.

Titanium carbide, TiC, is an extremely hard (Mohs 9–9.5) refractory ceramic material. It has an elastic modulus of approximately 400 GPa and a shear modulus of 188 GPa. Titanium carbide is used in the preparation of cermets, which are frequently used to machine steel materials at high cutting speed. TiC is also used as an abrasion-resistant surface coating on metal parts, and also used as a heat shield coating. Addition of 6%–30% of TiC to tungsten carbide-cobalt increases the resistance to wear, corrosion, and oxidation of the WC-Co material.

Tungsten carbide, WC, contains equal parts of tungsten and carbon atoms. In its most basic form, tungsten carbide is a fine gray powder, but it can be pressed and formed into shapes for use in industrial machinery, cutting tools, abrasives, armor-piercing rounds, surgical tools, and jewelry. Tungsten carbide has a high melting point at 2,870°C (5,200°F), a boiling point of 6,000°C (10,830°F). Tungsten carbide is extremely hard (9 Mohs, Vickers number of 2,600). It has Young's modulus of approximately 530–700 GPa, and ultimate tensile strength of 344 MPa. With a low electrical resistivity of ($\sim 2 \times 10^{-7}$ Ωm), tungsten carbide's resistivity is comparable with that of some metals. WC is readily wetted by both molten cobalt and nickel. Indeed, the most common use of WC as an engineering material is in the form of WC-Co composite.

Hardmetal is normally referred to as a composite material with hard particles of Tungsten Carbide embedded in a softer matrix of metallic cobalt. In general, however, hardmetals include several are other metallic carbides. They consist of finely divided tungsten, titanium, tantalum, or vanadium mixed with carbon and molten cobalt or nickel. They are also known as cemented carbides or sintered carbides. Cemented carbides combine the high hardness and strength of metallic carbides

(WC, TiC, TaC) or carbonitride (e.g., TiCN) with the toughness and plasticity of a metallic alloy binder (Co, Ni, Fe), in which the hard particles are evenly distributed to form a *metallic* composite. Significant increases in toughness are achieved by a higher amount of metallic binder but at the expense of hardness. Sintered carbides have a high elastic coefficient and a high elastic limit. Therefore, even when they are subjected to a large compressive force, deformation and wear are small. These qualities make them a suitable material for a high-precision forging die that will work for a long time. However, it is difficult to machine with a usual machine tool because of its hardness. EDM which is often used is problematic in that the machining speed is slow and that it makes microcracks on the machined surface. While ECM is a possible alternative, the process is still in the investigation stage. In the following section, the anodic behavior of these materials and some of the issues related to their high-rate dissolution under ECM conditions are discussed.

6.4.1 Anodic Behavior of Ti and W Carbides

Anodic behavior of titanium carbide: Investigation of anodic polarization behavior and identification of dissolution products of titanium carbide in 2 N H_2SO_4 was conducted by Cowling and Hintermann [28]. They measured the potentiostatic anodic polarization curve for TiC in 2 N H_2SO_4 which showed that the anodic dissolution of titanium carbide begins at about 0.8 V (vs. SHE). The species formed during dissolution can be speculated based on the equilibrium potential-pH diagram for the system titanium-H_2O [29], which indicates that the most stable species at pH $= 0$ and at potentials in the region of 1 V is the tetravalent ion TiO^{2+}; and the equilibrium potential for the redox couple Ti^{3+}/TiO^{2+} at pH $= 0$ is about 0.46 V. Since the potential of the titanium carbide is above 0.8 V during its dissolution, it is evident that any Ti^{3+} formed would be quickly oxidized to TiO^{2+}. The net result is its oxidation to the tetravalent state. Passivation of the electrode, which occurs between 1.2 and 1.7 V when it is polarized potentiostatically, is due to Ti(IV) oxide. At potentials above 1.75 V, a sharp rise in current occurs indicating the dissolution of TiC. During the initial dissolution of the carbide, both CO and CO_2 were detected. Based on these data they proposed that the dissolution of TiC proceeds according to the following reactions:

$$TiC + 3H_2O = TiO^{2+} + CO_2 + 6H^+ + 8e^- \tag{6.1}$$

and

$$TiC + 2H_2O = TiO^{2+} + CO + 4H^+ + 6e^- \tag{6.2}$$

In a subsequent study, Cowling and Hintermann [30] examined the anodic dissolution of titanium carbide in 2 N H_2SO_4 by coulometric experiments and by the application of a rotating ring-disk electrode (RRDE). The RRDE technique showed that the dissolution of TiC proceeds through the intermediate formation of Ti(III) ions, which undergo immediate oxidation to Ti(IV) at the electrode surface. Coulometrically derived data indicated the dissolution valence in the initial stage of dissolution to be around 6.5.

Corrosion behavior of hardmetals: Technical applications of hardmetals are limited by their poor corrosion resistance. Considering the pH of the environment, the two major alloying components, Co and W have opposite behaviors [29]. In alkaline solutions, Co shows stable passivity, whereas W readily dissolves; the situation is opposite in acidic electrolytes. Therefore, hardmetals typically fail by corrosion of the less resistant phase, which is defined by the environment. Since WC is significantly nobler than Co, galvanic interactions between the phases of different nobility take place. Additions of alloying elements such as Cr_3C_2 and Ni in the Co binder phase can generally enhance the corrosion resistance [31]. Moreover, small additions of TiC or TaC were found to significantly improve the corrosion performance [32]. Diffusion of W and C into the binder phase during the sintering procedure [3] also affects the corrosion resistance [33,34]. The addition of small amounts of Cr_3C_2 and VC, inhibits the growth of the WC grains [32,35], thus improving its mechanical properties. A decrease in the grain size of WC can be expected to show also an influence on the corrosion behavior. A small-grain size increases the surface area of interfaces and could enhance the formation of galvanic couples on the material surface. Both these factors are expected to lead to an impairment of the corrosion behavior. Human and Exner [36,37] found almost no influence of the grain size on the corrosion behavior in acids, while Tomlinson and Ayerst [38] reported an increase of the passive current densities by increasing grain size in acidic solutions. Kellner et al. [39] investigated the corrosion behavior of hardmetals with different grain sizes in alkaline solutions. They observed that the smaller the grain size, the higher was the corrosion resistance. Analysis by angle-resolved Auger electron spectroscopy revealed that significant amounts of W and C diffuse into the Co binder matrix during the sintering process and that the W and C concentration in the Co phase increases with decreasing grain size. They attributed the higher corrosion resistance of the small-grained hardmetals to the higher amount of fcc Co, which has better corrosion stability than hcp Co.

Anodic behavior of WC-6% Co: The WC-Co alloys are dispersions of micron-size WC particles in a binder phase. These alloys are prepared by a liquid phase sintering process, during which some WC grains dissolve in Co, producing a Co-W-C ternary binder alloy [40]. These alloys are used in many applications and under a variety of corrosive environments. Examples of some of these environments are geothermal brines in drilling applications, acid extracts from wood in the cutting of lumber, and lubricants in metal finishing industries. Because cemented carbide alloys are nonhomogeneous, the different components would be expected to dissolve at different rates. The anodic behavior of WC-6% Co alloy, WC, and Co in phosphoric acid solution was studied by Ghandehari [41]. Figure 6.8 shows the potentiostatic, steadystate polarization curves for anodic dissolution of WC-6% Co, WC, and Co. The open-circuit potential for WC-6% Co was typically about −300 mV which is between the open-circuit potential for Co (−500 mV) and WC (0–200 mV). The WC+6% Co electrode shows three distinctive regions of active, passive, and transpassive dissolution. In the active region, the logarithm of the current density increases approximately linearly with the potential. The solution analyses and surface morphology lead to the conclusion that the Co-W binder phase selectively dissolves in the active region, leaving behind the WC grains. Because most of the current is carried by

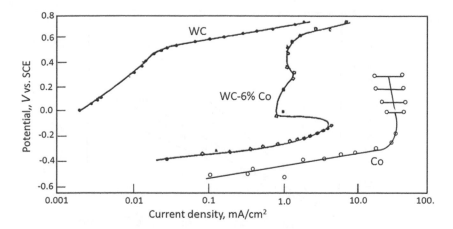

FIGURE 6.8   Steady-state anodic polarization curves for WC, Co, and WC-6% in 1.2 M H_3PO_4 solution [41].

cobalt dissolution, it is not surprising that the polarization curve for the composite is qualitatively the same as the cobalt polarization curve. In the passive region, the binder dissolution undergoes an abrupt decrease and then attains a somewhat constant value. Somewhat similar behavior is obtained for cobalt. The composition of the passive film could conceivably be caused by the formation of either tungsten oxide or cobalt phosphate. The current appears to be limited by the dissolution of a film of one of these substances in the passive region. As the potential is increased further, current increases corresponding to transpassive dissolution.

6.4.2   HIGH-RATE ANODIC DISSOLUTION OF TiC, WC, AND HARDMETALS UNDER ECM CONDITIONS

The feasibility of the ECM of TiC, and TiC/10% Ni composite was investigated by Coughanowr et al. [42]. They also compared the data to that of the components, TiC, and nickel. Experiments were conducted in an ECM system in which the cathode tool was advanced at a constant rate toward the anode (workpiece). Two electrolytes used were 2 M KNO_3 and 3 M NaCl. The TiC/10% Ni specimens were fabricated by using standard cold-pressing and sintering techniques. Experiments were aimed at determining the valence of dissolution and dissolved surface morphology. Dissolution valence was determined from a plot of machining rate, r, as a function of current density and applying Faraday's Law:

$$r = \frac{Mi}{nF\rho} \tag{6.3}$$

where M is the molecular weight, i is the current density, n is the dissolution valence, F is the Faraday constant, and ρ is the density of the material. ECM experiments were conducted by varying applied voltage between 10 and 31 V and current densities of 15–115 A/cm^2. Figure 6.9 shows a plot of machining rate vs. current density

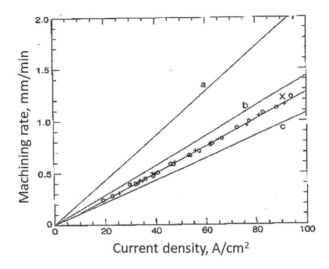

FIGURE 6.9   Machining rate vs. current density for TiC in 2 M KNO$_3$ (+) and 3 M NaCl (O). Line X: experimental correlation using both data sets. Theoretical lines: (a) $n=4$, (b) $n=6$, (c) $n=8$ [42].

for TiC. The dissolution data in each electrolyte fall on one line, the slope of which corresponds to a value of $n=6.6$. This number agrees well with the dissolution valence values reported by Cowling and Hintermann [28], although their results were obtained at much lower currents. The surface brightening was not observed on TiC for the range of ECM operating conditions employed. The post-ECM surfaces looked very similar to the original surfaces. The ECM behavior of a TiC/10% Ni composite and Ni 200 was investigated in 2 M KNO$_3$. TiC/10% Ni dissolved with an apparent valence of 6.5; the value is very close to that for TiC. For Ni 200, the dissolution valence was found to be 2.8. Since nickel dissolution leads to the formation of Ni^{2+} [43], the observed value of 2.8 indicates concurrent anodic oxygen evolution during ECM under these conditions. SEM micrographs of typical post-ECM surfaces are shown in Figure 6.10. In the case of TiC, the surface after dissolution appeared extremely rough (Figure 6.10a) while a finely disperse surface texture resulted from the dissolution of nickel 200 (Figure 6.10c). The post-ECM TiC/10% Ni surface (Figure 6.10b) was somewhat in between the pure TiC and Ni 200 surfaces. The surface morphologies of Figure 6.10 indicate that polishing conditions were not achieved under the employed current density range.

Schubert et al. [44] studied the anodic dissolution behavior of tungsten carbide in alkaline electrolyte under ECM conditions. They used pure, binder-free WC samples that were produced by sintering of two different tungsten carbide powders of different particle size. The potentiodynamic experiments under ECM conditions were carried out in a flow channel cell coupled with a microscope. Polarization curves of tungsten carbide in sodium hydroxide electrolyte, in situ investigations of the development of the surface structure during the ECM-process, and analytical investigations of reaction products such as oxygen were carried out. They claimed that during ECM of WC, an adherent, supersaturated, viscous film of poly-tungstate is formed close to

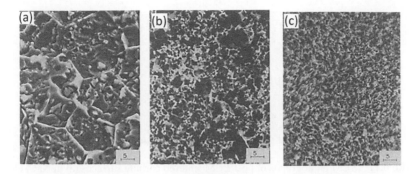

FIGURE 6.10   SEM micrographs of typical post-ECM anode surfaces: (a) TiC; (b) TiC/10% Ni; (c) Ni-200. Machining conditions: $V = 15$ V; tool feed rate $= 0.5$ mm/min [42].

the interface which continuously dissolved and reproduced. The dissolution proceeds in the active state up to 30 A/cm^2. At current densities >30 A/cm^2, an additional layer forms reflecting a passive state and high-field oxide films with thicknesses around 10 nm. The formation of an oxide film and, therefore, the active/passive transition is also indicated by the onset of oxygen evolution which takes place on the oxide films. The active/passive transition resulted in a current plateau the value of which varied with experimental parameters such as electrolyte flow rate and cell geometry. Oxygen evolution which commences at around 30 A/cm^2, consumes about 20% of the anodic charge and remains constant at that number at higher current densities.

ECM of sintered carbide using commonly used ECM electrolytes poses several issues. One of the important problems is the preferential dissolution of Co, which is used as the binder of the material. When too much Co is dissolved, the strength of the material is deteriorated, and it cannot be used as a die. Goto et al. [45] reported that prolonged immersion (8 hours) of sintered carbide (particle size: 0.8 μm, WC: 87 wt%, Co: 13 weight %) resulted in co-elution and the material deteriorated. They conducted experiments in different electrolytes and concluded that co-elution can be prevented by the addition of CoCl$_2$ in the electrolyte. They arrived at this conclusion from the immersion data for 110 hours in NaCl with and without CoCl$_2$ solutions and comparing the SEM and EDS data of each sample. Goto et al. [45] also performed machining experiments in an ECM tool using the AC power supply. The gap distance between the electrode and the workpiece was set at 0.3 mm at the beginning of machining and the electrode was advanced at a speed of 0.3 mm/min. The machining time was 2 minutes. The two electrolytes used consisted of (i) 150 g/L NaCl in water and (ii) 150 g/L NaCl + 50 g/L CoCl$_2$ in water. Figure 6.11 shows the surfaces resulting from machining of sintered carbide in these electrolytes. The photographs clearly show that machining in NaCl solution yields a highly porous surface due to co-elution. The surface machined in CoCl$_2$ containing NaCl electrolyte is largely devoid of such pores indicating that co-elution is indeed prevented by the presence of CoCl$_2$ in the electrolyte. The machining speed remained almost the same in both the electrolytes [45].

Schubert et al. [46] addressed the problem of nonhomogeneous anodic dissolution of cemented carbides from the thermodynamic point of view. They considered

FIGURE 6.11 SEM photographs of post-ECM sintered carbide surfaces (a) in NaCl electrolyte and (b) in $NaCl + CoCl_2$ electrolyte [45].

WC6Co composite, which mainly consists of a hard phase (WC) enclosed in a binder phase (Co). The electrochemical behavior of both is completely different and complicates a homogenous dissolution process of the material. According to Pourbaix's diagram [29], tungsten carbide dissolves to a soluble product only at pH > 7, whereas soluble cobalt species are stable at pH < 6.

The anodic dissolution of tungsten carbide can be given as:

$$WC + 7H_2O \rightarrow WO_4^{2-} + CO_3^{2-} + 14H^+ + 10e^- \tag{6.4}$$

In neutral and acidic solutions, tungsten carbide forms stable passive layers according to:

$$WC + 5H_2O \rightarrow WO_3 + CO_2 + 10H^+ + 10e^- \tag{6.5}$$

In neutral and acidic electrolyte, dissolution of cobalt is given by:

$$Co \rightarrow Co^{2+} + 2e^- \tag{6.6}$$

And in alkaline pH range, hydroxide layer is formed according to:

$$Co + 2H_2O \rightarrow Co(OH)_2 + 2H^+ + 2e^- \tag{6.7}$$

However, since an effective ECM requires a homogeneous dissolution of both phases, the authors used an alkaline electrolyte with complex agents, which suppresses the passivity of cobalt. It is well known, that cobalt easily forms complexes with ammonia [29]. In the ammoniacal solution, cobalt hydroxides are formed at pH ≥ 11 [47]. Addition of ammonia leads to the dissolution of hydroxide by the formation of hexamine cobalt complex:

$$Co(OH)_2 + 6NH_3 \rightarrow [Co(NH_3)_6]^{2+} + 2OH^- \tag{6.8}$$

The soluble cobalt complex is easily removed by the electrolyte flow during machining. An additional advantage of the usage of ammonia as an additive is its function as a pH buffer. The buffer effect guarantees that the pH keeps in the alkaline range during the machining process.

Schubert et al. [46] conducted electrochemical dissolution experiments using a microcapillary flow through cell, which enabled experiments to be conducted under near-ECM conditions using high current densities. Using an electrolyte consisting of a mixture of 2 M NH_3 and 2.9 M $NaNO_3$ at pH 12, they demonstrated that a homogenous dissolution of the hard phase, as well as the binder phase, is achieved. The current efficiency was found to be 100% at low current density (5 A/cm²) at higher current densities the current efficiency decreased and stabilized at around 70% between 20 and 30 A/cm². Lohrengel et al. [48] using an electrolyte consisting of a mixture of 1.2 M $NaNO_3$ and 0.6 M NaOH for ECM of sintered WC/Co reported that reduction of surface resulted from dissolution at higher voltages.

REFERENCES

1. E. Krasicka-Cydzik. Anodic Layer Formation on Titanium and Its Alloys for Biomedical Applications, Titanium Alloys. *in Towards Achieving Enhanced Properties for Diversified Applications*, A. K. M. Nurul Amin ed., pp. 175–200, InTech, (2012), ISBN: 978-953-51-0354-7.
2. J. L. Trompette, L. Massot, L. Arurault, S. Fontorbes, *Corros. Sci.*, 53, 1262 (2011).
3. I. Dugdale, J. B. Cotton, *Corros. Sci.*, 4, 397 (1964).
4. T. R. Beck, *J. Electrochem. Soc.*, 120, 1310 (1973).
5. T. R. Beck, *J. Electrochem. Soc.*, 120, 1317 (1973).
6. G. T. Burstein, R. M. Souto, *Electrochim. Acta*, 40, 1881 (1995).
7. J. B. Mathieu, *Ph.D. Thesis No 314*, EPFL, Lausanne, (1978).
8. J. B. Mathieu, D. Landolt, *Oberflache Surf.*, 20, 24 (1979).
9. D. Landolt, P.-F. Chauvy, O. Zinger, *Electrochim. Acta*, 48, 3185 (2003).
10. C. Madore, D. Landolt, *J. Micromech. Microeng.*, 7, 270 (1997).
11. J. Bannard, *J. Appl. Electrochem.*, 6, 477 (1976).
12. D. Baehre, A. Ernst, K. Weißhaar, H. Natter, M. Stolpe, R. Busch, *18th CIRP Conference on Electro Physical and Chemical Machining (ISEM XVIII), Procedia CIRP*, vol. 42, p. 137, (2016).
13. M. M. Lohrengel, C. Rosenkranz, I. Klüppel, A. Moehring, H. Bettermann, C. Van den Bossche, J. Deconinck, *Electrochim. Acta*, 49, 2863 (2004).
14. Y. Liu, N. Qu, *J. Electrochem. Soc.*, 166 (2), E35 (2019).
15. T. Heumann, N. Stolica, *Electrochim. Acta*, 16, 643 (1971).
16. T. Heumann, N. Stolica, *Electrochim. Acta*, 16, 1635 (1971).
17. R. D. Armstrong, K. Edmondson, R. E. Firman, *J. Electroanal. Chem.*, 40, 19 (1972).
18. M. Anik, K. Osseo-Asare, *J. Electrochem. Soc.*, 149, B224 (2002).
19. S.-J. Jaw, *High Rate Anodic Dissolution and Jet Electrochemical Micromachining of Tungsten, Ph.D. Thesis*, University of Connecticut, (1996).
20. S.-J. Jaw, M. Datta, J. M. Fenton, *Proceedings of the Symposium on High Rate Dissolution Processes*, M. Datta, B. R. MacDougall, J. M. Fenton eds., vol. 95–19, p. 236, Electrochemical Society, Pennington, NJ, (1995).
21. P. J. Bryant, H. S. Kim, Y. C. Zheng, R. Yang, *Rev. Sci. Instrum.*, 58, 1115 (1987).
22. S. Kerfriden, A. H. Nahle, S. A. Campbell, F. C. Walsh, J. R. Smiths, *Electrochim. Acta*, 43 (12–13), 1939 (1998).
23. O. L. Guise, J. W. Ahner, M.-C. Jung, P. C. Goughnour, J. T. Yates, *Nano Lett.*, 2 (3), 191 (2002).
24. B.-F. Ju, Y.-L. Chen, Y. Ge, *Rev. Sci. Instrum.*, 82, 013707 (2011).

25. W.-T. Chang, I.-S. Hwang, M.-T. Chang, C. Y. Lin, W.-H. Hsu, J.-L. Hou, *Rev. Sci. Instrum.*, 83, 083704 (2012).
26. M. A. H. Mithu, G. Fantoni, *Int. J. Precis. Technol.*, 2 (4), 301 (2011).
27. T.-H. Duong, H.-C. Kim, *Int. J. Precis. Eng. Manuf.*, 16 (6), 1053 (2015).
28. R. D. Cowling, H. E. Hintermann, *J. Electrochem. Soc.*, 117, 1447 (1970).
29. M. Pourbaix, *Atlas of Electrochemical Equilibria in Aqueous Solutions*, National Association of Corrosion Engineers (NACE), Houston, (1974).
30. R. D. Cowling, H. E. Hintermann, *J. Electrochem. Soc.*, 118, 1912 (1971).
31. J. Zackrisson, B. Jansson, G. Uphadyaya, H. Andrén, *Int. J. Refract. Met. Hard Mater.*, 16, 417 (1998).
32. S. Sutthiruangwong, G. Mori, R. Kösters, *Int. J. Refract. Met. Hard Mater.*, 23, 129 (2005).
33. S. Hochstrasser, Y. Müller, C. Latkoczy, S. Virtanen, P. Schmutz, *Corros. Sci.*, 49, 2002 (2007).
34. A. M. Human, H. E. Exner, *Mater. Sci. Eng. A*, 241, 202 (1998).
35. W. Lee, B. Kwon, S. Jung, *Int. J. Refract. Met. Hard Mater.*, 24, 215 (2006).
36. A. M. Human, H. E. Exner, *Mater. Sci. Eng. A*, 209, 180 (1996).
37. A. M. Human, H. E. Exner, *Int. J. Refract. Met. Hard Mater.*, 15, 65 (1997).
38. W. Tomlinson, N. Ayerst, *J. Mater. Sci.*, 24, 2348 (1989).
39. F. J. J. Kellner, H. Hildebrand, S. VirtanenInt, *Int. J. Refract. Met. Hard Mater.*, 27, 806 (2009).
40. P. Schwarzkopf, R. Kietter, Chap. 5, *in Cemented Carbides*, The Macmillan Company, New York, (1960).
41. M. H. Ghandehari, *J. Electrochem. Soc.*, 127, 2144 (1980).
42. C. A. Coughanowr, B. A. Dissaux, R. H. Muller, C. W. Tobias, *J. Appl. Electrochem.*, 16, 345 (1986).
43. M. Datta, D. Landolt, *Electrochim. Acta*, 25, 1263 (1980).
44. N. Schubert, M. Schneider, A. Michaelis, M. Manko, M. M. Lohrengel, *J Solid State Electrochem.*, 22, 859 (2018).
45. A. Goto, A. Nakata, N. Saito, *Procedia CIRP*, 42, 402 (2016).
46. N. Schubert, M. Schneider, A. Michaelis, *Int. J. Refract. Met. Hard Mater.*, 47, 54 (2014).
47. C. Vu, K. N. Han, *Met. Trans. B*, 10 (1), 57 (1979).
48. M. M. Lohrengel, K. P. Rataj, N. Schubert, M. Schneider, S. Höhn, A. Michaelis, M. Hackert-Oschätzchen, A. Martin, A. Schubert, *Powder Metall.*, 57 (1), 21 (2014).

7 Anodic Dissolution of Metals in Electropolishing Electrolytes

7.1 INTRODUCTION

Electropolishing is the process of improving the surface finish of a metallic workpiece by making it an anode in an electrolytic cell. The use of appropriate electrolyte and operating conditions produces surfaces that are smooth, bright, and free of defects, stress, and contamination. Because of these inherent features, electropolishing is widely employed in metal finishing, vacuum technology, preparation of samples for metallography and electron microscopy, and in technologies requiring surfaces of controlled quality and cleanliness. Electropolishing involves removal of both macro surface roughness, i.e., leveling, and micro surface roughness, i.e., surface brightening. Smooth and highly reflecting surfaces require the elimination of surface roughness comparable in size to the wavelength of light.

Electropolishing is normally carried out in aqueous solutions of concentrated acids such as phosphoric acid, sulfuric acid, and their mixtures or in nonaqueous, mainly alcoholic, acid solutions. Several books and reviews provide a list of electropolishing electrolytes for different metals and alloys [1–3]. The original work on electropolishing was conducted by Jacquet [4,5], who postulated that a viscous layer formed on the metal surface was responsible for microsmoothing. Later work by Elmore [6], Wagner [7], Hoar et al. [8–11], and Kojima and Tobias [12] emphasized the role of mass transport and the presence of compact solid films for polishing effects. Edwards [13,14] performed computation of smoothing efficiencies and introduced the terms macrosmoothing and microsmoothing for anodic leveling and anodic brightening, respectively. Landolt et al. [15–18] compared the numerical simulation and experimental results for the dissolution of triangular macro and microprofiles. Under electropolishing conditions, the predictions for the rate of leveling for microprofiles compared very well with the experimental data. A comprehensive analysis of the above literature data was presented by Landolt [19] in a review paper published in 1987. Since then, many research papers on electropolishing of a variety of materials and their new-found applications have appeared in the literature.

7.2 ELECTROPOLISHING OF SELECTED METALS AND ALLOYS: RATE-CONTROLLING SPECIES

Most of the research work established that the electropolishing takes place in the transpassive potential region and the occurrence of mass transfer-controlled dissolution is generally a prerequisite for electropolishing. The value of the limiting current is governed by the rate of transport of dissolution products from the anode into the bulk. This is certainly the case for high-rate dissolution in aqueous salt solutions where, as described in the earlier chapter, the estimated surface concentration of dissolution products agrees well with the saturation concentration of the corresponding metal salt [20–23]. Under such conditions, the formation of a salt film leads to surface brightening. In most of the electropolishing systems also the limiting current corresponds to the salt-film formation at the anode surface. On the other hand, since many electropolishing electrolytes contain a limited amount of water, the possibility of its role as a limiting reactant has been claimed in the literature. In this chapter, we present a discussion on the anodic dissolution behavior of some selected metals and alloys in different electropolishing solutions to identify the mechanism leading to microfinishing in these systems.

7.2.1 COPPER

The copper-phosphoric acid system has been studied extensively to understand the phenomena involved in the electropolishing process [6–14,24–28]. Elmore [6] was one of the early researchers who conducted anodic polarization studies to understand the mechanism of electropolishing of copper in phosphoric acid. He concluded that under polishing conditions, the concentration of dissolved copper ions at the anode reaches its highest value. Accordingly, the concentration gradient which exists at the anode limits the rate at which copper can dissolve and diffuse into the bulk electrolyte. A rough surface, thus, becomes progressively leveled since there is greater diffusion from raised areas of the anode. Under these conditions, the surface etching is suppressed since the rate of dissolution is controlled by the concentration gradient yielding a polished surface.

Similar to Elmore's findings, Edwards [13,14] observed that the anodic dissolution of copper in phosphoric acid is diffusion controlled. Edwards, however, disagreed with Elmore's hypothesis that the onset of polishing coincides with the attainment of the solubility limit of copper in the electrolyte. Instead, the influence of additions of copper to the electrolyte was interpreted to be due to the modification of the viscosity. He advanced a hypothesis according to which the controlling process is depletion of the anode layer concerning those ions or molecules with which the copper ions are combined in the solution, in this case, the phosphate ions. Thus, according to Elmore, the principal feature of electropolishing is the attainment of zero concentration of an ion-acceptor (phosphate ions) at the anode. This represents a modified version of Elmore's hypothesis in which the solubility limit of copper is replaced by the effective concentration of phosphate ions in the bulk of the solution.

Hoar and Rothwell [9] studied the influence of linear solution flow during electropolishing of a horizontal copper anode in different concentrations of aqueous phosphoric acid. Surface etching resulted from dissolution at low current densities until a limiting current density was observed at which polishing set in. The value of the limiting current density increased with the 0.5 power of the flow rate. The results were interpreted in terms of the limitation of the overall rate of dissolution mainly by the diffusion layer in the solution and partly by several types of compact solid film on the dissolving metal surface.

Kojima and Tobias [12] conducted unsteady-state anodic dissolution of copper in 5–10 M phosphoric acid in the absence of convection, under galvanostatic and potentiostatic conditions. They evaluated different possible transport mechanisms considering the variation of transport properties with electrolyte composition. Based on their data as well as the data available in the literature, they ruled out the possibility of anion diffusion mechanisms being responsible for electropolishing. Dissolved metal cation was concluded to be the transport limiting species. They contended that the solubility limit of copper phosphate is exceeded while the concentration of phosphoric acid remained quite appreciable at the anode surface. The limiting current density could be predicted by considering the transport of copper phosphate away from the anode surface.

Voltammetry measurements for copper dissolution in phosphoric acid conducted by Glarum and Marshall [24,25] using a rotating disk electrode indicated that the dissolution current depended only upon the viscosity of the electrolyte and is otherwise indifferent to the electrolyte composition. This is the same as the viscosity effect reported by Elmore. However, Glarum and Marshall dismissed Elmore's ion-acceptor mechanism instead they proposed a water acceptor mechanism and suggested the formation of a highly viscous, dehydrated surface layer. Under certain conditions, current oscillations were observed during electropolishing. Glarum and Marshall [25] also reported results of an AC impedance study which further supported a water-acceptor mechanism. They also concluded that the region depleted of water behaves as a thin viscous film with a space charge. During the initial phase of layer growth, observed current oscillations reflect a negative impedance relaxation at low frequencies.

Vidal and West [26] conducted anodic polarization measurements using a copper rotating disk electrode in phosphoric acid at different temperatures. The results are summarized in Figure 7.1. Potentiodynamic anodic polarization curves for Cu in 85% H_3PO_4 at different rotation speeds at a scan rate of 5 mV/s and an electrolyte temperature of 30°C are shown in Figure 7.1a. The open circles in the figure correspond to steady-state potentiostatic measurements. The data show that the limiting current plateaus extend over a range of 1 V and that the charge-transfer controlled region is insensitive to rotation speed. Polarization beyond the plateau results in an oxygen-evolution reaction. The valence of dissolution determined by weight-loss measurements was found to be 2 at potentials below and at the limiting current. It was therefore concluded that copper oxidizes into its divalent state along the plateau (limiting current) with a current efficiency of 100%.

Based on the polarization curves shown in Figure 7.1a, the limiting current densities were measured at an applied potential of 0.6 V at different temperatures and different rotation speeds. Figure 7.1b shows the Levich plot obtained from polarization measurements performed at different temperatures. These results are consistent with a process controlled by mass transfer. Vidal and West [27] also used AC-impedance and electrohydrodynamic (EHD)-impedance spectroscopy to further investigate the copper electropolishing mechanism. High-frequency AC-impedance measurements

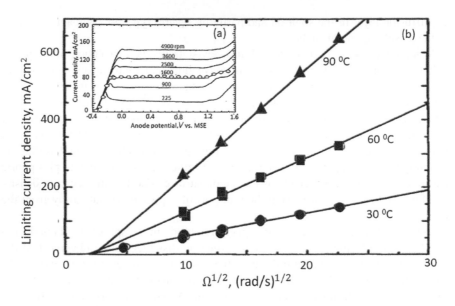

FIGURE 7.1   (a) Potentiodynamic anodic polarization curves for Cu in 85% H₃PO₄ at different rotation speeds. Scan rate: 5 mV/s; electrolyte temperature: 30°C. The open circles correspond to potentiostatic measurements. (b) Levich plot at three temperatures. The limiting current densities were measured at an applied potential of 0.6 V [26].

were used to estimate the ohmic resistance, capacitance, and polarization resistance of the system. EHD-impedance experiments, in which the disk-rotation speed instead of the electrode potential is modulated, were used to probe the convective-diffusion phenomena. Assuming a water-acceptor type of mechanism and with a known viscosity and diffusion coefficient, the Levich analysis of the variation of the limiting current with rotation speed was used to estimate the bulk concentration of acceptor. The value obtained was found to agree with the known water content in phosphoric acid thus indicating that water is the acceptor species [27].

Mendez et al. [28] employed potential transient techniques to characterize copper electropolishing in 85% (14.7 M/L) phosphoric acid. They first performed potentiodynamic polarization experiments at different rotation speeds of the RDE which provided curves the same as those of Figure 7.1a [26]. From these data, the authors then chose two different current densities to perform potential transient experiments: one below the limiting current density (6.3 mA/cm²) and the other at the limiting current density (19.6 mA/cm²). The potential response obtained at the limiting current indicated a two-step process. Consistent with the potential transient behavior, they proposed a two-step mechanism for the copper electropolishing process [28]. According to the authors, a flux imbalance between the dissolving copper ions and their transport out of the boundary layer region causes a buildup of copper ions near the anode surface. This concentration increases until the solubility limit of the copper salt is reached, and a resistive film is formed. During the second stage, the resistive film increases in thickness due to the imbalance between the current-dependent dissolution and the removal of copper ions by transport. Under the constant current, the film thickness increases linearly with time, giving rise to a corresponding increase in the potential.

7.2.2   Fe, Cr, and Their Alloys

7.2.2.1   Fe

Electropolishing of steels and stainless steel is widely practiced in industry. Phosphoric acid-based electrolytes are commonly employed for electropolishing these materials. Gabe [29] measured the anodic polarization curves for mild steel in phosphoric acid and phosphoric-sulfuric acid mixtures. Anodic passivation was observed in these solutions and the value of the passivation current was found to depend on the electrolyte composition. A phosphoric acid solution containing sulfuric acid gave a better surface finish than a pure phosphoric acid.

Datta et al. [30–32] investigated the anodic dissolution behavior of iron in concentrated (14 M) phosphoric acid solution using potentiodynamic and potentiostatic techniques (Figure 7.2). The influence of temperature on the polarization behavior of iron is shown in Figure 7.2a. At 90°C, the current density rises sharply in the active region until a limiting plateau is observed which is followed by a region of the current fluctuations. At higher potentials, a second limiting current plateau is observed. The magnitudes of both limiting currents increase with increasing rotation speed. In Figure 7.2b, the inverse of the limiting currents below and above fluctuations are plotted as a function of $\omega^{-1/2}$. It is interesting to note that for limiting current density below the current fluctuations, $1/i_l$ vs. $\omega^{-1/2}$ passes through the origin following the Levich equation indicating that the anodic reaction below fluctuations is mass-transport controlled. On the other hand, the limiting current density values above the current fluctuations show an intercept, indicating the presence of a kinetic step during the dissolution process. For a first-order reaction proceeding under mixed

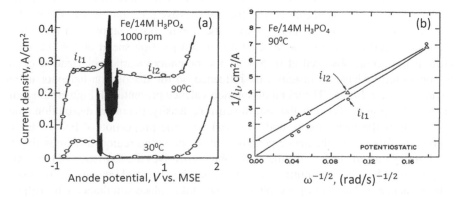

FIGURE 7.2   (a) Anodic polarization curves for Fe in 14 M H_3PO_4 at two different temperatures; RDE at 1,000 rpm. (b) Plots of the inverse of limiting currents below and above current fluctuations as a function of $\omega^{-1/2}$ [33].

(kinetic as well as mass transport) control, the limiting current density measured by a rotating disk electrode can be expressed as:

$$\frac{1}{i_l} = \frac{1}{i_k} + \frac{1}{B\omega^{1/2}} \tag{7.1}$$

where i_k is the kinetic current density that depends on the charge transfer or chemical reaction and is independent of the rotation speed, and B is a constant, which for a rotating disk electrode is equal to $0.62 \, nF \, D^{2/3} \, v^{-1/6} \, \omega^{1/2} \Delta C_s$. Equation 7.1 allows one to evaluate the relative importance of kinetic and mass transport-controlled steps in the anodic dissolution process. For a mixed controlled reaction, a linear plot of $1/i_l$ vs. $\omega^{1/2}$ is obtained but with a positive intercept at extrapolation to $\omega^{1/2} = 0$, thus yielding the value of i_k.

The dissolution valence determined from weight loss measurements indicated that below the current fluctuations (−04 V), iron dissolution leads to Fe^{2+} formation, at the fluctuation region (+0.05 V) the valence was measured to be 2.46 indicating that both Fe^{2+} and Fe^{3+} are formed. At potentials above the fluctuation range (+0.75 V), the valence was measured to be close to 3 indicating the formation of Fe^{3+}. Finally, at still higher potentials, a sharp increase in current occurs, accompanied by oxygen evolution.

At 30°C, the polarization behavior at low anodic potentials is qualitatively similar to that at 90°C, except for the magnitude of the anodic current, which is relatively very low. Beyond the current fluctuation region, anodic passivation occurs, as indicated by a sharp drop in the current. At higher potentials, oxygen evolution occurs. Metal dissolution at 30°C is insignificant over a wide range of potential. The polarization curves of Figure 7.2a help to explain why electropolishing of iron and iron-based alloys in phosphoric acid is best achieved at elevated temperatures.

Anodic polarization of Fe in a concentrated phosphoric-sulfuric acid mixture (2:1) is shown in Figure 7.3, which shows that the addition of sulfuric acid to the phosphoric acid completely suppresses the active dissolution of iron [33]. Metal dissolution in this solution occurs only in the transpassive potential region. Figure 7.3a compares potentiodynamic and potentiostatic anodic polarization data of iron in phosphoric-sulfuric solution at 90°C at a constant rotation speed of 1,000 rpm. A current minimum observed at 0.45 V in the potentiodynamic polarization curve is considered to have an insignificant influence on the dissolution behavior under steady-state conditions. The metal dissolution valence presented in Figure 7.3b indicates that at potentials below the limiting current density, the metal dissolution reaction involves the formation of both Fe^{2+} and Fe^{3+}, the proportion of Fe^{3+} being an increasing function of the anode potential. In the potential region corresponding to the limiting current density, the metal dissolution involves the formation of Fe^{3+} only. At potentials beyond the limiting current region, the observed increase in the dissolution valence is due to oxygen evolution which takes place simultaneously with the metal dissolution reaction. Figure 7.4 shows the surface roughness of iron samples dissolved at different potentials. Also included in the figure are the SEM pictures of surfaces dissolved at potentials below the limiting current, and at or above the

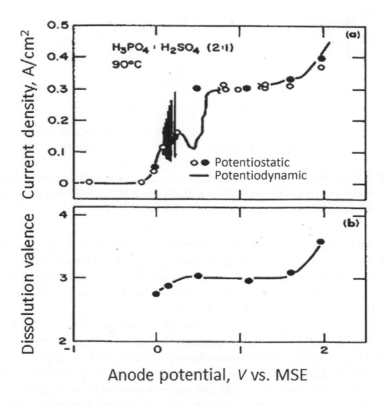

FIGURE 7.3   (a) Potentiodynamic and potentiostatic anodic polarization curves for Fe in phosphoric and sulfuric acid solution and (b) dissolution valence as a function of potential. Electrolyte temperature: 90°C; RDE speed: 1,000 rpm [33].

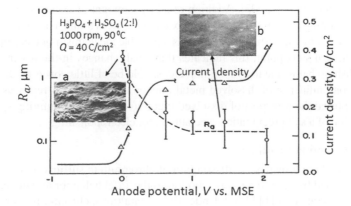

FIGURE 7.4   Average surface roughness, R_a, of iron anodes dissolved at different potentials in a mixture of phosphoric-sulfuric acids at 90°C. SEM photographs of tow surfaces (a) below and (b) at the limiting current [33].

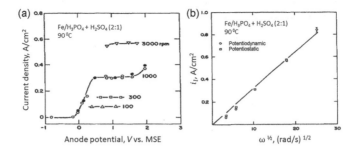

FIGURE 7.5 (a) The influence of rotation speed on the anodic limiting current density for iron dissolution in phosphoric acid-sulfuric acid mixture at 90°C. (b) Levich plot of the data showing the dissolution process to be mass transport controlled [33].

limiting current. Dissolution below the limiting current led to pitting, grain boundary, and other forms of localized attack thus yielding a dull appearing rough surface (Figure 7.4a). On the other hand, dissolution at potentials corresponding to the limiting current or into the oxygen evolution region yielded highly reflecting microscopically flat electropolished surfaces (Figure 7.4b).

Figure 7.5 shows the strong influence of the rotation speed on the limiting current density; the higher the rotation speed, the higher is the limiting current density, indicating that the anodic process under these conditions is mass transport controlled. Furthermore, the data of Figure 7.5a show that potentiostatic and potentiodynamic techniques give the same results. A straight-line relationship between the measured limiting current density and $\omega^{1/2}$ presented in Figure 7.5b conclusively demonstrates that under the present experimental conditions, electropolishing is governed by convective mass transport controlled anodic reactions. To determine the rate-limiting diffusing species, the role of water and dissolved metal ions on the measured limiting current was investigated. For a quantitative estimation of their influence, viscosity and density of the electrolytes were measured and the diffusion coefficients were determined by employing stokes – Einstein equation [33]. Addition of a large amount of water to the electrolyte altered the limiting current only to a small extent which was found to be in contradiction to that predicted by the diffusion of water as a possible mechanism of electropolishing. From a practical standpoint, this indicates that small changes in the water content of electropolishing baths due to evaporation etc. should be of little concern. On the other hand, a strong influence of dissolved metal ions on the limiting current was observed which supported the transport of dissolved metal ions as the rate-limiting species and the formation of a salt precipitation mechanism for electropolishing.

7.2.2.2 Cr and FeCr Alloys

The kinetics of the transpassive dissolution of chromium in sulfuric acid electrolytes has been studied by several authors who found the Tafel behavior and dependence of the dissolution rate on pH [34,35]. Under these conditions, chromium goes into solution in the hexavalent state, according to

$$2Cr + 7H_2O \rightarrow Cr_2O_7^{2-} + 14H^+ + 12e^-. \tag{7.2}$$

Several electropolishing electrolytes and corresponding operating conditions are described in the literature [1] according to which chromium can be electropolished in phosphoric acid at its concentrations exceeding 30 wt.%. Improvement is obtained by adding sulfuric acid, glycerol, or citric acid to the solution. Ponto and Landolt [36] studied the anodic dissolution behavior of chromium in phosphoric acid-sulfuric acid electrolytes using a rotating disc electrode. They investigated the influence of electrolyte composition on the resulting surface finish and observed that electropolishing of chromium occurs in the transpassive potential region under mass transport control. The transport limiting species is the anodically generated hexavalent chromium ion. The formation of anodic film at the anode surface is potential dependent, and film instabilities leading to current oscillations occur in a certain potential region of the limiting current plateau.

Ponto et al. [37] investigated the electropolishing of Fe13Cr, Fe24Cr, and 304 (10Ni18Cr) stainless steel in H_3PO_4–H_2SO_4 electrolytes using rotating disk electrodes. Their experimental data indicated that the electropolishing of these alloys takes place in the transpassive region and is mass transport controlled. Electropolishing is favored by high temperatures. The anodic polarization of the FeCr alloys in 65% H_3PO_4 – 20% H_2SO_4 – 15% H_2O is insensitive to the elemental composition of the alloy and is the same for the austenitic-type 304 stainless steel. At 70°C, electropolishing is obtained over a wide concentration range of H_3PO_4-H_2SO_4, but in the absence of H_3PO_4, iron passivity prevents the establishment of electropolishing conditions. Dissolved Fe^{3+} was found to be the most likely rate-controlling species responsible for electropolishing of the different FeCr alloys and 304 stainless steel. A similar mechanism involving the transport of dissolved nickel species was found for electropolishing of nickel in sulfuric acid [38].

Datta and Vercruysse [39] studied the anodic polarization behavior of 420 (Fe13Cr) stainless steel in concentrated phosphoric acid, sulfuric acid, and their mixtures to identify the processes that lead to microsmooth surfaces. The influence of temperature on the anodic polarization behavior of 420 stainless steel in these concentrated acid electrolytes was studied. In these electrolytes, the anodic polarization curves showed active, passive, transpassive transitions at all temperatures, but at 25°C the transpassive reactions were little influenced by the rotation speed of the RDE. The anodic polarization curves in different electrolytes at 60°C are given in Figure 7.6. The anodic polarization curve in concentrated phosphoric acid (Figure 7.6a) shows a well-defined rotation-speed-dependent limiting current in the transpassive potential region. The anodic polarization behavior in sulfuric acid (Figure 7.6b) shows that metal dissolution is insignificant in this electrolyte even at very high anode potentials. Concentrated sulfuric acid is, therefore, unsuitable for electropolishing of 420 stainless steel. In a mixture of the phosphoric and sulfuric acid solution, a rotation speed-dependent limiting current is obtained in the transpassive potential region (Figure 7.6c). An increase in the temperature of the phosphoric-sulfuric electrolyte from 25°C to 90°C decreased the active current peak values but increased the transpassive limiting currents as shown in Figure 7.7. At 60°C and 90°C, the measured limiting current densities as a function of the rotation speed follow the Levich behavior as shown in Figure 7.7b. Estimation of surface concentration of dissolution product at the limiting current and its dependence on temperature pointed toward dissolved metal ions to be the rate-controlling species.

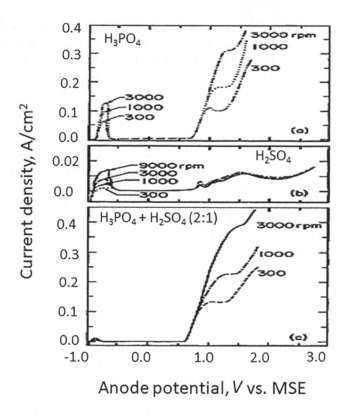

FIGURE 7.6 The influence of electrode rotation speed on the anodic polarization behavior of 420 (Fe13Cr) stainless steel in (a) concentrated phosphoric acid, (b) concentrated sulfuric acid, and (c) a mixture of two parts by volume of phosphoric acid and one part of sulfuric acid. Electrolyte temperature: 60°C scan rate: 5 mV/s [39].

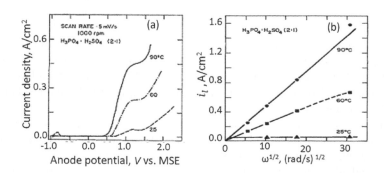

FIGURE 7.7 (a) The influence of electrolyte temperature on the polarization behavior of 420 (Fe13Cr) stainless steel in a mixture of two parts by volume of phosphoric acid and one part of sulfuric acid. (b) The influence of temperature on the i_l vs. $\omega^{1/2}$ relationship [39].

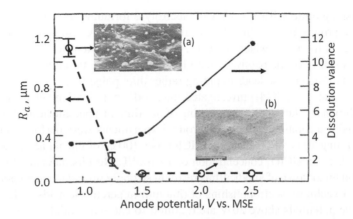

FIGURE 7.8 Average surface roughness and dissolution valence of 420 stainless steel samples dissolved at different potentials in a mixture of two parts by volume of phosphoric acid and one part of sulfuric acid at 90°C. Electrode rotation speed: 1,000 rpm. Also included are the SEM photographs (a and b) showing the influence of the anode potential on the surface microroughness [39].

On the other hand, Matlosz et al. [40] reported water acceptor to be the rate-controlling species in the same system.

Potentiostatic experiments at a constant charge were performed to determine the influence of applied potential on dissolution valence, and surface morphology [39]. The extent of microsmoothness and reflectivity of surfaces was strongly dependent on the electrolyte composition and temperature. At 25°C, both phosphoric acid and phosphoric acid-sulfuric acid electrolytes produced gray-appearing dull surfaces over a wide potential range beyond the transpassive current plateau. Increasing temperature of the electrolyte led to improved microfinishing, and highly reflecting surfaces were obtained at 90°C. Figure 7.8 shows the measured average roughness of surfaces dissolved at different potentials in a mixture of phosphoric acid and sulfuric acid (2:1) at 90°C. The surface dissolved at 0.9 V shows a severe localized attack (Figure 7.8a) and the measured surface roughness value is very high. At 2 V a highly reflecting surface is obtained (Figure 7.8b). Except for some micropits at random locations, the surface is almost flat, with an average surface roughness value of 0.05 μm. The apparent valence of dissolution at the limiting current is close to 3.4, indicating that under these conditions, Fe and Cr dissolve in their highest oxidation states of Fe^{3+} and Cr^{6+}, respectively.

7.2.3   Ti AND NiTi ALLOYS

7.2.3.1   Titanium

Titanium and titanium alloys are lightweight, high-strength, corrosion-resistant materials. They are used in a variety of applications including aerospace, photovoltaics, chemical process industry, and biological implants. In many of these devices, it is critical for the metals to have defect-free, smooth surfaces. For such applications, the electrochemical polishing method provides a reliable, flexible, and cost-effective microfinishing option. An overview of electrolyte formulations for electropolishing of titanium proposed in the literature is given in Ref. [3]. A common feature of these electrolytes is their low water content. A high water content tends to increase the chemical stability of the passive oxide film formed on the titanium surface. Anodic dissolution in such a case leads to pitting rather than polishing.

Mathieu and Landolt [41] investigated the anodic polarization behavior and reaction stoichiometry during electropolishing of titanium in a perchloric acid-acetic acid solution using a rotating cylinder electrode. Experiments were performed in a solution containing 10 cc of concentrated $HClO_4$ per 100 cc of glacial acetic acid which corresponds to an $HClO_4$ concentration of 1.1 mol/L. Results indicated that the formation and breakdown of anodic films result in strongly time-dependent polarization behavior. To achieve electropolishing of titanium in perchloric acid-acetic acid solutions anode potentials above 20 V are required to cause a breakdown of the anodic film of TiO_2 which protects the metal from dissolution at lower potentials. The onset of electropolishing coincides with a current plateau in the potentiostatic polarization curve. Its value depends little on hydrodynamic conditions because a gelatinous matrix consisting of solid reaction products covers the anode and masks the influence of convection on anodic mass transport processes. During anodic dissolution under electropolishing conditions perchlorate ion is reduced to chloride which leads to an apparent valence of dissolution of 1.3–1.6 independent of applied current or potential.

Mathieu et al. [42] also investigated the thickness and chemical nature of anodic films formed on titanium in perchloric acid-acetic acid electropolishing electrolyte using AES. Oxide film thickness determined by AES depth profiling is shown as a function of anode potential in Figure 7.9. The oxide film thickness increased to about 40 nm before the breakdown occurred at 20 V at which point the film thickness abruptly dropped to very low values independent of potential at higher potentials. For samples polarized under electropolishing conditions beyond 20 V, the measured film thicknesses scattered considerably but they were smaller than 10 nm, of the same order as that of the natural oxide films formed in air on titanium. The concentration depth profiles showed a chlorine maximum in the film formed at 28 V which corresponded to electropolishing conditions. No chlorine signal was observed on air formed film. These data indicated that chlorine species such as perchlorate or chloride ions could accumulate at the metal-oxide interface due to anodic polarization in perchloric acid. XPS analysis of the anodically polarized sample showed the presence of only Cl^- but ClO_4^- could not be detected [42], suggesting that the ClO_4^- ions are reduced at the metal surface. Based on these results, Mathieu et al. [42] suggested that the passive film breakdown on titanium is initiated at discrete sites leading to dissolution by pitting.

Piotrowski et al. [43] studied the mechanism of electropolishing of titanium in methanol-sulfuric acid electrolytes by steady-state and AC-impedance spectroscopic measurements using a rotating disk electrode. They investigated the effect of applied potential, electrode rotation rate, temperature, and electrolyte composition on anodic polarization curves. Figure 7.10a shows the potentiodynamic polarization curves for titanium measured at 0°C and different rotation speeds of the RDE in an electrolyte containing methanol and 6 M fuming sulfuric acid. Similar polarization measurements were also performed in electrolytes containing different sulfuric acid

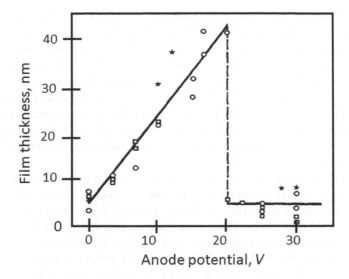

FIGURE 7.9 The estimated thickness of anodic films formed on titanium in perchloric acid–acetic acid solution as a function of anode potential. □, o, Stationary electrolyte, ★ two independent sets of experiments; RDE at 1,200 rpm [41].

concentrations. Figure 7.10b shows a plot of reciprocal of the limiting current density vs. the reciprocal of the square root of the rotation speed determined for titanium in methanol containing different concentrations of fuming sulfuric acid at 0°C. The linear relationship with an intercept at zero indicates purely mass transport-controlled anodic reaction at the limiting current. Experiments were performed in electrolytes containing different amounts of Ti^{4+} by dissolving $Ti(OC_2H_5)_4$ in a methanol-3 M H_2SO_4 solution. A linear decrease in current density with increasing Ti^{4+} concentration was observed, suggesting that the rate of reaction is controlled by mass transport of tetravalent titanium species from the anode surface to the bulk electrolyte.

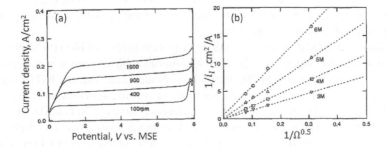

FIGURE 7.10 (a) Potentiodynamic polarization curves for titanium determined at 0°C in an electrolyte containing methanol and 6 M fuming sulfuric acid. (b) Reciprocal of the limiting current density vs. the reciprocal of the square root of the rotation speed determined for titanium in methanol containing different concentrations of fuming sulfuric acid at 0°C [43].

The impedance data confirmed the presence of a compact film at the anode surface during titanium dissolution at the limiting current. The presence of water at a concentration above a few percent in the electrolyte led to passivation of the surface thus suppressing electropolishing. Similar results were reported by Piotrowski et al. [44] for the electropolishing of tantalum in sulfuric acid-methanol electrolytes. Steady-state measurements performed at different temperatures showed that at the limiting current plateau the rate of dissolution is controlled by mass transport of Ta^{5+} species from the anode to the bulk electrolyte. The impedance data obtained under limiting current conditions suggested the presence of a compact film through which tantalum ions migrate by high-field conduction.

The beneficial effect of the presence of a small amount of water has been claimed by Huang et al. [45] in a sulfuric-ethanol system, which is exactly opposite to the results reported above for sulfuric-methanol system [43]. The experimental results of Huang et al. [45] showed that Ti could be electropolished in a 3 M H_2SO_4 – ethanol electrolyte, but inadequately electropolished in 0.5 and 1 M H_2SO_4 – ethanol electrolytes. However, the brightening effect on the Ti surface could be improved by the addition of a few volume percent of H_2O to the 0.5 and 1 M H_2SO_4 – ethanol electrolytes. An electropolished Ti surface with a surface roughness of a few nanometers could be achieved when the Ti was electropolished in a 3 M H_2SO_4 – ethanol electrolyte with 0.1 vol.% or lower of H_2O or in the 1 M H_2SO_4 – ethanol electrolyte with 1 vol.% H_2O.

7.2.3.2 NiTi (Shape Memory Alloy)

Nickel-titanium (NiTi) alloys possess unique properties such as shape memory effect and pseudo-elasticity. The binary nickel-titanium alloy, also known as shape memory alloy, transforms around human body temperature and is widely used in medical devices such as orthodontic archwires and coronary stents [46,47]. The biocompatibility and corrosion resistance of NiTi in physiological media is attributed to the formation of a harmless layer of TiO_2. Various studies on NiTi alloys report the dependence of its biocompatibility and corrosion behavior on surface conditions [48–50]. Several publications on the topic have been published by Hassels et al. [48–50] who conducted investigations to understand and enhance the surface stability of the NiTi surface. Among various methods, electropolishing was found to be one of the inexpensive and effective methods in treating implants which generally have intricate shapes and geometries.

Fushimi et al. [48] performed anodic polarization studies of NiTi (51Ni49Ti) alloy in various aqueous and methanolic-H_2SO_4 solutions with sulfuric acid concentration varying between 0.1 and 7 M. Their results indicated that whereas NiTi easily passivates in aqueous H_2SO_4, methanolic solutions of sulfuric acid prevents the passivity on NiTi. Figure 7.11a compares anodic polarization curves of NiTi in 0.1 M aqueous H_2SO_4 and 0.1 M H_2SO_4 solution in methanol using a potential sweep of 5 mV/s from open-circuit potential. In aqueous solution, very low anodic currents flow at potentials lower than 1.5 V, indicating the passive state of the NiTi. At potentials higher than 1.5 V, the anodic current increases rapidly with increasing potential leading to simultaneous transpassive dissolution of NiTi and oxygen evolution. In a methanolic solution, on the other hand, the passivity of the material is much less pronounced.

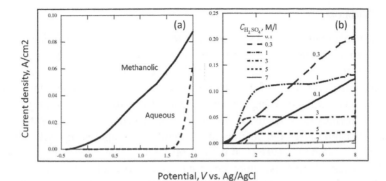

Potential, V vs. Ag/AgCl

FIGURE 7.11 (a) Anodic polarization curves of NiTi in aqueous 0.1 M H_2SO_4 and methanolic 0.1 M H_2SO_4 solutions at 298 K (25°C). The potential was ramped from open-circuit potential at a scanning rate of 5 mV/s. (b). Anodic polarization curves of NiTi in methanolic solution with 0.1–7 M H_2SO_4 at 263 K (−10°C). For experiments in (b), the potential was first held at 8 V for 100 seconds and was then swept down to 0 V at a scanning rate of 10 mV/s [48].

The onset of anodic dissolution current is at −0.35 V and increases with increasing potential. Figure 7.11b shows anodic polarization curves of NiTi in methanolic sulfuric acids with H_2SO_4 concentration ranging from 0.1 to 7 M H_2SO_4. In these studies, the potential was first held at 8 V for 100 seconds and then swept down to 0 V at a rate of 10 mV/s. For concentrations of up to 0.3 M H_2SO_4, a linear increase of current with potential is observed and electropolishing was not observed up to 8 V in this concentration range. For concentrations of 1 M or higher, the current shows a steep increase before reaching a plateau. The value of the limiting current decreases with an increase in acid concentration, showing a linear dependence on the logarithm of the acid concentration. The best results were obtained in a 3 M methanolic H_2SO_4 which at the limiting current gave a smooth mirror-like surface with the lowest root-mean-square (RMS) in a reasonable time.

Neelakantan and Hassel [49] investigated the influence of scan direction, rotation rate, process temperature and addition of ions to arrive at a mechanistic understanding of the electropolishing of NiTi (55Ni45Ti) alloy using a rotating disk electrode. During potentiodynamic experiments, they observed that the scanning direction produces a significant difference in the current vs. potential behavior, suggesting that the history of the NiTi surface plays a significant role. The polarization curves exhibited a limiting current, the value of which increased with an increase in the rotation speed. The Levich plot showed a mass transport-controlled reaction as the rate-limiting step during the electropolishing of NiTi. The temperature dependence on the electropolishing of NiTi showed a typical Arrhenius behavior with an activation energy of 19.2 (±1.33) kJ/mol. The addition of Ni and Ti ions to the electropolishing solution showed a corresponding decrease in the limiting current density, confirming that the mass transport of dissolved metal ion species from the anode surface to the bulk of the solution is the rate-determining step.

7.2.4 NIOBIUM

Niobium superconducting radiofrequency (SRF) cavities are used in advanced particle accelerators because of their high operating efficiency. Since the superconducting RF current flows in a surface layer about 40 nm deep, it is essential to assure good superconducting properties of the surface layer. Attainment of super-smoothness of the surface is critical to meeting the performance demands. The standard preparation of Nb SRF cavities includes the removal of a "damaged" surface layer, by buffered chemical polishing after the cavities are formed. Electropolishing with $HF-H_2SO_4$ electrolytes yields better surface finishing of cavities that meet SRF performance goals, but a less-hazardous, more environmentally friendly process is desirable. To develop an HF-free electropolishing system, Zhao et al. [50] studied the anodic polarization behavior of Nb in sulfuric acid-methanol electrolytes with H_2SO_4 concentrations ranging from 0.5 to 3 M and the temperature varying between 0°C and −30°C. Anodic polarization curves showed a current plateau, the value of which decreased with increasing H_2SO_4 concentration and decreasing temperature. The impedance spectroscopic studies pointed to a compact salt film being responsible for the presence of the current plateau. The metal removal rate under conditions that gave the best polishing was comparable to that obtained by $HF-H_2SO_4$ solution and improved microsmoothing was obtained. The temperature was determined to be a more critical parameter than H_2SO_4 concentration to achieve microsmoothing. With decreasing temperature, the surface quality improved substantially. At −30°C microsmoothing was achieved in all H_2SO_4 concentrations. With the desired material removal of 100 μm, the surface mechanical damage was completely removed, and nanometer-scale surface roughness was measured.

Barnes et al. [51] investigated the electropolishing of Nb and Ti in H_2SO_4-MeOH electrolytes at temperatures varying between 20°C and −70°C. Polarization experiments were performed to determine the potentials at which a current plateau corresponding to electropolishing occurred. The current plateaus decreased in value and the voltage range expanded when the acid concentration was increased, or the temperature was lowered. This behavior is explained by a polishing process that operates according to the salt-film mechanism. Further evidence for this mechanism was established once the temperature of the electrolyte was dropped to −70°C. A visible salt film that developed at this temperature on the surface of the metal was found to contain oxygen and sulfur, along with the respective metal species. Barnes et al. [51] obtained optimum electropolished conditions for Nb and Ti with an electrolyte of 2 M H_2SO_4, a constant voltage of 15 and 25 V, respectively, and a temperature of −70°C for 2 hours. This polishing technique is capable of creating mirror finishes on both metals with root mean square (R_a) values from atomic force microscopy analysis to be 1.64 nm for Ti and 0.49 nm for Nb. The authors claim that the established electropolishing procedure may be effective for other valve metals, such as vanadium, zirconium, hafnium, and tantalum [52].

Neelakantan et al. [52] studied the electro-dissolution behavior of a 30% Nb-Ti alloy in a non-aqueous methanolic sulfuric acid solution using a rotating disc electrode. The influence of sulfuric acid concentration, rotation speed, and process temperature on the dissolution kinetics was investigated. The observed limiting current density increased linearly with an increase in rotation rate and followed the Levich behavior, thus confirming a mass transport-controlled process. The dissolution rate showed a strong dependence on the sulfuric acid concentration between 1 and 5 M. The value of the limiting current density decreased with increasing sulfuric acid concentration. The dissolving cations were identified as the rate-limiting species, thus indicating a compact salt-film mechanism for electropolishing. A 3 M sulfuric acid was chosen as an optimum electrolyte owing to its better controllability of the material removal rate. The average RMS roughness value for an electropolished surface was approximately 10 nm, which is significantly lower than a mechanically polished surface.

7.3   CURRENT OSCILLATIONS

Periodic fluctuations of current or potential are encountered in many electropolishing and ECM systems [23,24,30–33,36,37,41,53–55]. The majority of the reported fluctuations have involved active-passive or passive-transpassive transitions during the anodic polarization of metals. The literature available on this topic has been reviewed and several schemes have been proposed for the periodic phenomena. For the dissolution of iron in sulfuric acid, Frank developed a general theory in which the local changes in pH are responsible for the periodic current fluctuations [53]. The changes in solution pH brought about by electrode reactions coupled with diffusion may shift the Flade potential for active-passive transitions. The sequential formation, oxidation, and chemical dissolution of a salt layer or an oxide layer may induce the transitions [54]. The depletion and accumulation of an acceptor anion necessary for active dissolution may also lead to such fluctuations [54]. The periodic current fluctuations may arise from changes in the stoichiometry of anodically formed solid layers. Dielectric or mechanical disruption of the resistive layer may alternate with periods of film growth and repair.

Periodic fluctuations have been observed in many ECM systems, including Fe/sodium chlorate [23] and Cu/sodium chlorate [53] systems. An extensive study of the periodic phenomena observed in a Cu/sodium chlorate system has been conducted by Cooper et al. [54], who interpreted the periodic oscillations as sequential periods of film growth, field collapse, and partial dissolution and removal of the film material. Field collapse was considered to follow the onset of resistive switching transitions, while film dissolution/removal took place at discrete sites on the anode. For high-rate anodic dissolution of iron in sodium chlorate solution, Datta and Landolt [23] observed potential fluctuations at limiting and higher current densities corresponding to surface brightening. The frequency of these oscillations increased with increasing current density varying between 5 and 100 Hz. The amplitude was typical of the order of 3–4 V. Solution analysis confirmed that the chloride ions were produced due to the decomposition of chlorate ions. Periodic formation and breakdown of surface films leading to pitting-type phenomenon and reduction of chlorate ions on the active surface within the pits produced during the high-rate anodic dissolution of iron were attributed to the current oscillations.

Ponto and Landolt [36] observed current oscillations during anodic polarization of chromium in phosphoric-sulfuric solution under certain conditions in the limiting current region. The oscillations were attributed to interfacial reactions such as film breakdown or adsorption processes. Auger electron spectroscopy performed on electrodes polarized at the limiting current showed the presence of sulfur in the film indicating that sulfate ions are adsorbed at the metallic surface when the potential is sufficiently high. Similar behavior was previously observed during electropolishing of titanium in a perchloric acid-acetic acid solution where chloride was found on the surface after dissolution at sufficiently high potentials [41]. This behavior resembles pitting, which requires a critical potential for film breakdown and involves inter-actions between electrolyte anions and the oxide film. According to Glarum and Marshall [24], current oscillations during electropolishing of copper in phosphoric acid may occur in the presence of a negative impedance, and the oscillation period is affected by the electrolyte resistance. The pitting-type mechanism suggested above could be responsible for the negative impedance behavior of the oscillating systems.

The periodic current fluctuations observed during the anodic polarization of iron in phosphoric acid, described above (Figure 7.2), have been analyzed further. Figure 7.12 shows a set of current fluctuations observed at various intervals during potentiostatic anodic polarization at 0.05 V. After an initiation time, the general characteristics of the individual fluctuating pulses are reproducible over an extended period. An expanded view of the current fluctuations observed is shown in the inset of Figure 7.13. After an abrupt jump to a peak value, i_{max}, the current slowly decays and reaches a plateau value, $i_{plateau}$, then drops to a minimum value, i_{min}. The influence of the applied potential, disk rotation speed, and temperature on the periodic characteris-tics has been investigated [30]. The potential range of current fluctuations under these conditions is from -0.1 to $+0.15$ V. The dissolution valence in this potential range is 2.5, which corresponds to the simultaneous formation of Fe^{2+} and Fe^{3+}. The ampli-tude of the current fluctuations has been found to increase with increasing rotation speed, indicating that mass transport effects play an important role. In Figure 7.13, the inverse of the amplitude has been plotted vs. the inverse of the square root of the disk rotation speed. The straight line and positive intercept indicate the mixed control nature of the reaction, as previously observed for the current plateau, i_{l2}, above the fluctuations (Figure 7.2b). The current fluctuations were observed in a potential range situated between two limiting current plateaus, one corresponding to the formation of a salt layer at the surface that is purely mass-transport controlled, and the other corre-sponding to a mixed (mass transport and kinetic) control. Periodic fluctuations in this

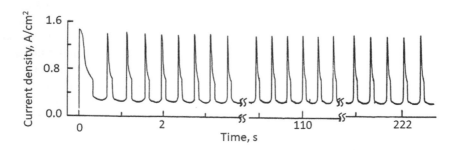

FIGURE 7.12 Periodic current fluctuations during anodic polarization of iron in 14 M phos-phoric acid at 0.05 V. Electrolyte temperature: 90°C; RDE rotation speed: 1,000 rpm [30].

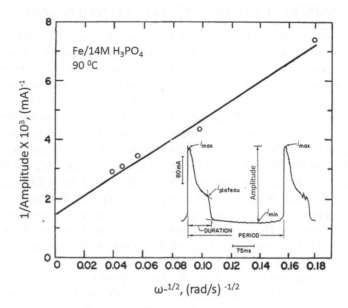

FIGURE 7.13 The inverse of current oscillation amplitude as a function of $\omega^{-1/2}$. The straight line and nonzero intercept are indicative of mixed (mass transport and kinetic) control of the anodic reactions. Inset: Expanded view of current fluctuations depicting amplitude and period [30].

system may, therefore, represent instabilities associated with the switching between these two film formation processes in the intermediate potential range. The oscillation period was found to increase with an increase in potential but remained independent of dissolution time. The amplitude of the fluctuations also showed a similar dependence. The dependencies imply that the charge associated with the fluctuations is an increasing function of potential. The charge per oscillating pulse, calculated by integrating the area under each trace of current vs. time, was found to increase with increasing potential, indicating that the film thickness increased with potential [55]. The electrolyte temperature had an insignificant influence on the fluctuation characteristics. The above results suggest that the periodic oscillations observed during the electropolishing of iron in phosphoric acid are associated with cyclic growth and removal of surface films, accompanied by periodicity in the anodic reaction stoichiometry that switches between the formation of Fe^{2+} and Fe^{3+}.

Current oscillations have also been observed during the electropolishing of niobium radiofrequency cavities. Diepers [56] and Crawford [57] reported that during the electropolishing of niobium in a mixture of sulfuric acid and hydrofluoric acid, the best results are obtained by operating under conditions of large current oscillations. In this system, sulfuric acid acting as an oxidizing agent forms an insulating layer of niobium pentoxide (Nb_2O_5), which gets dissociated by HF. Under the influence of the electrical potential, both processes occur simultaneously at a reasonably high rate. As electropolishing proceeds, the formation and breakdown of the dielectric layer of the niobium salt lead to current oscillations. The fluctuations repeat indefinitely, provided the bulk electrolyte solution can supply the required amount of HF.

7.4   TRANSPORT MECHANISM OF ELECTROPOLISHING

The above discussions adequately establish that the electropolishing of metals and alloys is achieved by operating at the mass transport-controlled limiting current. The observed limiting currents have been explained in the literature by two different mechanisms: a salt-film type or an acceptor type mechanism. The two mechanisms are schematically depicted in Figure 7.14. In the salt-film mechanism, dissolved metallic ions accumulate near the electrode surface until at the limiting current a thin salt film is formed on the surface and the concentration of the dissolving metal ions at the surface corresponds to the saturation concentration of the salt formed with the electrolyte anions. In the presence of the film, the dissolution rate depends on the mass transfer of the metallic ions from the film/electrolyte interface to the bulk solution. In an acceptor theory, the dissolution process is limited by transport of an acceptor species such as water or a complexing ion to the electrode surface. These species react with the dissolving metal ions to form complexed or hydrated species. At the limiting current, the surface concentration of the acceptor is zero.

Electropolishing studies related to salt-film mechanisms have generally involved steady-state measurements. The addition of a precipitating salt to the electrolyte is a commonly used method of testing for a salt-film mechanism. Taking into consideration the effect of salt addition on the electrolyte transport properties, such as diffusion coefficient, and kinematic viscosity, investigators have provided a quantitative interpretation of the observed mass transport controlled limiting currents. Electrochemical impedance spectroscopy (EIS) is another method that has proven useful. While the EIS measurements are readily performed, interpretation may not be easy. When using EIS to test a salt-film mechanism, the high-frequency range of the spectrum may be particularly valuable. A physical interpretation of the low-frequency part of the spectra is sometimes elusive. The proponents of the acceptor mechanism have mostly used the EIS technique. The application of flow modulation spectroscopy provides a measure of the diffusion time constant, thus allowing for

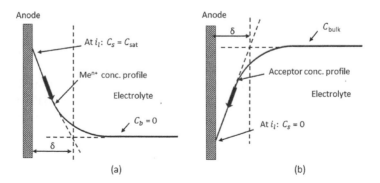

FIGURE 7.14   Schematic diagrams showing (a) transport of reaction products away from the anode surface (salt-film mechanism), and (b) transport of an acceptor to the anode surface (acceptor mechanism). δ is the Nernst diffusion layer thickness.

the determination of the diffusion coefficient of the acceptor independent of knowledge of its bulk concentration [27]. However, the experimental evidence in support of acceptor mechanisms is not straight forward. Table 7.1 summarizes some of the published work on different electropolishing systems and the proposed mechanism. While it is apparent that salt-film mechanisms have received overwhelming acceptance, for some systems such as copper-phosphoric acid, the rate-limiting species remains a controversial topic.

In an attempt to describe the role of mass transport-controlled limiting current and surface films on microsmoothing, we invoke the theories put forward by Edward [13], and Hoar [11] and elaborated by Landolt [19]. Mass transport-controlled dissolution results in surface leveling because protrusions of a rough surface are more easily accessible than recesses from a diffusion point of view. Furthermore, at the mass transport-controlled limiting current, the influence of charge-transfer kinetics

TABLE 7.1
Compilation of the Literature Data Showing the Proponents of the Salt Film vs. Acceptor Mechanism for Electropolishing

Author	Metal/Electrolyte	Year	Ref.
Rate-Limiting Species: Dissolved Metal Ions (Salt Film)			
W. C. Elmore	Cu/H_3PO_4	1939	[6]
T. P. Hoar, G. P. Rothwell		1964	[9]
K. Kojima, C. W Tobias		1973	[12]
J. Mendez, R. Akolkar, T. Andryuschenko, U. Landau		2008	[28]
L. Ponto, D. Landolt	$Cr/H_3PO_4 + H_2SO_4$	1987	[36]
L. Ponto, M. Datta, D. Landolt	$Fe_{13}Cr, Fe_{24}Cr, 304 SS/ H_3PO_4 + H_2SO_4$	1987	[37]
K. Kontturi, D. J. Schiffrin	Ni/H_2SO_4	1989	[38]
L. F. Vega, M. Datta	Fe/H_3PO_4	1990	[31]
M. Datta, D. Vercruysse	$Fe_{13}Cr/H_3PO_4, H_3PO_4 + H_2SO_4$	1990	[39]
M. Datta, L. F. Vega, L. T. Romankiw, P. Duby	$Fe/H_3PO_4 + H_2SO_4$	1992	[33]
O. Piotrowski, C. Madore, D. Landolt	$Ti/Methanolic H_2SO_4$	1998	[43]
O. Piotrowski, C. Madore, D. Landolt	$Ta/Methanolic H_2SO_4$	1999	[44]
L. Neelakantan, A. W. Hassel	$Ni_{55}Ti_{45}/Methanolic H_2SO_4$	2007	[49]
X. Zhao, S. G. Corcoran, M. J. Kelly	$Nb/Methanolic H_2SO_4$	2011	[51]
P. Barnes, A. Savva, K. Dixon, H. Bull et al.	$Ti, Nb/Methanolic H_2SO_4$	2018	[51]
L. Neelakantan, A. Pareek, A. W. Hassel	$Nb_{30}Ti_{70}/Methanolic H_2SO_4$	2011	[52]
Rate-Limiting Species: Acceptor (Water or Anion)			
J. Edwards	Cu/H_3PO_4	1953	[13]
S. H. Glarum, J. H. Marshall		1985	[25]
R. Vidal, A. C. West		1995	[27]
M. Matlosz, S. Magaino, D. Landolt	$Fe_{13}Cr/H_3PO_4 + H_2SO_4$	1994	[40]

is negligible, and therefore, differences due to grain orientation, grain boundaries, dislocations, or small inclusions do not play a significant role. Furthermore, the presence of an ionically conducting surface film suppresses the crystallographic effects on the dissolution process. As a result, random removal of atoms from the metal results in a very smooth surface topography leading to smooth and highly reflective surfaces.

It is interesting to discuss the similarities between the processes involved in surface brightening during high-rate dissolution in salt solutions under ECM conditions and electropolishing in concentrated acid solutions. Since salt solutions contain high water concentration, an acceptor mechanism can be easily ruled out. As we have noted in Chapter 5, there is no ambiguity about the presence of a salt film on the anode at the limiting current in aqueous salt solutions. With the knowledge of the physical properties of the electrolyte, and solubility of the reaction products in the electrolyte, the estimation of the surface concentration of dissolving metal ions at the limiting current yields values that agree with the saturation concentration. Electropolishing in concentrated acid solutions involves a low amount of water which creates confusion about it being the rate-limiting species. Despite this, the salt-film mechanism appears to be the favored mechanism in electropolishing electrolytes as well, as is evidenced in Table 7.1, where an overwhelming majority of researchers validate the salt-film mechanism. Based on this information, it may be concluded that electropolishing in concentrated acid solutions and in salt solutions under ECM conditions follow the same mechanism – the salt-film mechanism. The key difference between the two being the magnitude of the limiting current. In salt solutions, the high solubility of reaction products and low electrolyte viscosities lead to relatively high limiting currents. On the contrary, the limiting current density values in concentrated acid solutions are relatively small due to the high viscosity of the electrolyte and low saturation concentration of the dissolved products. Because of these reasons, concentrated acid solutions are preferred in most of the surface finishing processes which require only a small amount of material to be removed.

REFERENCES

1. W. J. McTegart, *The Electrolytic and Chemical Polishing of Metals*, Pergamon Press, London, (1956).
2. P. V. Shigolev, *Electrolytic and Chemical Polishing of Metals*, 2nd edition, Freund Pub, Tel-Aviv, Israel, (1974).
3. P. C. Madore, D. Landolt, *Plating Surf. Finish.*, 85 (5), 115–119 (1998).
4. P. A. Jacquet, *Trans. Electrochem. Soc.*, 69, 629 (1936).
5. P. A. Jacquet, *Met. Rev.*, 1, 157 (1956).
6. W. C. Elmore, *J. Appl. Phys.*, 10, 724 (1939).
7. C. Wagner, *J. Electrochem. Soc.*, 101, 225 (1954).
8. T. P. Hoar, *Modern Aspects of Electrochemistry*, J. M. O. Bockris ed., vol. 2, p. 262, Butterworths, London, (1959).
9. T. P. Hoar, G. P. Rothwell, *Electrochim. Acta*, 9, 135 (1964).
10. T. P. Hoar, D. C. Mears, G. P. Rothwell, *Corros. Sci.*, 5, 279 (1965).
11. T. P. Hoar, *Corros. Sci.*, 7, 341 (1967).
12. K. Kojima, C. W. Tobias, *J. Electrochem. Soc.*, 120, 1026 (1973).
13. J. Edwards, *J. Electrochem. Soc.*, 100, 189C (1953).
14. J. Edwards, *J. Electrochem. Soc.*, 100, 223C (1953).

15. R. Sautebin, H. Froidevaux, D. Landolt, *J. Electrochem. Soc.*, 127, 1096 (1980).
16. R. Sautebin, D. Landolt, *J. Electrochem. Soc.*, 129, 946 (1982).
17. C. Clerc, D. Landolt, *Electrochim. Acta*, 29, 787 (1984).
18. C. Clerc, M. Datta, D. Landolt, *Electrochim. Acta*, 29, 1477 (1984).
19. D. Landolt, *Electrochim. Acta*, 32, 1 (1987).
20. M. Datta, D. Landolt, *J. Electrochem. Soc.*, 122, 1466 (1975).
21. M. Datta, D. Landolt, *J. Appl. Electrochem.*, 7, 247 (1977).
22. M. Datta, D. Landolt, *Electrochim. Acta*, 25, 1255 (1980).
23. M. Datta, D. Landolt, *Electrochim. Acta*, 25, 1263 (1980).
24. S. H. Glarum, J. H. Marshall, *J. Electrochem. Soc.*, 132, 2872 (1985).
25. S. H. Glarum, J. H. Marshall, *J. Electrochem. Soc.*, 132, 2878 (1985).
26. R. Vidal, A. C. West, *J. Electrochem. Soc.*, 142, 2682 (1995).
27. R. Vidal, A. C. West, *J. Electrochem. Soc.*, 142, 2689 (1995).
28. J. Mendez, R. Akolkar, T. Andryushchenko, U. Landau, *J. Electrochem. Soc.*, 155 (1), D27 (2008).
29. D. R. Gabe, *Corros. Sci.*, 13, 175 (1973).
30. M. Datta, *IBM J. Res. Dev.*, 31 (2), 207 (1993).
31. L. F. Vega, M. Datta, *Electrochemical Society Extended Abstract No. 367, Fall Meeting, Seattle*, p. 531, (1990).
32. L. F. Vega, *Ph.D. Thesis*, Columbia University, (1990).
33. M. Datta, L. F. Vega, L. T. Romankiw, P. Duby, *Electrochim. Acta*, 37, 2469 (1992).
34. R. Knoedler, K. E. Heusler, *Electrochim. Acta*, 17, 197 (1972).
35. R. D. Armstrong, M. Henderson, *J. Electroanal. Chem.*, 32, 1 (1972).
36. L. Ponto, D. Landolt, *J. Appl. Electrochem.*, 17, 205 (1987).
37. L. Ponto, M. Datta, D. Landolt, *Surf. Coat. Technol.*, 30, 265 (1987).
38. K. Kontturi, D. J. Schiffrin, *J. Appl. Electrochem.*, 19, 76 (1989).
39. M. Datta, D. Vercruysse, *J. Electrochem. Soc.*, 137, 3016 (1990).
40. M. Matlosz, S. Magaino, D. Landolt, *J. Electrochem. Soc.*, 141, 410 (1994).
41. J. B. Mathieu, D. Landolt, *J. Electrochem. Soc.*, 125, 1044 (1978).
42. J. B. Mathieu, H. J. Mathieu, D. Landolt, *J. Electrochem. Soc.*, 125, 1039 (1978).
43. O. Piotrowski, C. Madore, D. Landolt, *J. Electrochem. Soc.*, 145, 2362 (1998).
44. O. Piotrowski, C. Madore, D. Landolt, *Electrochim. Acta*, 44, 3389 (1999).
45. C. A. Huang, F. Y. Hsu, C. H. Yu, *Corros. Sci.*, 53, 589 (2011).
46. T. Duerig, A. Pelton, D. Stockel, *Mater. Sci. Eng. A*, 273–275, 149 (1999).
47. N. B. Morgan, *Mater. Sci. Eng. A*, 378, 16 (2004).
48. K. Fushimi, M. Stratmann, A. W. Hassel, *Electrochim. Acta*, 52, 1290 (2006).
49. L. Neelakantan, A. W. Hassel, *Electrochim. Acta*, 53, 915 (2007).
50. X. Zhao, S. G. Corcoran, M. J. Kelley, *J. Appl. Electrochem.*, 41, 633 (2011).
51. P. Barnes, A. Savva, K. Dixon, H. Bull, L. Rill, D. Karsann, S. Croft, J. Schimpf, H. Xiong, *Surf. Coat. Technol.*, 347, 150 (2018).
52. L. Neelakantan, A. Pareek, A. W. Hassel, *Electrochim. Acta*, 56, 6678 (2011).
53. U. F. Frank, *Z. Electrochem.*, 62, 649 (1958).
54. J. F. Cooper, R. H. Muller, C. W. Tobias, *J. Electrochem. Soc.*, 127, 1733 (1980).
55. H. P. Lee, K. Nobe, A. J. Pearlstein, *J. Electrochem. Soc.*, 132, 1031 (1985).
56. H. Diepers, O. Schmidt, H. Martens, F. Sun, *Phys. Lett.*, 37A (2), 139 (1971).
57. A. C. Crawford, *Nucl. Instrum. Methods Phys. Res. A*, 849, 5 (2017).

8 Electrochemical Machining

8.1 INTRODUCTION

The demand for machining large and complex shaped high strength, heat-resistant super alloy products led to the development of noncontact manufacturing techniques. Electrochemical machining (ECM) is one of such nonconventional machining processes [1–5]. ECM is based on the principles of electrolysis, according to which the machining takes place by the anodic dissolution of the workpiece at high current densities. A schematic of the ECM process is shown in Figure 8.1. In the electrolytic cell, the piece of the metal that is to be shaped (the workpiece) is made the anode and the metal tool giving rise to shaping is made the cathode. The two electrodes face each other with a narrow space in between. A properly selected electrolyte is pumped through the narrow interelectrode space. The commonly used electrolytes consist of aqueous salt solutions of sodium chloride (NaCl), sodium nitrate ($NaNO_3$), or a mixture of the two. The electrolysis is initiated by applying a voltage between the two electrodes which results in the dissolution of the workpiece anode material, while hydrogen gas evolves at the cathode tool. Electrolyte flow at high velocities removes the dissolved material from the anode and the gaseous by-products produced at the tool surface. During machining, either

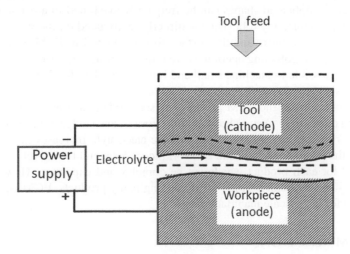

FIGURE 8.1 Schematic showing the working principle of ECM. During ECM, the tool cathode is moved toward the workpiece anode; an interelectrode gap of constant thickness is maintained, and the tool shape is reproduced on the workpiece. The dotted lines show the initial position of the tool and the workpiece.

the workpiece anode or the tooling cathode (commonly the tool) is fed toward the other. This ensures that a high material removal rate is maintained throughout machining. Over time, as the tool advances, the workpiece adopts a profile that is an approximate mirror image of the tool shape. Thus, an equilibrium state is reached, and the workpiece profile remains invariant. Since there is no contact between tool and workpiece there is no tool wear and no residual stress is imparted on the final workpiece.

ECM process is characterized by high current density, high electrolyte flow velocity, and narrow interelectrode spacing. The cell voltage is typically in the 5–20 V range. The high current densities of the order of 15–230 A/cm^2 encountered in ECM necessitate a rapid movement of the electrolyte through the narrow interelectrode gap to remove anodic and cathodic reaction products and to dissipate the heat generated. Electrolyte flow velocities through the gap usually range between 3 and 30 m/s. To pump the electrolyte at these velocities, high pressure of the order of 2–30 atmosphere is needed. A small electrode separation, between 0.1 and 1 mm, is essential to reproduce the contours of the cathode onto the anode workpiece and also to minimize the ohmic resistance in the gap thereby decreasing the electrical power consumption.

Compared to conventional metal removing processes, ECM has some unique features. Metal removal being electrochemical in nature is governed by Faraday's laws of electrolysis and depends on the electrochemical equivalent of the material independent of its hardness. So, a variety of materials, from hard metals to soft metals, can be machined with equal ease as long as they are electrically conducting. Since the tool never directly touches the workpiece, no mechanical deformation due to cold work is introduced during machining. The tool electrode used in the process does not wear; therefore, soft metals can be used as tools to form shapes on harder workpieces. Complicated shapes can be frequently machined in a single operation because, in principle, any tool shape is directly reproduced on the workpiece. The product is free from burrs and the surface finish obtained in ECM is such quality that no subsequent polishing operations are required even in quite demanding applications. It is possible to machine small and intricate work in hard or unusual metals and composite materials.

On the other hand, the ECM process has some inherent drawbacks. Equipment cost is high due mainly to complex pumping, storage, and sludge elimination systems required for the electrolyte. A saline or acidic electrolyte increases the risk of corrosion for the equipment, tool, and the workpiece. A special tool development effort is required for each product which can be complex and expensive. However, once made, the same tool can be used for producing many products. ECM is, therefore, best suited for large-scale manufacturing.

8.2   ECM REACTIONS

Atom by atom removal of metal by anodic dissolution is the basic principle underlying the electrochemical metal removal process. The anodic reaction is:

$$M \rightarrow M^{n+} + ne^- \tag{8.1}$$

where n is the valence of metal dissolution or the number of electrons removed from the dissolving metal atoms by anodic oxidation. At the cathode, electrolytic reduction of water takes place resulting in the formation of hydrogen gas and hydroxyl ions:

$$2H_2O + 2e^- \rightarrow H_{2\,(gas)} + 2OH^-. \tag{8.2}$$

Since only hydrogen gas is evolved at the cathode, its shape remains unaltered during the machining process. Depending on the nature of the material being machined and the pH of the electrolyte, dissolved metal ions and hydroxyl ions may combine to form hydroxide precipitates:

$$M^{n+} + nOH^- \rightarrow M(OH)_n \text{ (solid)}. \tag{8.3}$$

The hydroxide precipitates remain in suspended form in solution and need to be filtered.

The anodic reaction shown above is not the only reaction that may take place during machining. In electrolytes containing oxidizing anions, passivating oxide films are formed at the open-circuit or at low potentials. The oxide films so formed are electronically conducting such that at potentials exceeding that of oxygen potential (E_{O_2}), anodic oxygen evolution takes place. The oxygen evolution reaction is given by:

$$2H_2O \rightarrow 4H^+ + O_2 + 4e^-. \tag{8.4}$$

It is to be noted that while simultaneous oxygen evolution and metal dissolution take place at low current densities, metal dissolution rate increases with increasing current density leading to high-rate metal dissolution under ECM conditions.

Oxygen evolution reaction produces H^+ ions thus creating localized acidic conditions at the anode surface. However, the amount of H^+ ions produced is generally insignificant compared to the electrolyte volume used in the process. In any case, pH control of the electrolyte is one of the key operational items of the ECM process.

Anodic and cathodic gas evolution implies that during machining the interelectrode gap becomes filled with a two-phase, gas-liquid mixture. This, along with the electrolyte heating with the flowing current, leads to the variation of the electrolyte conductivity in the gap and, correspondingly, of the current density at the anode. These aspects are extremely important in the simulation of shape evolution and machining precision.

While hydrogen evolution by the decomposition of water is the sole cathodic reaction in nonpassivating electrolytes, in the nitrate electrolyte, the reduction of nitrate ions may take place on the cathode. The cathodic reduction of nitrate in the neutral and alkaline solutions yields nitrite ions or complete reduction yields ammonia:

$$NO_3^- + H_2O + 2e^- \rightarrow NO_2^- + 2OH^-, \tag{8.5}$$

$$NO_3^- + 7H_2O + 8e^- \rightarrow NH_4OH + 9OH^- \tag{8.6}$$

However, the degree of reduction depends on the prevailing electrochemical conditions. Since the limiting current for nitrate reduction is generally very low compared to the high currents employed in ECM, the rate of nitrate reduction is very insignificant, and such reactions do not contribute to the electrolyte composition changes.

8.3 PROCESS DESCRIPTION

8.3.1 ECM EQUIPMENT

Schematic of a manufacturing scale ECM installation is shown in Figure 8.2. It consists of four main components: (i) the electrochemical machine with a control unit for tool movement; (ii) the electrolyte circulation system consisting of pump, centrifuge, sludge removal, filter, heat exchanger, and electrolyte reservoir; (iii) the power supply with short-circuit protection; and (iv) the control system for feed rate, gap width, electrolyte temperature, pH value, pressure, and concentration. The process is conducted in an enclosed chamber which is connected to an exhaust to extract the fumes. The machining system consists of an electrolytic cell in which the workpiece and the tool electrode are positioned. The workpiece is firmly held by a fixture to provide stable operation at a constant gap. The workpiece is connected to the positive pole of a power supply, and the tool serves as the cathode. The tool movement is controlled by a tool feed system. The electrolyte is usually an aqueous solution of inorganic salt ($NaNO_3$ or $NaCl$) which is pumped through the interelectrode gap at a high velocity to remove the electrode reaction products and the heat generated. The applied voltage typically varies between 5 and 20 V. As the machining proceeds, the anode dissolves, and the tool is moved toward the workpiece. The equilibrium interelectrode distance is typically kept between 0.1 and 1 mm. A smaller interelectrode gap

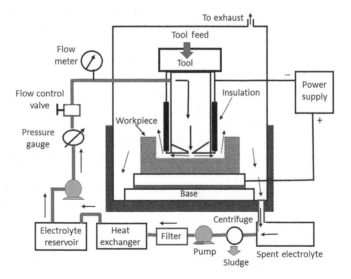

FIGURE 8.2 Schematic of a complete industrial ECM installation including electrolyte pumping and control systems.

width provides more accurate machining and a higher dissolution rate. The anodic and cathodic reaction products generated during ECM must be removed and a fresh electrolyte must be fed as the machining process proceeds. A centrifuge in the flow system separates the sludge from the spent electrolyte. The electrolyte is filtered and the heat is extracted by a heat exchanger before it is collected in an electrolyte reservoir from where the clean electrolyte is pumped into the interelectrode gap in the machining chamber, while the inlet pressure, the electrolyte flow velocity, pH, and temperature of the electrolyte are monitored and controlled.

8.3.2   ECM ELECTROLYTES

ECM electrolytes are categorized into two types: passivating (electrolytes containing oxidizing anions such as nitrates and chlorates) and nonpassivating (electrolytes containing aggressive anions such as chlorides, bromides, iodides, and fluorides). The electrolyte solutions are made up of 10%–20% salt in water and are generally in the neutral pH range. In many cases, a slightly acidic pH of the electrolyte may be essential to keep the reaction products dissolved in solution and to prevent their interference in the machining precision. There are some exceptional cases where either acidic or alkaline solutions are used for electrochemical machining. For example, for drilling of high aspect ratio small diameter holes, acid solutions are used. Acid electrolytes are essential in such cases to prevent sludge formation which may prematurely stop the machining process. Another case is the machining of tungsten and tungsten carbides where the use of alkali solutions is essential to anodically dissolve such materials. Employing strong acid or alkali solutions require extra precautions since they create hazardous operating and equipment corrosion conditions. In general, for ECM of most of the engineering materials, neutral aqueous salt solutions are commonly used.

8.3.3   THE INTERELECTRODE GAP

The ECM performance is directly associated with the processes within the interelectrode gap and its thickness distribution. To precisely reproduce the contours of the tool on the workpiece, the tool must be very close to the workpiece. However, maintaining the metal removal rate uniformity and thickness uniformity within the thin interelectrode gap is extremely challenging. During ECM, the reaction products produced at the electrodes remain in the electrolyte and accumulate in the flow direction within the gap. These products include metal hydroxide sludge and gas bubbles such as hydrogen and oxygen. Formation and accumulation of these products change the properties of the electrolyte in the machining gap. Furthermore, high currents encountered in ECM lead to a temperature rise of the electrolyte in the flow direction. Figure 8.3 shows a schematic of the variation of electrolyte temperature, the void fraction of bubbles, and electrolyte conductivity across the machining surface. Such changes of the electrolyte properties in the interelectrode gap lead to changes in the metal removal rate from the workpiece, which in turn affects the shape control accuracy. Removal of the reaction products and heat from the interelectrode gap by utilizing the flow of high-velocity electrolyte is one of the major activities in the case of conventional ECM.

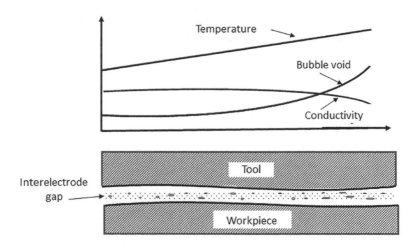

FIGURE 8.3   Electrolyte property changes in the interelectrode gap during ECM.

8.3.4   PROCESS MONITORING AND CONTROL

Monitoring and control of key parameters is essential for achieving precision machining and maintaining an environmentally friendly ECM process. From a machining precision point of view, the interelectrode gap is a key parameter that needs to be monitored and controlled. The reproduction of tool shape into the workpiece requires the tool to be as close as possible. Therefore, creating and maintaining a small interelectrode gap is very important to achieve precision dimensional control of the machined part. Current and voltage signals are used to control the machining rate and gap width. The voltage and current feedback signals from sensors and the waveform generated during the ECM process are used to predict the MRR, machining precision, and machining time. In many advanced ECM equipment, the machining process adjusts itself based on the feedback it receives. The process control agenda include short circuit detection and prevention which enables better process stability and precise machining.

Maintenance of electrolyte composition, pH, and temperature within the prescribed rage is another important aspect of ECM process control. During the ECM process, the electrolyte composition and its temperature in the interelectrode gap changes due to electrochemical reactions proceeding on electrodes and chemical reactions proceeding in the gap. Monitoring electrolyte pH continuously enables keeping its value within the prescribed limits. Any deviation from the prescribed range requires the addition of alkali or acid to bring the pH value within the range. The dissolved metal ions that get converted to insoluble metal hydroxides need to be separated to prevent their impact on the ECM performance. Separation of the insoluble precipitates (sludge) from the electrolyte is done by sedimentation or by using a centrifuge. In some cases, coagulation accelerating agents are used to make the separation process more effective.

8.4   CURRENT EFFICIENCY AND METAL REMOVAL RATE

Depending on the metal-electrolyte combination and operating conditions, different anodic reactions take place at high current densities. Rates of these reactions are dependent to a great extent on the ability of the system to remove the reaction products as soon as they are formed and supply fresh electrolyte to the anode surface. All of these factors influence the machining performance namely: dissolution rate, shape control, and surface finish of the workpiece. An understanding of the kinetics and stoichiometry of anodic reactions and their dependence on mass transport conditions is, therefore, essential to optimize the operating parameters.

Knowledge of anodic reactions that take place at high potentials is mostly derived from weight loss measurements and by applying Faraday's law [6–9]. The current efficiency for metal dissolution, θ is related to weight loss, ΔW, by

$$\theta = \frac{\Delta W n F}{ItM} \tag{8.7}$$

where I is the applied current, t is the time, F is the Faraday constant, and n is the valence of metal dissolution (the number of electrons removed from dissolving metal atoms by anodic oxidation and M is the atomic weight). For alloys, the atomic weight is calculated by using $M_{\text{alloy}} = \Sigma x_j M_j$, with x_j being the mole fraction of the component j and M_j its atomic weight.

The use of weight loss as a measure of dissolution valence is strictly applicable to anodic reactions involving metal dissolution only. In the presence of other reactions simultaneously occurring at the anode, such as oxygen evolution, it is essential to get a complete analysis of the reaction products including collecting the gas to determine the contribution of each reaction to the current efficiency. Accordingly, the weight loss measurements have been frequently used to determine an "apparent dissolution valence" [6–9].

The value of n provides information on the overall reaction stoichiometry and is useful for the calculation of the metal removal rate. Material removal rate, MRR, is related to the dissolution valence, n, according to Equation 8.8:

$$\text{MRR} = \frac{IM\theta}{nFA\rho} \tag{8.8}$$

where the material removal rate, MRR, is in cm/s, A is the surface area in cm^2, and p is density in g/cm^3.

Determination of current efficiency for metal dissolution in nonpassivating electrolytes, such as chloride solutions, is somewhat easier since there is no simultaneous anodic oxygen evolution occurring in these systems and the value of θ can be directly obtained from weight loss measurements. Figure 8.4 shows metal dissolution efficiency as a function of the current density for a few selected materials in nonpassivating and passivating electrolytes. For nickel dissolution in chloride solutions (Figure 8.4a), the current efficiency based on divalent nickel formation is 100% independent of current density and electrolyte flow rate [6]. Cr dissolution in chloride solution yields a value of $n = 6$ independent of current density [10]. For iron dissolution in chloride solutions, the current efficiency based on divalent iron (Fe^{2+}) is 100% at low current densities, but at current densities higher than the limiting current, the value of which depends on the electrolyte flow rate, the current efficiency based on Fe^{2+} decreases. Complete analysis of reaction products showed no evidence of anodic oxygen evolution [11]. The decrease in current efficiency, therefore, is attributed to the simultaneous production of Fe^{2+} and Fe^{3+}, with an average value of 2.5 [6]. This behavior is shown by curve b in Figure 8.4. The dissolution behavior of Fe13Cr alloys is similar to that of pure Fe with the measured values of n being 2.1 at low current densities and 2.7 at high current densities above the limiting current. The measured values for Fe13Cr are explained by postulating formation of Fe^{2+}, Cr^{2+}, and Cr^{3+} below the limiting current and formation of Fe^{2+}, Fe^{3+}, and Cr^{6+} (chromate) above the limiting current. Anodic dissolution of Fe24Cr alloy yields $n = 3.6$ independent of current density [12], which corresponds to the formation of Fe^{3+} and Cr^{6+}. For 304 stainless steel, the experimental data yields $n = 3.5$ independent of current density. This corresponds to the formation of Fe^{3+}, Cr^{6+}, and Ni^{2+} [10]. In summary, anodic dissolution of nickel, Cr, Fe24Cr, and 304 SS in NaCl (nonpassivating) electrolyte correspond to the curve (a) in Figure 8.4a where current efficiency is shown to be independent of current density. On the other hand, for anodic dissolution of Fe and Fe13Cr in NaCl, the current density-dependent dissolution stoichiometry leads to a decreased current efficiency at high currents (Figure 8.4b).

Anodic reaction stoichiometry during the high-rate dissolution of iron and nickel in passivating ECM solutions (sodium nitrate and sodium chlorate) has been extensively studied [7,10,12,13]. In these studies, oxygen evolution has been found to be the predominant anodic reaction at low current densities in the transpassive potential region. At higher current densities, the relative rate of metal dissolution increases

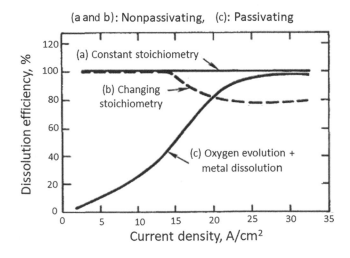

FIGURE 8.4 Current efficiency for metal dissolution as a function of current density in passivating and nonpassivating electrolytes: (a) nickel in 5 M NaCl, (b) iron in 5 M NaCl, and (c) nickel and iron in $NaNO_3$ and $NaClO_3$ electrolytes [6,7].

with increasing current density (Figure 8.4c). The transition from predominantly oxygen evolution to high-rate transpassive dissolution is governed by the specific metal-electrolyte interactions which influence passivation and depassivation processes. Different results have indicated that even in passivating systems, the metal dissolution current efficiency reaches close to 100% at high current densities under ECM conditions [7,10,12,13]. The most important aspect of passivating electrolytes is that the dissolution efficiency is an increasing function of current density which plays a significant role in achieving machining accuracy.

8.5 MASS TRANSPORT AND SURFACE FINISH

The technical feasibility of an electrochemical metal shaping process is determined by the metal removal rate, shape profile, and surface finish. These criteria are dependent on the ability of the system to provide desired mass transport, current distribution, and surface film properties at the dissolving workpiece surface. In most applications, mass transport and current distribution are intimately related. Mass transport conditions at the dissolving anode not only affect the current distribution on a macroscopic and a microscopic scale, but they are also crucial for the surface texture resulting from dissolution.

The rate of dissolution at any given point on the electrode surface is proportional to the local current density. The shape evolution of an anodically dissolving metal, therefore, depends on the current distribution. In ECM, the macroscopic current distribution on the workpiece is most important because it determines the final shape resulting from anodic dissolution. Because of the high current densities used, current distribution problems can be well described by the primary current distribution approximation. However, local conductivity changes resulting from the presence of gas bubbles in the interelectrode gap or from Joule heating create complications that need to be considered. Furthermore, in many systems, the current efficiency varies with the current density. For example, high-rate transpassive dissolution of nickel or iron in passivating (sodium nitrate and sodium chlorate) electrolytes, current efficiency increases with an increase in current density as discussed above [7]. In these systems, closer tolerances have been achieved in practice due to this effect. For surface finishing operations, both the macroscopic and the microscopic current distribution are of importance. While macroscopic current distribution depends on the potential distribution and on the hydrodynamic conditions, the current distribution on a microscopic scale is mostly governed by mass transport. From a mathematical point of view, the current distribution on a micro profile for a purely diffusion-controlled process is the same as the primary current distribution and, therefore, Laplace's equation can be used to simulate the rate of anodic leveling. However, although anodic leveling under primary current distribution conditions and under mass transport-limited conditions are mathematically the same, the resulting surface finish differs.

Metal dissolution at high current densities in both passivating and nonpassivating systems leads to an increase in the metal ion production at the anode. When the metal ion concentration at the surface exceeds the saturation limit, precipitation of a thin salt film occurs. The polarization curve under these conditions exhibits a limiting current plateau. The limiting current density has been found to increase with increasing electrolyte flow in a channel cell or with increasing rotation speed in an RDE system. The limiting current density is, therefore, controlled by convective mass transport. The formation of salt films on the anode influences the surface morphology of a dissolved workpiece [6,7,9,13]. Different studies have conclusively demonstrated that two distinctly different surface morphologies result from dissolution. At current densities lower than the limiting current density, extremely rough surfaces are obtained which, depending on the metal-electrolyte combination, reveals crystallographic steps and etch pits, preferred grain boundary attack or finely dispersed microstructure. On the other hand, dissolution at or higher than the limiting current leads to electropolished surfaces.

Surface finishing is a combination of leveling and brightening effects. Anodic dissolution results in surface smoothing because protrusions of a rough surface are more easily accessible than recesses. This is known as the leveling effect. At the mass transport-controlled limiting current, the influence of charge-transfer kinetics is negligible and differences due to grain orientation, grain boundaries, dislocations, or small inclusions, therefore, do not play a significant role. This leads to a brightening effect caused by the random removal of atoms from the metal surface. In ECM electrolytes, the surface smoothing is identified to be the dissolution product limited transport known as the salt-film mechanism in which the rate of transport of dissolving metal ions from the anode surface into the bulk is rate-limiting. At the limiting current, a thin salt film is present on the surface and the concentration of the dissolving metal ions at the surface corresponds to the saturation concentration of the salt formed with the electrolyte anions. The plateau current, due to the resistive salt film, may extend over several volts, depending on the salt-film properties. Surface brightening of nickel, Fe, and stainless steel under ECM conditions in $NaCl$, $NaClO_3$, and $NaNO_3$ electrolytes has been shown to occur when the limiting current leading to salt-film formation is reached or exceeded [6,7]. Similar behavior has been observed for copper [14], titanium [15], and stainless steel [10]. The solubility of the dissolving metal ions in ECM electrolytes is generally high and, in addition, high flow rates and/ or pulsating current lead to a small diffusion layer thickness at the anode. As a consequence, the limiting current densities needed to achieve surface brightening under ECM conditions are usually high, typically on the order of several A/cm^2.

Bright and polished surfaces often exhibit pits and pits with tail in the flow direction. These phenomena have been observed with many metals and alloys, and it has been shown that these tails may give rise to so-called flow streaks mentioned in the ECM literature [3]. It has been established that pit tailing and flow streak formation are observed only during transpassive dissolution with the direct current when anodic dissolution is entirely, or partially mass transport controlled [16]. The formation of pit tails as an initial step to flow streak formation is therefore due to a local disturbance of hydrodynamic conditions introduced by the presence of a pit. This is possible if flow separation [17] caused by the presence of pits creates small vortices and eddies that increase the local mass transport rate at the downstream end. However, flow streaks formation is restricted to direct current ECM under brightening conditions where the anode surface is covered with a salt film that leads to surface brightening. On the other hand, surfaces dissolved in the active mode or during transpassive dissolution with pulsating current do not show pit tails. The fact that under pulsating current no tails are observed even under transpassive dissolution conditions is also consistent with the fact that the effective diffusion layer thickness during dissolution with the pulsating current is much smaller than the steady-state diffusion layer thickness and is essentially determined by the pulse parameters. Therefore, local variations of hydrodynamic conditions have little influence on the dissolution rate. This aspect is one of the important virtues of using pulsating current in metal shaping and finishing operations.

8.6 MACHINING ACCURACY

In ECM, the distribution of metal dissolution rate on the workpiece determines its final shape in relation to the tool [18,19]. From a machining accuracy point of view, two types of metal removal rate (MRR) distribution need to be considered. The first type relates to MRR in the machining area versus the stray current area which mainly acts to define the initial machining profile against the tool profile. This is mainly dependent on the choice of electrolyte. The second type is the MRR distribution within the interelectrode gap. Variation of the thickness of the interelectrode gap and the physical properties of the electrolyte within the gap affects the entire machining process. A combination of the two types of MRR distribution affects the final shape of the workpiece.

8.6.1 PASSIVATING ELECTROLYTES

For a given metal, the anodic reactions occurring at different current densities under ECM conditions depend mainly on the electrolyte used and the machining performance is significantly influenced by the current density dependence of the anodic reactions. This is illustrated in Figure 8.5 for commonly used engineering materials such as Ni, Fe, and their alloys in passivating and nonpassivating electrolytes. The reaction rate distribution at the beginning of a shaping operation for the three cases discussed above are shown in Figure 8.5a [18,19], and the corresponding machined profiles during electrochemical drilling (ECD) are shown schematically in Figure 8.5b. In the machining area where the workpiece directly faces the cathode tool, the anodic reaction rate is constant for a constant interelectrode gap and conductivity. Away from the machining area, current density on the workpiece decreases asymptotically to zero with increasing distance. Current efficiency for metal dissolution, which is a function of the current density and local flow conditions, varies as a function of distance from the tool. Three different metal-electrolyte systems exhibiting different types of anodic behavior are considered in Figure 8.5. Two among the three types are in nonpassivating electrolytes such as chloride electrolytes: (case A) where the dissolution stoichiometry, hence the current efficiency, is independent of current density, and (case B) where due to changes in the dissolution stoichiometry, the dissolution efficiency is higher at low current densities but lower at high current densities. And the third case is in passivating electrolytes such as nitrate or chlorate solutions (case C), where at low current densities, oxygen evolution occurs simultaneously with metal dissolution, and the dissolution efficiency is an increasing function of the current density.

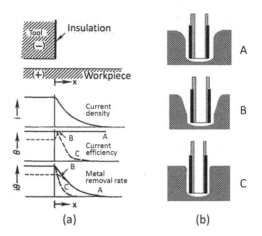

FIGURE 8.5   (a) Schematic of reaction rate distribution on a flat surface: the variation of current density, current efficiency, and metal removal rate as a function of the distance from the machining area for three different types of anodic behavior of Figure 8.4: (A) nickel in NaCl, (B) iron in NaCl, and (C) nickel and iron in nitrate and chlorate electrolytes. (b) Machined profiles during ECD corresponding to anodic behaviors (A–C) [18,19].

As shown in Figure 8.5a, the metal removal rate at any location is proportional to the product of current density and metal dissolution efficiency. Figure 8.5b shows shapes of drilled holes for the three cases considered. It is evident that machining accuracy is much better in passivating electrolytes since in the stray current region there is very low to no metal dissolution while high-rate metal dissolution takes place in the machining area (case C). The worst case is that of a system where dissolution efficiency is higher in the stray current region compared to the machining area (for example, Fe or Fe13Cr in NaCl; case B). It is therefore evident that shape prediction in ECM requires knowledge not only of local current distribution but also of the functional dependence of metal dissolution efficiency on current density and electrolyte flow conditions. In agreement with the above discussions, passivating metal-electrolyte systems are known to give better ECM precision because of their ability to form oxide films and evolve oxygen in the stray current region [18]. Similar results have been confirmed during electrolytic jet electrochemical micromachining where passivating electrolytes have been found to yield minimized stray cutting [20].

8.6.2   Pulse ECM

As described above, a small interelectrode spacing is required for the precise reproduction of the tool shape onto the workpiece. A small interelectrode gap in ECM results in improved dimensional accuracy control and simplified tool design. However, using a small gap in ECM makes it difficult to evacuate the reaction products which lead to process instability. Application of pulsating voltage/current with alternating pulse-on and pulse-off times permits the system to relax and refresh the electrolyte within the machining gap thus enhancing the accuracy of the ECM process.

Datta and Landolt investigated the feasibility of using pulsating current in ECM [21,22]. An important aspect of the use of a pulsating current is the possibility of varying mass transport conditions at the anode by independently varying pulse parameters as opposed to direct current ECM where mass transport is defined by hydrodynamic conditions [21]. Mass transport in pulse electrolysis is a combination of steady-state and non-steady-state diffusion processes. The presence of a pulsating current leads to the formation of a time-dependent diffusion layer close to the anode surface within which the dissolved metal ion concentration is a periodic function of time. In pulse ECM, anodic dissolution takes place during the pulse-on time while the system relaxes during the pulse-off time. Thus, by suitable choice of the magnitude of the pulse parameters, it should be possible to apply extremely high peak current densities and achieve high instantaneous mass transport rates even at low electrolyte flow rates. An analysis aimed at predicting electrolyte heating in narrow gaps under pulsed ECM conditions indicated that the instantaneous and average electrolyte heating can be minimized by working with a small "pulse-on" time and a low duty cycle [21]. On the other hand, one of the serious drawbacks of pulsed dissolution is that the average current, and hence the average metal removal rate, is very low. Experimental results obtained under well-controlled hydrodynamic conditions in a flow channel cell using passivating and nonpassivating systems indicated that a good surface finish and high current efficiency for metal dissolution can be obtained even at low average current densities [21,22]. Since local variations of hydrodynamic conditions have little influence on the dissolution rate, the formation of flow streaks is largely absent in pulse ECM as explained above. Another important advantage of pulse ECM is the localization and improved directionality of dissolution due to its ability to employ high instantaneous current densities.

Rajurkar et al. [23] investigated PECM taking into consideration the movement of the bubble-mixed electrolyte layer in the interelectrode gap. They concluded that both the current density and the pulse on-time have significant effects on the effective volumetric electrochemical equivalent. Short pulses are preferable for the enhancement in the localization of the anodic dissolution, which leads to a significant improvement in dimensional accuracy. Proper setting of parameters such as the distance of electrolyte flow path in the gap, electrolyte flow velocity, and pulse on-time, is important to reduce the variation in the anodic removal rate along the electrolyte flow direction in the gap. Kozak et al. [24] proposed a mathematical model for PECM processes that considered the non-steady physicochemical phenomena in the interelectrode gap. Based on the model analysis, they concluded that the gap distribution in PECM is more uniform as compared with the direct current ECM. This fact permits the tool electrode shape to be designed as an equidistant surface from the required anode shape. The gradients of temperature, gas concentration, and conductivity along the electrolyte flow direction are small during PECM, particularly when shorter pulse on-times are used. For uniform physicochemical conditions of anodic dissolution along the flow path, the machining should be performed with pulse on-times as short as possible. Shorter pulse on-times bring about both higher accuracies and metal removal rates due to the significant effect of pulse on-time on the dissolution localization factor. PECM machining with small gap sizes should be accompanied by shorter pulse-on times.

Chen and Han [25] investigated the possibility of using the polarization voltage during pulse off time to enable gap detection and avoid short circuits in PECM. The short circuit between the tool electrode and the workpiece in ECM is one of the main factors affecting process stability, machining efficiency, and the workpiece surface quality. Most short-circuit detection methods identify a large current after the occurrence of a short circuit, by which time burns and other damage caused by a short circuit have already occurred. Chen and Han found that there is a polarization voltage that does not corrode the workpiece in PECM, and the voltage varies with the nature of the electrolyte. By measuring the polarization voltage during the pulse-off time, they presented a new method to enable ECM gap detection and machine tool feed servo feedback control. This enabled the automatic discharge of a short circuit and recovery of the machining process, which greatly improves its stability and efficiency.

Based on the above discussion it is evident that PECM provides an attractive alternative to the direct current ECM. The improved electrolyte flow condition in the interelectrode gap, enhanced localization of anodic dissolution, and small and stable gaps found in PECM lead to higher dimensional accuracy, better process stability, and relatively simpler tool design. It is well known that pulse parameters such as pulse on-time, pulse off-time, and pulse waveform, and inlet electrolyte parameters such as temperature, pressure, and concentration have significant effects on the gap distribution. A proper selection of these parameters greatly reduces the shaping errors associated with the electrolyte conductivity variations and also increases the localization of anodic dissolution.

8.6.3 OTHER METHODS

Davydov and Volgin [5,26] recommended the application of air-electrolyte mixtures as a working medium to increase accuracy in ECM. The air-electrolyte mixture enables one to raise the localization of metal dissolution in places within the smallest gaps and, thus, to enhance the accuracy of electrochemical reproduction of the tool on the workpiece. In order to achieve the highest efficiency, the air-electrolyte mixture should be formed in the immediate vicinity of the gap inlet and the flow rate should be adequately high. A ratio between gas and liquid amounts from 3.0 to 3.5 is considered to be most effective.

Sectional tool electrodes (STE) have been considered to improve ECM accuracy [5,26]. STE is particularly suitable for ECM of large surfaces where several difficulties are encountered that include nonuniformity of the electrolyte conductivity in the direction of flow within the interelectrode gap and the necessity to employ extremely powerful power source that may go up to a few tens of thousands of amperes. One way to overcome these difficulties is to apply cathodes comprising several sections separated from one another by insulating bushings and connected to the power source in a designed order (Figure 8.6). Figure 8.6a is a schematic of the sectioned tool assembly. The assembly was used for electrolytic shaping of nickel in a 15% NaCl solution [26]. The metal removal rate was determined during ECM with and without the sectioned tool. Figure 8.6b shows how shaping with a sectioned tool makes the voltage distribution more uniform and simultaneously enhances the process efficiency. A serious drawback of the sectioned tools is that sections next to the working

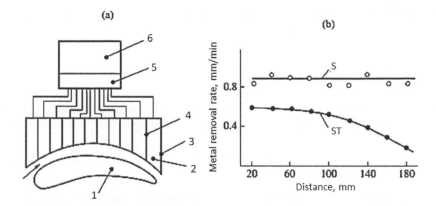

FIGURE 8.6 (a) ECM with sectioned tool: (1) workpiece, (2) section, (3) sectioned tool, (4) intersection insulation, and (5 and 6) process controlling device; and (b) distribution of metal removal rate over the interelectrode gap during ECM of nickel in 15% NaCl using (S) solid and (ST) sectioned tools [26].

section dissolve, due to a bipolar effect, the dissolution being more pronounced at smaller gaps and thinner insulators. One of the two methods proposed for solving this problem was the application of corrosion-resistant materials for sectioned tools, for example, titanium. The other involved electrochemical protection of nonoperating tool sections by applying a relatively low voltage to them. Electrochemical protection may also be employed for improving the copying accuracy.

8.7 SHAPE PREDICTION AND TOOL DESIGN

Accurately reproducing the tool shape onto the workpiece is one of the key challenges of the ECM process. Predicting the workpiece shape evolution for a given tool shape is known as the direct problem in ECM, whereas the inverse problem deals with designing the shape of the tool capable of producing a desired workpiece shape under specified machining conditions. For the prediction of workpiece shape, it is essential to estimate the anode material removal thickness at a given time increment. The material removal thickness is a function of current density distribution at the gap including the varying electrical conductivity of the electrolyte. The electrolyte properties depend on the temperature and gas bubble formation, which in turn depends on the velocity and pressure fields besides current density. Other parameters affecting the current distribution at the anode include geometry, anodic reaction kinetics, and hydrodynamic conditions.

The basic equations required for the numerical analysis of the shape prediction are discussed below. In the absence of concentration gradients, the potential distribution in the electrolyte obeys Laplace's equation,

$$\nabla^2 \phi = 0 \qquad (8.9)$$

The solution to Equation 8.9 gives the potential at any point in the workpiece. The current density is given by Ohm's law, Equation 8.10:

$$i = - \kappa \nabla \phi. \tag{8.10}$$

where κ is the electrolyte conductivity. Since the electrolyte in the interelectrode gap is filled with reaction products, the conductivity expression in the gap is expressed as:

$$\kappa = \kappa_e (1 - \alpha)^n \left[1 + v (T - T_e) \right] \tag{8.11}$$

where κ_e is the electrolyte conductivity at the entrance, α is the void fraction, v is the conductance constant representing temperature dependence of electrolyte conductivity, n represents a generalization of heterogeneous conduction mechanism and is taken to be 1.75 [27] and T_e is the electrolyte temperature at the entrance.

The local current density at the electrode is proportional to the potential gradient normal to the surface. Considering that the electrode potential does not vary appreciably with current density, the current distribution depends only on the geometry of the electrochemical system. Since the kinetic resistances at the electrodes are neglected, the electrodes are treated as equipotential surfaces. At the anode, the potential, ϕ, is given by:

$$\phi = V_a, \tag{8.12}$$

and at the cathode:

$$\phi = 0 \tag{8.13}$$

The gradient of the potential at all lines of symmetry and insulators is zero:

$$\frac{\partial \phi}{\partial n} = 0. \tag{8.14}$$

For the computation of shape evolution in ECM, the current distribution at the anode is combined with Faraday's law according to which the rate, r, at which the anodic surface recedes is determined as:

$$r = i \frac{M\theta}{nF\rho} \tag{8.15}$$

where i is the current density, M is the molecular weight of the metal, n is the metal dissolution valence, ρ is the density of the metal film, F is Faraday constant, and θ is the current efficiency of metal dissolution. For the simulation of shape evolution, the current distribution is solved for an initial anode profile. The surface of the anode is then moved proportionately to the current distribution using Faraday's law. This process is repeated at several time steps to predict the shape evolution at the anode.

Several researchers have attempted to predict the workpiece shape machined by a given shaped tool. Different numerical techniques including the finite difference method, the finite element method (FEM), finite difference method (FDM), and the boundary element method (BEM) have also been employed [28–30].

Thorpe and Zerkle [27] proposed a one-dimensional, two-phase, fluid flow model and showed that most ECM can be treated as a quasi-steady process. The electrolyte flow is two-dimensional in the transition region, where the gap between the electrodes changes rapidly. A two-dimensional flow model is needed in predicting the workpiece shape machined by a tool of complex geometry, since the electrolyte flow may recirculate or separate around corners.

Narayanan et al. [31] employed BEM programs using linear and quadratic elements to predict workpiece shapes during ECM. Accuracy is influenced by element size, time step size and the type of iso-parametric element used. Mesh refinement and small-time steps gave more accurate solutions but with a substantial increase in computational time. Results obtained using quadratic elements showed an increase in accuracy and a reduction in computational time compared with equivalent linear elements.

Jain et al. [32] employed the finite element technique to analyze EC bit drilling operations by considering one- and two-dimensional problems. The models account for the effects of temperature, void fraction, feed rate vectors inside and transition zones, and voltage history on the predicted anode profiles.

Hourang and Chang [33] applied a one-phase, two-dimensional fluid model to predict the flow and thermal fields between electrodes in an ECD process. A body-fitted curvilinear coordinate transformation was used such that a moving, irregular surface of the workpiece in the physical domain becomes a fixed, regular boundary in the computational domain. Some transport properties were found to vary abruptly in the transition region. In the transition region, the rate of change of pressure in the streamwise direction was higher near the tool than near the workpiece. The electrolyte temperature near the surfaces of the workpiece and the tool in the transition and side regions was higher than in other places.

The determination of the tool shape to produce a desired workpiece shape is more challenging. Therefore, tool design forms a major part of the modeling efforts of the ECM process [34–38]. Jain and Rajurkar [34] reviewed different models proposed for the analysis of tool design in ECM. These models include methods based on the $\cos\theta$ method, complex variable approach, empirical and nomographic approach, and numerical simulation methods such as finite differences, finite element, and BEMs. The numerical simulation approaches vary in the methodology employed to solve the Laplace equation. The assumptions and applications are also somewhat different, and they have their own merits and limitations. The interdependence of different parameters was suggested to be the main cause of the low success in tool design for ECM [34]. Das and Mitra [35] considered the problem of cathode shape determination for a given anode shape in ECM by developing an algorithm based on the boundary integral equation technique and nonlinear optimization. An additional flux condition at the anode was used as the constraint. Through an iterative process, the shape of the cathode was determined. Based on the testing of the algorithm, they claimed their technique to be superior to the numerical techniques based on the embedding method or the method of lines. Sun et al. [37] investigated the application of FEM to tool design for ECM freeform surface. Pattavanitch et al. [38] developed a 3-D boundary element model to simulate the ECM process. The BEM has the advantage that it does not require the solution to current density at points inside the domain but only at discrete points on the workpiece surface. They applied the model to EC drilling, turning, and milling and concluded that the accuracy of prediction is strongly dependent on the magnitude of the time step and mesh density. Since small time steps are involved, the simulation of the ECM process is very time-consuming. They could reduce the computing time by classifying the nodes as active or passive.

The generic aspects of tool design for ECM were presented by Westley et al. [39]. Since the controlling parameters of the process are difficult to predict when designing a new electrode for a newly designed workpiece, they propose to produce a primary tool from a soft material, such as brass, to facilitate any subsequent alterations. When the electrode performs satisfactorily, a hard material, such as copper-tungsten, should be used to manufacture the production electrode, as this has a high melting point and will resist damage from electrical sparks. In order to avoid problems due to short circuit, the electrode should be manufactured slightly less than the length of the casting gate. The dimensions of the finished component drawing are entered into a CMM machine to produce a digital profile of the required workpiece and a brass tool is then machined in accordance with the digitized data. The brass tool is then altered as needed in order to achieve the best shape. The knowledge gained from developing the electrode can then be used in developing electrodes for subsequent components.

8.8 ECM TECHNIQUES AND APPLICATIONS

ECM is a high precision contact-free shaping process applicable to electrically conducting materials regardless of their hardness. The process does not induce any thermal or mechanical stresses. A high standard of surface finish is achieved with high reproducibility. With typical tolerances down to 5 μm, complex macro, and microfeatures can be machined by ECM in a variety of materials including exotic superalloys that are difficult to machine by conventional machining methods. Diverse machining operations such as turning, grinding, milling, profiling, drilling, structuring, deburring, and polishing can be performed by the ECM process. Due to these capabilities, the ECM process has found cross-industry applications including aerospace, automotive, machine tooling, biomedical, energy, and general engineering. A showcase of different complicated shaped parts machined by ECM is presented in Figure 8.7. In the following, a brief discussion of a few selected ECM techniques and their application is presented.

8.8.1 DIE SINKING AND COMBINED TOOL MACHINING

Die sinking is the simplest form of ECM in which the tool shape is reproduced on the workpiece as shown in Figure 8.8. During this process, the tool and the workpiece are positioned in a high-pressure cell with an electrolyte pumped between the two electrodes. A voltage is applied, and the tool is fed toward the workpiece resulting in material removal from the surface of the workpiece. As the tool moves closer to the

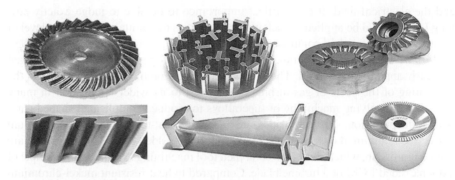

FIGURE 8.7 A showcase of parts machined by ECM. With typical tolerances down to 5 μm, complex macro, and micro features can be machined by ECM in a variety of materials independent of their hardness including exotic superalloys that are difficult to machine by conventional machining methods. (Reproduced with permission from Scott Kowalski, President, RM Group, PEM Technologies Americas.)

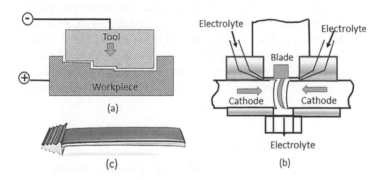

FIGURE 8.8 (a) Schematic of a die-sinking ECM process, (b) turbine blade shaping by combined tool ECM, and (c) a turbine blade.

workpiece, the material removal rate will tend toward a rate that equals the feed rate and an equilibrium condition is reached.

Combined tool machining is commonly associated with the manufacture of turbine blades. It is similar to the die-sinking approach described above, except in this method two tooling electrodes are used simultaneously to machine the workpiece from both sides. Figure 8.8b shows the schematic of a combined tool machining system. Due to the nature of this style of machining, it is impossible to reach an equilibrium workpiece shape – the leading and trailing edges of the blade will be continually machined. The simulation of this type of process is a more complicated problem than that of simulating die-sinking ECM due to the high level of curvatures of the leading and trailing edges. These regions are defined by the complex spatial electrostatic field in this machining area. In addition to this, equilibrium machining will not be reached at the leading and trailing edge of the workpiece. In situations such as this, the dominant factors are the initial form of the workpiece, the feed rate, and the electrical field. It is of critical importance to be able to judge exactly how long ECM should be applied to such workpieces so as to maintain the correct level of accuracy for the finished component.

Turbine blades: The machining of turbine blades is perhaps the most important and early application of ECM. The successful application of the ECM process in the machining of the parts of gas-turbine engines led to its wider application in many other fields involving machining of superalloys to produce complicated shaped parts with high precision. Turbine blades are made of heat-resistant nickel–chromium alloys and titanium alloys. Both one-sided and double-sided modes of machining are used. Figure 8.8b, which shows the combined tool machining system, is an example of a double-sided ECM of a turbine blade. Compared to heat-resistant nickel-chromium alloys, the titanium alloy blades are difficult to machine because of the presence of more resistant oxide films on the surface. Application of higher voltage is necessary and a mixture of nonpassivating (NaCl, NaBr) and passivating ($NaNO_3$) is commonly used for ECM of titanium blades [26,40]. Some of the methods employed for improvement of the performance of ECM of titanium blades include the use of STEs and the use of gas-liquid mixtures [26,40]. The mixtures of electrolyte solution and air are formed immediately before their injection into the interelectrode space. Sectioned tools are used to overcome the difficulties encountered during ECM of large surfaces of turbine blades. The problems include the nonuniformity of the electrolyte conductivity within the interelectrode gap in the direction of electrolyte flow and the necessity to employ extremely powerful power supplies ranging up to few tens of thousands of amperes. The use of a sectioned tool makes the metal removal rate distribution more uniform as shown in Figure 8.6 [26].

8.8.2 ELECTROCHEMICAL DRILLING

ECD is one of the principal applications of ECM. A tubular electrode is used as the cathode-tool through which an electrolyte is pumped. During the machining process, the electrolyte flows down the central bore of the tool, across the main machining gap, and comes out between the side gap that forms between the wall of the tool and the machined hole. The main machining action is carried out in the gap formed between the leading edge of the drill tool and the base of the hole in the workpiece. As the tool is fed toward the workpiece, anodic dissolution also proceeds laterally between the sidewalls of the tool and the workpiece component due to stray current. The overall effect of the stray current dissolution is to increase the diameter of the hole produced due to overcutting. Several methods are applicable to reduce the amount of overcut. As shown in Figure 8.9, a common method is to use an insulation layer on the external walls of the tool, which minimizes the stray current effect. The other method is to use a passivating electrolyte such as sodium nitrate solution, in which current efficiency is an increasing function of current density with high current efficiency at high current densities. In hole drilling, these high current densities occur between the leading edge of the drill and the base of the workpiece. If nonpassivating electrolyte such as sodium chloride is used, the overcut is much greater. Reversal of the electrolyte flow can often produce a considerable improvement in machining accuracy. Another drilling process, known as capillary drilling, is shown in Figure 8.9b. The drill tube is a glass

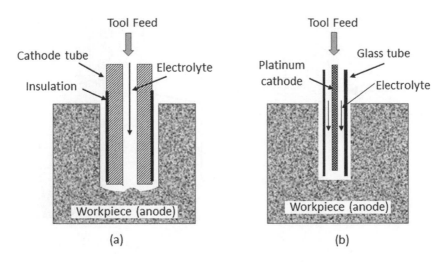

FIGURE 8.9 Working principles of (a) ECD and (b) capillary drilling.

capillary through which the electrolyte flows. The cathode is a platinum wire which is sized to suit the fine tube bore without restricting the electrolyte flow through the tube. The wire is positioned 1–2 mm back from the tube tip to ensure minimal influence on the integrity and the direction of electrolyte flow at the tip. Employing this method, holes are drilled on production components within 0.05 mm of nominal position and to a diametral tolerance of 0.05 mm [41].

Drilling of small deep holes with a depth-to-diameter ratio sometimes up to 100:1 and even higher in the difficult-to-machine materials (high-strength, super-alloy materials) is a very important area of ECD application [5,42–50]. Compared to mechanical drilling where only round holes can be commonly drilled, the ECD operations enable the production of different shaped holes such as rectangular, triangular, oval, and other sections. It also permits drilling of several holes of varying diameters and/or shapes simultaneously. ECD also enables production curved holes or slots, holes that are not perpendicular to the surface, and to perform other unique drilling operations in the difficult-to-machine materials [5,42–50].

Air cooling holes: The ECD of small deep holes was developed predominantly for producing air-cooling holes in gas turbine buckets and vanes to generate higher power output of jet engines for the same unit size and with improved fuel efficiency. The turbine blades are cooled by passing cool air through radial passages. These holes are also used to surround the turbine blades with a film of air that is at a lower temperature than the air entering the same section of the blade. A thermally insulating layer then effectively surrounds the blade. The radial passage is normally been formed in some turbine blades when they are cast, and the film cooling holes produced by EDM or ECM. Such blades often need holes with a depth to diameter ratio of 160:1. Even EDM and laser machining have limitations and an ECM-based process, shaped tube electrolytic machining (STEM) drilling is commonly employed. The configuration of electrodes is similar to that of ECM. The cathode drill electrode is made of titanium tube, its outer wall having an insulating coating. During drilling,

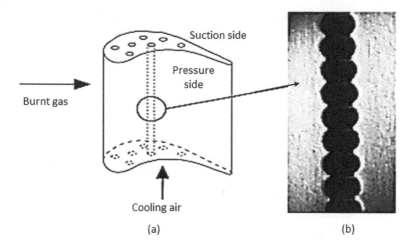

FIGURE 8.10   (a) Turbine blade with holes and (b) contoured cooling holes [50].

the periodic reversal of polarity is employed to prevent an accumulation of undissolved machining products on the surface of the cathode drill.

The cooling efficiency can be further increased by incorporating contoured holes (turbulators) at regular intervals along the length of the cooling channel [45,46]. Turbulators are formed by local changes in diameter superimposed, at regular intervals, on the core diameter of the cooling channel. Figure 8.10 shows the picture of a turbine blade with contoured cooling holes [50]. These turbulators with contoured holes not only increase the heat transfer area but also induce turbulent flow as the cold air flows past them. Whereas cooling channels with aspect ratios greater than 100 and diameters as small as 0.8 mm are successfully machined using STEM drills, the machining of contoured holes poses an enormous challenge. Jain et al. [46] produced a simple turbulator shape by alternately changing the feed rate from its initial value to a lower value and vice versa. They demonstrated the feasibility of contoured hole drilling using shaped tube electrolytic drilling and stainless steel and Inconel superalloy (IN 738 LC) as workpiece material. Experimentally obtained profiles were compared with the profiles derived theoretically from the basic ECM equations. The quality performance factor was evaluated for the machined holes to find the best hole profile. Various types of hole profiles were generated using the shaped tube electrolytic process [46]. This involved variation of process parameters such as voltage, the variation of feed rate from fast to slow, and vice versa during drilling and fixing different step lengths.

8.8.3   ELECTROCHEMICAL DEBURRING

Electrochemical deburring (ECDB) is an electrochemical dissolution-based process to remove burrs that are created by traditional machining operations. ECDB is one of the largest applications of electrochemical metal removal technology. The advantages of the method include the ability to finish hard-to-reach areas, such as inside of bores and grooves that are normally inaccessible for ordinary mechanical methods; high productivity due to short machining times and the possibility of deburring of several workpieces simultaneously; high repeatability; and a relatively low-cost process. Other advantages include stress relieving of the surface; removal of impurities embedded in the workpiece during its manufacturing; and improving surface finish and enhancing the overall appearance. All these advantages may extend the service life of the parts. One of the key technical problems is to localize the dissolution solely on the burrs. Since surrounding areas of the burr are also exposed to the electrolyte, metal dissolution of the surface adjacent to burr areas takes place. Properly designed tool insulation, use of passivating electrolyte, and proper design of the electrolyte flow focusing on the burr are some of the measures that are needed to alleviate the problems due to unwanted dissolution from surrounding areas. Some other problems include certain environmental and safety concerns that should be addressed when using this process. The burrs are electrochemically dissolved into the typically salt electrolytes forming a sludge that needs to be replaced and properly disposed of according to environmentally friendly practices. Depending on the materials involved, the sludge may include heavy metals.

ECDB has evolved as an attractive deburring process in many areas of engineering including for tribological sensitive components such as gears, bearings, bushes, etc. [51]. ECDB is an effective burr removal technique for flank surfaces of gear teeth that are simultaneously fine finished by the process. A preshaped cathode is fixed at a predetermined distance from the gear edge such that the burrs are located in the path of the current flow without touching the cathode as shown in Figure 8.11 [51]. A short-circuit detection system is used to ensure that no burrs touch the cathode before the DC power is switched on. The voltage and amperage settings are a function of the burr size and finishing requirements. The cycle time may then be estimated based on the selected process parameters. Deburring commences due to current flow at the workpiece gear. Deburring occurs selectively by arranging the process such that the areas to be finished are preferentially found in high current density areas. This limits the dissolution of the stock material. Modern ECDB machine tools are designed with the ability to fully control all applicable process parameters to ensure repeatable part quality. The duration of an ECDB process is a function of the burr size and surface finish requirements and may, therefore, range from as little as 5 seconds to few minutes.

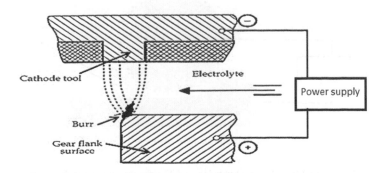

FIGURE 8.11   Schematic showing the working principle of ECDB [51].

8.9   ASSISTED ECM

Assisted machining is a combination of two or more processes to remove material. In assisted ECM, while the key technique remains electrochemical metal removal, the inclusion of additional mechanical/physical means to the process is aimed at enhancing the machining performance such as rate, precision, and surface finish. In the following, some of the assisted ECM processes are discussed that have been proven to be effective.

8.9.1   ELECTROCHEMICAL GRINDING

Electrochemical grinding (ECG) utilizes both mechanical and electrochemical actions to remove materials. For electrochemical action, an electrolyte solution is pumped into the workpiece which acts as an anode. The ECG process uses a conducting grinding wheel in which an insulating abrasive, such as diamond particles, is embedded as shown in Figure 8.12. The rotating grinding wheel works as a cathode tool. The abrasive particles of the grinding wheel contact the workpiece and the gap between the wheel and workpiece makes a passage for electrolyte circulation. The gap voltages range from 2.5 to 14 V [52]. The material removal is achieved by the action of the electrochemical process which leads to the formation of an anodic layer on the workpiece surface. The abrasive grains help in the removal of the surface layer. The dissolution process stabilizes by the continuous supply of the fresh electrolyte and exposing of the fresh workpiece surface. The wheel rotation is an important parameter that governs the accuracy and surface quality in ECG. The use of pulsating current/voltage in ECG offers better control of the machining process. By using pulsed power, the balance between electrochemical and mechanical removal can be adjusted by setting optimum pulse on-time and duty cycle. The contribution of the electrochemical versus the mechanical portion of the metal removal rate (MRR) in ECG depends on the current applied current density. With the application of high currents, the mechanical portion of the metal removal rate decreases [53]. At high current densities, the metal removal rate due to electrochemical dissolution increases, which results in an increased interelectrode gap. Hence, there is a reduction in mechanical force on the workpiece exerted

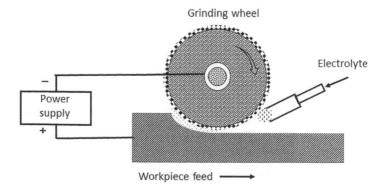

FIGURE 8.12   Schematic of ECG. During the ECG process, the electrolyte is fed in the gap between the rotating grinding wheel with abrasive and the workpiece.

by the ECG wheel. In addition, the bubble volume fraction also increases due to an increase in current density. This contributes to a decrease in MRR caused by erosion.

One of the key concerns of ECG is the wear of the cathode wheel. The wear of the abrasive wheel is affected by several mechanisms. In the interelectrode gap, the temperature rise of the zone affects the wear mechanism of abrasive particles. Cracking of the abrasive particles and spalling of top layers of particles occurs under high stress due to friction. Chemical wear is caused by the corrosive environment and high current density. Hydrogen gas generated due to electrochemical reactions destructs the abrasive particles and metallic binder, leading to the spalling of particles fragments and metallic binders. Other concerns include the generated overcut due to uncontrolled ECM caused by stray currents from the side face of the wheel which affects the process accuracy. Maintaining a small interelectrode gap requires very precise control of the electrode feed.

ECG yields a high metal removal rate, better surface finish, and accuracy compared to conventional grinding, particularly when working with high-strength, and temperature-resistant alloys, and other hard-to-machine materials. In conventional grinding, these materials tend to cause extreme wear and rapid loading of grinding wheels. A wide application of ECG is in the production of tungsten carbide cutting tools. ECG is also useful in the grinding of fragile parts such as hypodermic needles and honeycomb seal rings in aircraft engine components [54].

Several ECG investigations have focused on increasing the material removal rate (MRR), improving the machining accuracy, and enhancing the tool working life. Tehrani and Atkinson [55] found that the use of pulsed voltage was effective in diminishing the overcut, while the wheel grain type has no significant influence. Sapre et al. [56] analyzed the electrolytic flow effects in micro ECG. The higher metal removal rate was achieved at high electrolyte flow rates. Levinger and Malkin [57] investigated the ECG of WC-Co cemented carbides, with particular consideration of their heterogeneity. An analysis of the electrochemical removal process indicated that the electrochemical dissolution of the cobalt phase occurs initially at a much faster rate than the carbide phase. The selective removal of cobalt weakens the composite material, thereby reducing the power requirements for mechanical material removal during ECG.

8.9.2 Ultrasonic-Assisted ECM

Ultrasonically assisted ECM is the method of improving the performance of ECM by the introduction of electrode tool ultrasonic vibration. In general, the application of ultrasound to fluid causes a chaotic, turbulent flow. Propagation of the sound wave through the ultrasonically irradiated media creates a pressure drop, which can break forces holding the liquid molecules together and generates microbubbles [58]. In ECM, the ultrasonic vibrations improve the transport of heat and removal of reaction products from the machining area. Since ultrasound has a direct mechanical influence on electrolytes, electrodes, and the interelectrode gap, the use of ultrasonic vibrations during the ECM process enhances convective diffusion and decreases the rate of the passivation process. It also results in a decrease in potential drops in the layers adjacent to the electrodes and enables the creation of optimal hydrodynamic conditions for good surface finish.

The use of ultrasonic vibration provides the possibility for creating cavitation microbubbles near the workpiece and the tool surface. In the interelectrode gap, the presence of dissolution products, evolved gases, and increased electrolyte temperature during machining cause favorable conditions for cavitation bubbles to grow in the gap. The process of the microbubbles collapses in the area adjacent to electrode gives the possibility for increasing the intensification of mass and charge transport. Perusich and Alkire [59,60] conducted a theoretical and experimental analysis of ultrasonically induced cavitation studies of electrochemical passivity and transport mechanism. Their study indicated that ultrasonic vibrations have a significant influence on the kinetics of electrode processes and increase the rate of electrochemical dissolution. Ruszaj et al. [61] investigated the influence ultrasonic vibrations on the surface roughness during ECM of NC6 steel in 15% $NaNO_3$ solution using ultrasonic vibration power between 20 and 120 W. Their results indicated that the electrode ultrasonic vibrations change conditions of the electrochemical dissolution process and make it possible to decrease surface roughness parameters in comparison to classical and pulse electrochemical machining. For each case of machining, there is an optimal value of the amplitude of electrode ultrasonic vibrations for which the minimal value of R_a is obtained.

8.9.3 ULTRASONIC-ASSISTED ECG

The incorporation of ultrasonic vibrations extends the working life of the mechanical grinding tool and improves its grind ability. The ultrasonic-assisted grinding (UAG) has attracted attention due to features such as smaller grinding force, higher material removal rate, longer grinding wheel working life, and lower grinding heat generation due to a fundamental alteration in process kinematics [62]. Lower feed forces and process temperatures are achieved when ultrasonic vibration is superimposed on the drilling samples of Ti-6Al-4V. Similar results are obtained for several brittle materials including zirconia ceramics [63]. A combination of ECG and UAG, known as ultrasonic-assisted electrochemical grinding (UAECG), further reduces the tool wear and improves the MRR in the grinding of Ti-6Al-4V and similar materials.

The principle of operation of UAECG is similar to the ECG as explained above, except in the case of UAECG, the grinding wheel is ultrasonically vibrated. A typical UAECG system consists of a metal-bonded cubic boron nitride grinding wheel, ultrasonically vibrating in its axis at a predetermined frequency and amplitude, which is rotated clockwise [62]. The surface grinding operation is performed by setting up a depth of cut between the wheel and the workpiece and simultaneously feeding the workpiece in the right direction at the desired feed rate. The grinding wheel is made the cathode and the workpiece is made the anode. An electrolyte is supplied to the grinding zone between the wheel and the workpiece. A passivating electrolyte such as $NaNO_3$ is preferred. The depth of cut, the rotation speed, and the feed are so adjusted as to ensure that the maximum grain depth of cut is less than the protrusion height of the grain on the wheel working surface. This is to prevent the metal bond from contacting the work material; otherwise, electric short may occur between the workpiece and the grinding wheel.

Using a UAECG system described above, Li et al. [62] investigated the machining characteristics of the grinding of Ti-6Al-4V. The work-surface roughness in UAECG decreased with the increasing voltage, the surface finish is better than the conventional ECG without ultrasonic vibration. Plastic deformation and cracks were not observed in UAECG. The grinding forces in UAECG were significantly smaller than those in conventional grinding, UAG and ECG.

8.9.4 ELECTROCHEMICAL HONING

Electrochemical honing (ECH) is the coupling of ECM and mechanical honing processes to provide controlled functional surface generation and fast material removal capabilities in a single operation. The honing is a finishing process that creates the desired finish pattern on the interior of tubing or cylinder bores and provides the final sizing. Finishing is accomplished by expanding abrasive stones of suitable grit and grade against the work surface. The stones are rotated and reciprocated in the part with hone abrasive under controlled pressure. Combining rotation and reciprocation produces a crosshatch pattern on the surface of the part being honed. Honing machines are metal abrading tools and processes utilizing hard tooling and perishable abrasives stones for the correction of diameter, shape, surface finish, and positional tolerances of bores. The major limitations of the conventional honing process are difficulty in finishing a hard and hardened workpiece, the possibility of mechanical damage by inducing microcracks, and plastic deformation to the workpiece material, and low productivity. These limitations are overcome by the electrochemical dissolution process making ECH an ideal choice to explore as an alternative, superior, economical, and sustainable process for finishing. ECH is used in the finishing of complex-shaped products, such as helical and bevel gears, external and internal cylindrical surfaces. ECH offers a range of benefits to the machined surface which cannot be obtained by either of the processes when applied independently.

A schematic diagram of the ECH process is shown in Figure 8.13 [64]. It consists of a tooling system loaded with honing stones, a tool motion system, a machining

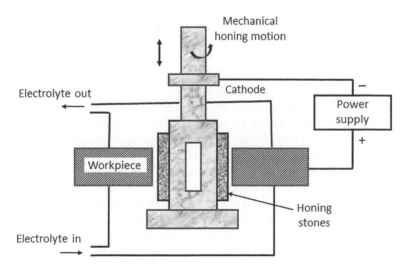

FIGURE 8.13 Schematic of the ECH process [64].

chamber with the properly held workpiece, an electrolyte supply system, and a power supply system. The tooling system is enclosed in a machining chamber, which includes provisions for supply and removal of the electrolyte, and for the escape of gases generated during the ECH process. The electrolyte supply system supplies supply the filtered electrolyte with a controlled flow rate and pressure to the machining zone. The power supply system consists has provision for operating at both continuous and pulsating condition. During ECH operation the tool rotates and reciprocates while the workpiece is stationary fixed in the machine chamber. This rotary and reciprocation movement of the ECH tool is the critical parameter in achieving the closer tolerances desired surface finish. Honing stones are embedded abrasive particles (Al_2O_3, SiC, CBN) bonded in vitreous bonding material with particular grit size honing stones. The choice of the abrasive type and size of the honing stones depends on the type of workpiece and the desired rate of surface finish. The honing stones are mounted on a stainless steel tool holder which is simultaneously given a rotary motion as well as reciprocating (or oscillatory) motion to perform a complete cycle. The reciprocating motion is along the axis of the workpiece to bring the entire work surface in contact with the honing stone.

The material removal in ECH is the sum of volumetric material removal due to electrochemical action and mechanical honing. While the significant part of the material removal is due to electrochemical dissolution, the action of honing further assists in the removal process. The electrochemical material removal in the ECH process is governed by the controlled electrolytic dissolution of the anodic workpiece which faces the cathode tool while maintaining an interelectrode gap in the range of 0.1–1.0 mm and an appropriate electrolyte flowing through the gap. DC or pulsating voltage is applied between the two electrodes. An aqueous solution of the salt-based electrolyte such as NaCl, $NaClO_3$, $NaNO_3$, or their mixture is supplied through the interelectrode gap at a flow rate in the range of 10–50 L/min [65]. During the pulsating ECH process, a layer of dissolution products forms on the dissolving surface, the thickness of this layer is more at surface valleys compared to that on the peaks. The honing tool selectively removes this layer enabling more effective electrolytic dissolution.

ECH combines the capabilities and advantages of ECM and mechanical honing. The ECH process is five to eight times faster than conventional honing and four times faster than grinding. It can provide a surface finish up to 0.05 μm. It provides fine surface generation by honing and fast material removal by ECM in a single operation. The ECH process offers advantages of high metal removal rate and extreme accuracy of 0.001 mm in a wide variety of hard to cut materials. Other advantages include the ability to correct out of roundness, cylindricity, circularity, and axis straightness in relatively round cylindrical workpieces. The ability of ECH to apply for these benefits productively has the potential of its widespread use in industries, especially in aerospace, automobiles, petrochemical reactor, roller, and gear manufacturing industries. However, like most of the assisted ECM processes, ECH is also in the development phase and further research efforts are needed for its successful industrial acceptance as a mature finishing process for a wide variety of advanced materials and complicated shaped engineering components [64].

8.10 ENVIRONMENTAL AND SAFETY ISSUES

Environmental issues in ECM arise from the spent electrolyte, ECM slurry, and rinse water while safety issues are posed mainly by the exposure to gaseous and aerosol products produced during processing, exposure to electrolyte and by the electrical short circuit [66–68]. Fresh ECM electrolytes, in general, do not pose environmental concerns. However, during ECM processing, the reaction products including heavy metals accumulate in the electrolyte. Chemical products such as toxic chromate, nitrate, etc., may get incorporated with the hydroxides in the slurry, which may be hazardous for air, water, or soil pollution. Waste disposal of slurries, if not managed properly, can contribute significantly to the processing cost. On the other hand, sludge reuse and electrolyte neutralization are an effective way to make the ECM a low-waste manufacturing process.

A large amount of sludge is produced during ECM. The sludge consists of a mixture of oxides and hydroxides of the metals. Most often the weight of the sludge exceeds that of the metal dissolved from the workpiece. Filtration of the sludge is a common method to separate the metal hydroxides from the electrolyte. In many cases where the filtration rate may be too slow to be practical, the sludges are treated by physical or chemical means so as to agglomerate the metal hydroxide particles and increase the filtration rate. The ECM sludge is used in the manufacturing of building materials and catalysts and also in pyrometallurgy and hydrometallurgy for recovery of the valuable components such as nickel, cobalt, tungsten, molybdenum, etc. Indeed, effective treatment of the sludge is to smelt the ECM wastes to produce new melts of the original components such as superalloys or stainless steels. The sludge produced by the ECM of titanium alloys may be used after suitable processing as a starting material for paints, pigments, and varnishes. Another promising use of ECM sludge is in the preparation of catalysts and adsorbents that are employed in the manufacture of ammonia, in the desulphurization of natural gas, and in the petrochemical industry.

Chromium alloys subjected to ECM anodically dissolve and pass into solution as hexavalent $Cr(VI)$. Due to its toxicity and environmental impact, the removal of $Cr(VI)$ is of utmost importance. Many methods that have been developed to reduce and remove chromium include electrodialysis, ultrafiltration, solution extraction ion exchange membrane, but the most common method to reduce hexavalent chromium is the chemical reduction and precipitation [69]. To transform dissolved chromium, substances converting $Cr(VI)$ into $Cr(III)$ are added to the spent electrolyte. Trivalent chromium is stable at low pH, so most reductions take place at low pH, and then the pH is raised to precipitate trivalent chromium hydroxide. Elemental iron can be used to reduce $Cr(VI)$; iron powder with high contact area is the most suitable form to use. This method creates lots of sludge waste and requires an excess of elemental iron to reduce chromium concentrations to acceptable levels. Ferrous sulfate heptahydrate ($FeSO_4 \cdot 7H_2O$) and sodium borohydride ($NaBH_4$) are added together; sodium borohydride reduces the ferrous sulfate to nanoparticles of zero-valent iron. The elemental iron particles produced then reduce the $Cr(VI)$. However, this method also produces large amounts of solid waste which can be costly to dispose of correctly. Another iron-based method for reducing $Cr(VI)$ is the use of ferrous sulfate in an acid environment followed by precipitation. Yet another method is the introduction of barium salts into the electrolyte whereby Cr(VI) can be directly precipitated in the form of insoluble $BaCrO_4$.

From a safety point of view, hydrogen gas generated during the ECM process can be explosive. Local exhaust ventilation must be installed to remove hydrogen gas from the working areas. Exposure to chromium compounds through inhalation, ingestion, and eye or skin contact can affect the skin, liver, and kidneys in humans and cause contact dermatitis. Several health effects can be related to the nitrates and nitrites in the electrolyte solution. Good work practices must be in place to avoid skin contact with the electrolyte. All pieces of ECM equipment should be reliably grounded. The power units should include an overload protection feature, which will interrupt the machining circuit in the case of an overload. Interlocks should also be built into machines so that power and electrolyte supply could be discontinued automatically in an emergency. Active technical safeguarding and personal protective equipment can be effective in managing the health and safety of workers.

REFERENCES

1. A. E. De Barr, D. A. Oliver eds., *Electrochemical Machining*, Macdonald, London, (1968).
2. J. F. Wilson, *Practice and Theory of Electrochemical Machining*, Wiley, New York, (1971).
3. J. McGeough, *Principles of Electrochemical Machining*, Chapman and Hall, Wiley, London, (1974).
4. K. P. Rajurkar, D. Zhu, J. A. McGeough, J. Kozak, A. De Silva, *CIRP Ann.*, 48 (2), 567 (1999).
5. A. D. Davydov, V. M. Volgin, Electrochemical Machining, *in Encyclopedia of Electrochemistry*, D. D. Macdonald, P. Schumki eds., vol. 5, Wiley-VCH Verlag GmbH & Co. KGaA, Weinheim, (2007).
6. M. Datta, D. Landolt, *Electrochim. Acta*, 25, 1255 (1980).
7. M. Datta, D. Landolt, *Electrochim. Acta*, 25, 1263 (1980).
8. M. Datta, D. Vercruysse, *J. Electrochem. Soc.*, 137, 3016 (1990).
9. M. Datta, *IBM J. Res. Dev.*, 37, 207 (1993).
10. E. Rosset, M. Datta, D. Landolt, *J. Appl. Electrochem.*, 20, 69 (1990).
11. K. W. Mao, *J. Electrochem. Soc.*, 118, 1870 (1971).
12. R. D. Grimm, D. Landolt, *Corros. Sci.*, 36, 1847 (1994).
13. M. Datta, D. Landolt, *J. Electrochem. Soc.*, 122, 1466 (1975).
14. D. Landolt, R. H. Muller, C. W. Tobias, *J. Electrochem. Soc.*, 116, 1384 (1969).
15. C. Madore, D. Landolt, *J. Micromech. Microeng.*, 7, 270 (1997).
16. M. Datta, D. Landolt, *J. Electrochem. Soc.*, 129, 1889 (1982).
17. P. K. Chang, *Separation of Flow*, Pergamon Press, Elmsford, New York, (1970).
18. M. Datta, D. Landolt, *Proceedings of the 2nd International Symposium on Industrial and Oriented Basic Electrochemistry*, SAEST, Madras, India, p. 4.3.1, (1980).
19. M. Datta, R. V. Shenoy, L. T. Romankiw, *J. Eng. Ind., Trans. ASME*, 118, 29 (1996).
20. M. Datta, L. T. Romankiw, D. R. Vigliotti, R. J. von Gutfeld, *J. Electrochem. Soc.*, 136, 2251 (1989).
21. M. Datta, D. Landolt, *Electrochim. Acta*, 26, 899 (1981).
22. M. Datta, D. Landolt, *Electrochim. Acta*, 27, 385 (1982).
23. K. P. Rajurkar, J. Kozak, B. Wei, J. A. McGeough, *Ann. ClRP*, 42 (1), 231 (1993).

24. J. Kozak, K. P. Rajurkar, B. Wei, *Trans. ASME, J. Eng. Ind.*, 116, 316 (1994).
25. W. Chen, F. Han, *Int. J. Adv. Manuf. Technol.*, 102 (5–8), 2531 (2019).
26. A. D. Davydov, V. M. Volgin, V. V. Lyubimov, *Russ. J. Electrochem.*, 40 (12), 1230 (2004).
27. J. F. Thorpe, R. D. Zerkle, *Int. J. Mach. Tool Des. Res.*, 9, 131 (1969).
28. V. K. Jain, P. C. Pandey, *Precis. Eng.*, 2, 195 (1980).
29. O. H. Narayann, P. G. Yogindra, S. Murugan, *Int. J. Mach. Tools Manuf.*, 26 (3), 323 (1980).
30. V. K. Jain, P. C. Pandey, *Int. J. Mach. Tool Des. Res.*, 22 (4), 341 (1982).
31. O. H. Narayanan, S. Hinduja, C. F. Noble, *Int. J. Mach. Tool Des. Res.*, 26 (3), 323 (1986).
32. V. K. Jain, P. G. Yogindra, S. Murugan, *Int. J. Mach. Tools Manuf.*, 27 (1), 113 (1987).
33. L. W. Hourng, C. S. Chang, *J. Appl. Electrochem.*, 24, 1170 (1994).
34. V. K. Jain, K. P. Rajurkar, *Precis. Eng.*, 13 (2), 111 (1991).
35. S. Das, A. K. Mitra, *Int. J. Numer. Methods Eng.*, 35, 1045 (1992).
36. S. Bhattacharyya, A. Ghosh, A. K. Mallik, *J. Mater. Process. Technol.*, 66, 146 (1997).
37. C. Sun, D. Zhu, Z. Li, L. Wang, *Finite Elem. Anal. Des.*, 43 (2), 168 (2006).
38. J. Pattavanitch, S. Hinduja, J. Atkinson, *CIRP Ann.*, 59 (1), 243 (2010).
39. J. A. Westley, J. Atkinson, A. Duffield, *J. Mater. Process Technol.*, 149 (1), 384 (2004).
40. A. D. Davydov, T. B. Kabanova, V. M. Volgin, *Russ. J. Electrochem.*, 53 (9), 941 (2017).
41. D. A. Glew, *Proceedings of ISEM-6*, Cracow, Poland, p. 309, (1980).
42. Y. Zhou, J. J. Derby, *Chem. Eng. Sci.*, 50 (17), 2679–2689 (1995).
43. J. Kozak, K. P. Rajurkar, R. Balkrishna, *J. Manuf. Sci. Eng. Trans. ASME*, 118, 490–498 (1996).
44. M. Uchiyama, T. Shibazaki, *J. Mater. Process. Technol.*, 149, 453–459 (2004).
45. J. Pattavanitch, S. Hinduja, *CIRP Ann. – Manuf. Technol.*, 61, 199 (2012).
46. V. K. Jain, A. Chavan, A. Kulkarni, *Int. J. Adv. Manuf. Technol.*, 44, 133 (2009).
47. S. Sharma, V. K. Jain, R. Shekhar, *Int. J. Adv. Manuf. Technol.*, 19, 492 (2002).
48. D. S. Bilgi, V. K. Jain, R. Shekhar, S. Mehrotra, *J. Mater. Process Technol.*, 149, 445 (2004).
49. D. S. Bilgi, V. K. Jain, R. Shekhar, A. V. Kulkarni, *Int. J. Adv. Manuf. Technol.*, 34, 79 (2007).
50. M. J. Noot, A. C. Telea, J. K. M. Jansen, R. M. M. Mattheij, *Comput. Visual Sci.*, 1, 105 (1998).
51. K. Gupta, N. K. Jain, R. Laubscher, Conventional and Advanced Finishing of Gears, *in Advanced Gear Manufacturing and Finishing*, K. Gupta, N. K. Jain, R. Laubscher eds., pp. 127–165, Academic Press, Cambridge, Massachusetts, (2017).
52. K. K. Saxena, M. Bellotti, D. V. Camp, J. Qian, D. Reynaerts, Electrochemical Based Hybrid Machining, *in Hybrid Machining*, X. Luo, Y. Qin eds., pp. 111–129, Elsevier, Berkeley, (2018).
53. D. Patel, V. K. Jain, J. Ramkumar, Electrochemical Grinding, *in Nanofinishing Science and Technology: Basic and Advanced Finishing and Polishing Processes*, V. K. Jain ed., CRC Press, Boca Rotan, (2019).
54. J. A. McGeough, Electrochemical Machining, *in Kirk-Othmer Encyclopedia of Chemical Technology*, vol. 1, John Wiley & Sons, Inc., Hoboken, New Jersey, (2016).
55. A. F. Tehrani, J. Atkinson, *Proc. Inst. Mech. Eng., B*, 214 (4), 259–268 (2000).
56. P. Sapre, A. Mall, S. S. Joshi, *ASME J. Manuf. Sci. Eng.*, 135 (1), 011012 (2013).
57. R. Levinger, S. Malkin, *J. Eng. Ind.*, 101 (3), 285 (1979).
58. S. Skoczypiec, *Int. J. Adv. Manuf. Technol.*, 52, 565 (2011).
59. S. A. Perusich, R. C. Alkire, *J. Electrochem. Soc.*, 138, 700 (1991).
60. S. A. Perusich, R. C. Alkire, *J. Electrochem. Soc.*, 138, 708 (1991).
61. A. Ruszaj, M. Zybura, R. Zúrk, G. Skrabalak, *J. Eng. Manuf.*, 217, 1365 (2003).

62. S. Li, Y. Wu, M. Nomura, T. Fujii, *J. Manuf. Sci. Eng. Trans. ASME*, 140, 071009-1 (2018).

63. X. Xiao, K. Zheng, W. Liao, H. Meng, *Int. J. Mach. Tools Manuf.*, 104, 58 (2016).

64. P. S. Rao, P. K. Jain, D. K. Dwivedi, *Procedia Eng.*, 100, 936 (2015).

65. S. Pathak, N. K. Jain, *Trans. IMF, Intl. J. Surf. Eng. Coatings*, 95 (3), 147 (2017).

66. G. E. Hunt, *JAPCA*, 38 (5), 672 (1988).

67. L. G. Twidwell, D. R. Dahnke, *Eur. J. Miner. Process. Environ. Prot.*, 1 (2), 1303–0868, 76 (2001).

68. J. Halder, M. S. J. Hashmi, *8.02 Health and Environmental Impacts in Metal Machining Processes*, pp. 7–33, Elsevier, Berkeley, (2014).

69. R. J. Leese, T. P. Noronha, *IJERT*, 6 (11), 29 (2017).

9 Electrochemical Micromachining
Maskless Techniques

9.1 INTRODUCTION

Micromachining refers to material removal from dimensions that range from several micrometers to a millimeter. Many advanced technical products rely on manufacturing techniques that produce structures in the micrometer range with a high degree of precision. Micromachining with a mechanical cutting tool is, in principle, capable of producing microstructures with high profile accuracy and the conventional mechanical machining processes have been extended to micromachining for low volume or customized single-piece production. However, unreliable tool life and early tool failure are the most common problems of employing microtools for micromachining applications. Micro-EDM also has the capability to machine microdimensional features irrespective of the material hardness but it suffers from surface integrity issues due to thermally induced defects. Its capability is restricted to features bigger than 100 µm. Other processes such as laser beam machining and electron beam machining are not effective for drilling in thick workpiece materials due to the limited working range of lasers. On the other hand, development of electrochemical material removal processes for microfabrication demonstrated that the electrochemical machining (ECM) concepts can be effectively employed for processing of thin films and micromachining of complicated structures that are of interest in a variety of applications including microelectronics and micro, nanomechanical systems (MEMS/NEMS). The application of ECM in microfabrication is referred to as electrochemical micromachining (EMM). Compared to conventional mechanical micromachining techniques, EMM offers several advantages such as no tool wear, stress, and deformation-free microsmooth surface, and the ability to machine complex shapes on a variety of materials including hard to machine high-strength superalloys and conducting ceramics. During a period of over three decades, enormous research work has been conducted on the understanding and development of EMM processes and the topic has been reviewed in several publications [1–5].

Microfabrication by EMM may involve maskless or through-mask material removal. Machining by Maskless EMM requires highly localized material removal which may be accomplished by the impingement of a fine electrolytic jet or by using a microwire as the cutting tool. Other examples of massless EMM include microdrilling and micromilling. In through-mask EMM (TMEMM) process, localization of the material removal is accomplished by the application of properly designed photoresist mask to insulate the surface that needs to be protected. The TMEMM process will be described in a separate chapter. In this chapter, maskless EMM processes will be described.

9.2   JET ELECTROCHEMICAL MICROMACHINING

In jet electrochemical micromachining (JEMM) a free jet of electrolyte locally removes material from an electrically conductive workpiece via anodic dissolution. This leads to an extremely localized dissolution at the impingement region. Due to the benefits offered by JEMM over other micromachining methods, JEMM has found applications in several manufacturing sectors, including but not exclusive to aerospace, microelectronics, medical and biomedical applications. Thus, JEMM has become an attractive topic for academic and industrial researchers [3,6–8,13–16]. An up-to-date review of the development of experimental and numerical simulation work on JEMM has been presented in a publication by Kendall et al. [8].

The micromachining performance in JEMM is determined by the ability of the jet to remove anodic reaction products from the machining area and supply fresh electrolyte at an extremely fast speed. Mass transfer studies at a free-standing jet as well as a submerged jet on a barrier substrate have been reported in the literature [9–12]. Figure 9.1 shows the flow pattern at the impingement region for a free-standing circular jet. After exiting the nozzle, the velocity profile prevailing in the pipe becomes uniform in the free jet flow [11]. At the impingement area, there is a stagnation region where the boundary layer thickness is relatively independent of the radial position. The diameter of this region of uniform mass transfer is roughly equal to twice the jet diameter. A flow boundary layer starts to develop in the wall jet region outside the impingement region. When the thickness of the boundary layer becomes equal to the thickness of the wall jet, there is a sudden increase in the thickness called hydraulic jump [10]. This region is followed by a region of undisturbed flow [10].

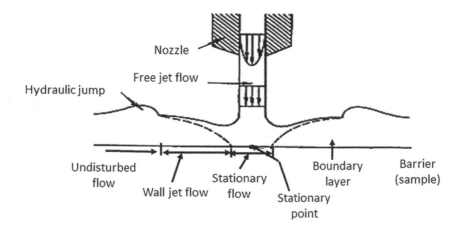

FIGURE 9.1   Flow pattern at the impingement region for a free-standing circular jet [11].

The selectivity in JEMM is governed by the current density dependence of the anodic reactions. In the machining area where the workpiece directly faces the electrolytic jet, the anodic reaction rate is constant. Away from the machining area, current density on the workpiece decreases asymptotically to zero. In a passivating metal-electrolyte system, in which oxygen evolution takes place at low current densities, the current efficiency for metal dissolution varies as a function of the distance from the machining area. In nonpassivating systems, on the other hand, current efficiency may remain independent of the current density. Therefore, a better selectivity is expected in a passivity system because of its ability to form oxide films and evolve oxygen in the stray current region.

A high-precision JEMM tool shown in Figure 9.2 was employed by Datta et al. for maskless micromachining/patterning [3,6,7]. The tool consists of a movable sample anode assembly, a jet assembly, and an electrolyte pumping system. All of these components are assembled on a vibration-free X–Y table. The sample holder is mounted on the arm of an X–Y table and moves in a plane perpendicular to the electrolyte jet while the nozzle is mounted on a rail-table moving in the Z-direction thus allowing the adjustment of the interelectrode distance. Two power supplies provide the required current or cell voltage and a multimeter is used to accurately measure the current. The electrolyte is pumped through a filter to remove corrosion products. A three-way valve directs the flow either directly in the electrolyte tank or through the nozzle. The temperature of the electrolyte is measured in the tank and at the nozzle. The hydraulic pressure is measured at the pump exit and just before the nozzle. A pH-meter is dipped in the electrolyte tank. The nozzle is electrically insulated from the remaining part of the flow system with a plastic pipe. The nozzle, shown in Figure 9.3, is a quartz capillary (microglass) mounted on a stainless-steel holder serving as the cathode. A systematic study was conducted to determine the influence of jet dimension (varying between 50 and 175 pm) and applied voltage (varying between 100 and 300 V for a constant interelectrode gap of 3 mm) on the selectivity and the rate of micromachining. Samples consisted of foils of copper,

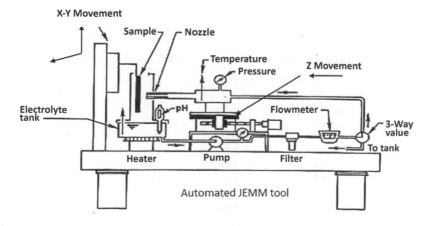

FIGURE 9.2   Schematic diagram of an automated JEMM setup [3].

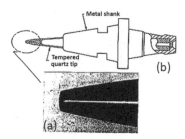
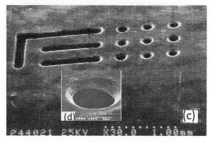

FIGURE 9.3 Details of the nozzle consisting of a quartz capillary (a) mounted on a stainless-steel holder (b). Microphotographs of through-holes and patterns (c and d) micromachined by JEMM [3].

molybdenum, and stainless steel of thickness varying between 50 and 250 μm. These studies demonstrated that neutral salt solutions can be effectively used for high-speed micromachining of many metals and alloys. An electrolytic jet of nitrate solution gave high current efficiency for metal dissolution and minimum stray cutting at high current densities. A 5 M sodium nitrate solution was used in experiments consisting of micromachining arrays of holes and slots and a preprogrammed pattern as shown in Figure 9.3. For EMM of holes, the dissolution time varied between 1 and 8 seconds. Results are summarized in Table 9.1, which shows that an increase in the cell voltage leads to an increased micromachining rate but has little effect on the diameter/width. The pattern diameter/width is approximately twice the nozzle diameter indicating that machining is mainly concentrated in the stagnation area of the impingement region.

Kozak et al. [14] conducted theoretical and experimental investigations of the effect of voltage, flow velocity, standoff distance, and electrolyte properties on the material removal rate (MRR) during JEMM. The performance of the JEMM process

TABLE 9.1
Micromachining Performance of JEMM for Selected Materials

Material	Nozzle Dia. (μm)	Voltage (V)	MRR (μm/s)	Hole Dia. (μm)	Ref.
302 SS	50	300	12.5	100	[5]
	100	100	–	173	
	100	200	12.5	195	
	100	300	16.7	200	
	175	300	16.7	304	
Copper	100	100	08.3	200	[5]
	100	200	16.7	210	
Molybdenum	100	100	04.5	194	[7]
	100	200	08.3	204	
	100	300	11.0	205	

was found to be governed by the heating of the electrolyte. The maximum value of the MRR was found to be limited by the boiling of the electrolyte. The authors proposed a mathematical model for the JEMM process which incorporated the variation of electrolyte conductivity due to temperature effects. The mathematical model enabled the determination of the maximum MRR and electrolyte flow velocity for given values of applied voltage and distance between the nozzle and the workpiece. Their experimental results showed good agreement with the theoretical values calculated from the proposed mathematical model.

Hackert-Oschätzchen et al. [15] investigated the capabilities of JEMM for drilling of blind holes and through-holes, cutting of a plate and milling of a microreactor using stainless steel plates. A 100 μm nozzle was used through which a 30% sodium nitrate solution was pumped at a constant flow velocity of 20 m/s. The working gap was varied between 10 and 100 μm, and the cell voltage ranged from 5 to 100 V. An increase in cell voltage increased the removal rates but the accuracy decreased due to stray currents. A voltage of 56 V was considered the optimum voltage. Reducing the working gap led to a rising current flow which caused a higher removal rate. However, working gaps smaller than half the nozzle radius caused an undefined widening of the dissolved area due to flashovers. A working distance of 100 μm was found to be appropriate. A series of blind holes were machined which showed highly reproducible geometries with their width being 220 ± 10 μm, approximately 2.2 times the nozzle diameter. With the surface profiler, the profiles of the blind holes were measured and plotted for selected process cycles as shown in Figure 9.4. The depths of the holes increased with increasing process time. Depths of about 37 μm at 0.5 seconds to approximately 90 μm at 2 seconds of process time were achieved. The widths ranged from 173 to 220 m. Figure 9.4 also includes the profiles obtained by FEM simulation. It is apparent from the figure that the geometries of experimental and simulation results compare well. The suitability of JEMM for cutting of foils and plates was demonstrated by using a 250 μm thick stainless-steel plate. With a cutting

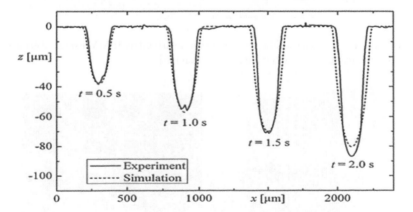

FIGURE 9.4 Simulated and experimentally measured profiles of blind holes micromachined by JEMM [15].

length of 4,300 μm and a width of 180 μm, a complete cut was realized after 13 crossings with a nozzle velocity of 200 m/s. Contrary to mechanical cutting technologies there was no deformation of the sample. To demonstrate the possibility of generating through-holes with JEMM, a 250-μm stainless steel plate was machined using a stationary nozzle. With process times above 8.5 seconds through-holes could be realized. The result for JEMM drilling after a processing time of 9 seconds is shown in Figure 9.5. The diameter of the hole which faced the nozzle widened to about 250 μm. On the outlet side, the hole is approximately 100-μm wide, similar to the dimension of the nozzle diameter.

Lu and Lang [16] developed a JEMM process to machine high aspect ratio micro-holes on titanium surfaces for biomedical applications. A pressurized tank connected to compression air was used to generate electrolyte jet flow through a metallic nozzle. The nozzle with a diameter of approximately 300 μm, which was made by embedding a metal tube in epoxy, controlled the jet flow toward the surface of titanium specimen placed about 3 mm away. The work stage holding the specimen provided vertical, horizontal movement and rotation in the horizontal plane. 50 mm long and 12 mm in diameter Ti6Al4V alloy cylinders were used. The electrolyte used was 5 M NaBr aqueous solution. The experimental results indicated that the JEMM can effectively machine titanium cylinder surfaces at high speed. Figure 9.6 shows arrays of through-holes on a Ti6Al4V cylinder fabricated by JEMM. The JEMM process was conducted with a voltage of 200 V and an average current of 45 mA. Each hole only

FIGURE 9.5 Front and back sides of a through-hole drilled by JEMM using a 100 μm diameter nozzle and using a 30% sodium nitrate solution [15].

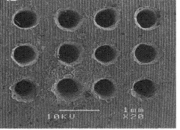

FIGURE 9.6 Arrays of through-holes micromachined on a Ti6Al4V cylinder using JEMM. Nozzle diameter: 300 μm; electrolyte: 5 M NaBr [16].

required approximately 2 minutes of machining time. The processing parameters such as applied voltage and jet pressure determined the surface finish and precision of the holes.

9.3   ELECTROCHEMICAL MICRODRILLING

Electrochemical microdrilling (EMD) is a controlled rapid electrolytic dissolution process on a micrometer scale in which a tubular-shaped tool, preferably made of brass, copper, or stainless steel, or a tungsten wire is used as the cathode. The tool is fed towards the workpiece at a predetermined speed. The electrolyte which is generally a concentrated salt solution is pumped at high pressure through interelectrode gap in order to allow high rate of dissolution by removing the reaction products and dissipating the heat generated. Commonly preferred electrolytes are NaCl, $NaNO_3$, $NaClO_3$, and their mixtures. Compared to conventional microdrilling techniques, EMD is one of the cost-effective techniques, which provides a better alternative in drilling microholes with reasonably accurate dimensions and good surface finish. The published literature on EMD has been reviewed by Sen and Shan [17] and Rahman et al. [18].

The major limitations of EMD are the failure of the tool insulation and stray removal [19–21]. Insulation failure in ECD occurs mainly due to the clogging of the holes on account of the reaction products. The stray removal that usually occurs on the internal sidewalls of the hole affects the process reliability significantly. The reduction of stray removal has been attempted by the use of good quality insulation such as ceramic layers of silicon carbide or nitride.

Wang et al. [19] proposed a new electrode sidewall insulation method. The double insulating layers which consisted of TiO_2 ceramic coating and organic film were fabricated on the electrode sidewall combining the microarc oxidation and cathodic electrophoresis coating techniques. EMD experiments were performed to investigate the influence of electrolyte jetting pressure, machining voltage, and the thickness of the TiO_2 ceramic coating and the organic film on the durability of the double insulating layers on the tool. The results showed that the double insulating layers performed well in terms of durability and a series of fine-shaped holes were fabricated by an electrode with such insulation layers. For further minimizing stray cutting, a dual pole tool has been implemented [20]. In a dual-pole tool, an insulated microtool electrode is covered with an anodic bush of metal. It has been found that the use of a dual-pole tool reduces the hole taper as compared to the insulated tool thereby improving the machining accuracy and process stability.

The shape of the tool is critical in obtaining the desired microhole. Using a flat-ended microtool for machining of a circular microhole, the bottom central part of the hole was not machined well due to erratic electrolyte supply. On the other hand, a round-ended tool provides good machinability because of the ease in electrolyte supply. The machining speed, machining stability, and microhole accuracy improve by using a round tool as compared to a cylindrical tool. Wang et al. [21] compared the results of microdrilling operation by cylindrical and disk-shaped microtools. They conducted the experiments on stainless-steel workpiece using $NaNO_3$ solution with and without vibration. Results shown in Figure 9.7 indicated that the taper of

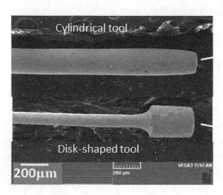
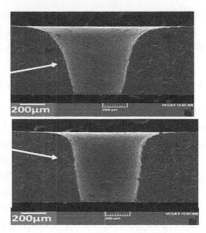

FIGURE 9.7 Holes microdrilled by cylindrical and disk-shaped microtools: radial over-cut and microhole taper reduced by using disk-shaped microtool as compared to cylindrical tool [21].

microhole was reduced by the disk-shaped microtool as compared to the cylindrical tool. Also, the radial overcut was minimized, and the surface finish improved by using ultrasonic vibration.

Accumulation of reaction products within the microgap between the tool and the workpiece is one of the key concerns of the effective operation of EMD processes. Fan and Hourng [22] employed a rotational system to extract insoluble sludge from a deep hole thereby minimizing the difficulty of filling a deep hole with electrolyte. A pulsed power supply was used to provide intermittent machining which led to an increase in the precision of the EMD. The influence of pulsed on-times, applied voltages, electrolyte concentrations, pulsed frequencies, tool feeding rates, tool diameters, tool rotational rates, and hole depth, on the hole overcut and tapering in an EMD, was investigated. A microhole with an aspect ratio (depth/diameter) of 8.6 was drilled on a 1,000 µm thick 304 stainless steel using a 50 µm diameter tungsten carbide pin tool. The results demonstrated that a rotational tool can be effectively used in the fabrication of deep holes by EMD [22].

EMD is also a feasible technique for producing multiple small holes in one run. In ECD of multiple small holes, side-insulated tube electrodes are arranged in an array and connected to an electrode holder. However, uneven electrolyte volume divided into each tube electrode would lead to different machining performance of each hole. Xiaolong et al. [23] designed an electrolyte dividing manifold for use in electrochem-ical drilling of multiple holes. An electrolyte is pumped into the manifold through an inlet and distributed to each tube electrode, as shown in Figure 9.8. The tube elec-trodes were horizontally arranged in a row and the electrolyte was pumped into the manifold inlet. The experimental multiple hole EMD set-up consisted of the X–Y–Z motion stage, a power supply, an electrolyte recycle system, a multiple tube electrode holder, and other fixtures. A tube electrode holder was designed for machining of 100 holes in an array, with a neighboring hole interval of 2 mm. The tube electrode

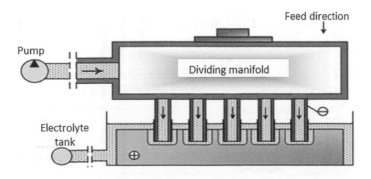

FIGURE 9.8   Schematic of a setup for electrochemical drilling of multiple holes [23].

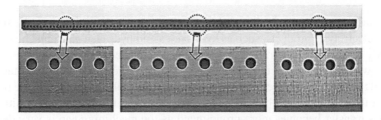

FIGURE 9.9   An array of simultaneously machined 100 holes by EMD [23].

length was 15 mm. An aqueous solution of 120 g/L of sodium nitrate was used as the electrolyte. The machining conditions used were as follows: applied DC voltage 12 V, electrode feed rate 0.9 mm/min, electrolyte pressure 0.6 MPa. The workpiece was made of 2 mm thick 304 stainless steel. Figure 9.9 shows a series of 100 microholes machined by EMD. The hole diameters were measured to be 1.08 ± 0.03 mm.

9.4   WIRE ELECTROCHEMICAL MICROMACHINING

In wire electrochemical micromachining (WEMM), a wire is used as a tool to cut the workpiece material. The wire tool cathode is fed towards the workpiece until the machining gap is suitable to initiate the required electrochemical dissolution. Compared to wire electrodischarge machining (EDM) the absence of thermally induced material removal makes the WEMM process promising as there is no recast layer and heat-affected zone [24]. The wire does not undergo dimensional change or wear during WEMM and can be reused. The electrolyte can be conveniently supplied to the machining zone without needing a complex electrolyte supply system. The potential candidate materials for wire for the WEMM process are tungsten, copper, and platinum [4]. The accuracy in WEMM depends mainly on the machining gap.

Figure 9.10 shows a schematic diagram of multiwire EMM in which the wire electrodes are arranged in a row and electrically connected to the pulse generator [25]. In the machining process, the wire electrodes are fed along the programmed tool trajectory simultaneously. The machining stability and accuracy are significantly affected

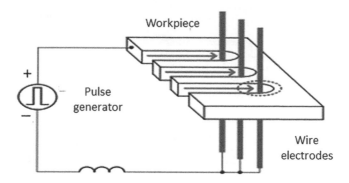

FIGURE 9.10 A schematic diagram of multiwire EMM in which the wire electrodes are arranged in a row and electrically connected to the pulse generator [25].

by the mass transport rate in the machining gap. A fresh electrolyte must be supplied and the electrolysis products, such as hydroxides and hydrogen gas, must be removed promptly to keep the electrolyte conductivity unchanged. Otherwise, electric short circuits will frequently occur as electrolysis products accumulate in the machining gap, and the machining process will become very unstable. Especially in the processing of a thick workpiece (thickness more than a few millimeters), the limitation of the mass transfer becomes more pronounced as the machining gap is very deep and narrow. An acidic solution is usually used avoid insoluble sludge formation in the gap. The enhancement of mass transport in the machining gap is essential during WEMM to improve the machining stability and the ability to cut thick workpieces. Application of low-frequency and small-amplitude tool vibration is an effective way to remove electrolysis products and renew the electrolyte in WEMM. Traveling wires improve the surface integrity and MRR markedly in an abrasive WEMM process.

Linear motion of the wire electrode along its axis during the machining process is an effective way to remove the electrolysis products and renew the electrolyte in the machining gap. Figure 9.11 shows a WEMM system with a monodirectional

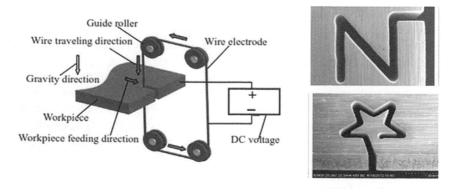

FIGURE 9.11 Schematic diagram of WEMM using a monodirectional traveling wire and photographs of structures machined on 5 mm thick 304 stainless steel plate [26].

traveling wire [26]. During the machining process, both the wire electrode and the workpiece are immersed in the electrolyte and then the workpiece profile can be formed by controlling the workpiece feed trajectory. Compared to electrolyte flushing and a reciprocated traveling wire, the monodirectional traveling wire provides more uniform flow in the machining gap as the wire is always traveling in one direction and there is no directional error. Using the monodirectional traveling WEMM process, structures with uniform slit width have been fabricated in of 5 mm thick 304 stainless steel plates. Figure 9.11 shows examples of structures machined in a 6 g/L solution of sodium nitrate using a 200 μm diameter wire with a traveling velocity of 0.02 m/s, a feeding rate of 1.2 μm/s, and an applied voltage of 9 V.

9.5 ASSISTED ELECTROCHEMICAL MICROMACHINING

9.5.1 LASER-ASSISTED JET ELECTROCHEMICAL MICROMACHINING

In laser jet electrochemical micromachining (LJEMM), the main purpose of employing a laser is to improve the precision by improving the process localization. While the predominant mechanism of material removal is electrochemical dissolution, the role of the laser is to assist and localize the electrochemical energy. The thermal energy of the laser transmitted to the workpiece enhances the kinetics of electrochemical reactions and hence enables the localization of dissolution to a specific area. The application of laser causes a higher MRR in the axial rather than in the lateral direction and thereby dimensional precision is improved.

Datta et al. [6] investigated the effectiveness of a laser jet employing neutral salt solutions for high-speed maskless micromachining of difficult-to-machine metals. The feasibility studies involved investigating metal removal rate and stray cutting during micromachining in passivating and nonpassivating electrolytes as a function of applied current density and laser power. The experimental setup shown in Figure 9.12 consists of two electrolyte reservoirs, a pump, the jet cell, Argon laser

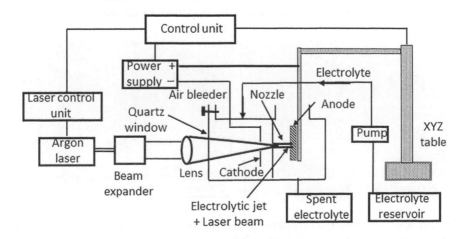

FIGURE 9.12 Experimental setup for LJEMM [6].

with beam expander, power supply, and the control system. The electrolyte from the reservoir is pumped to the jet cell where it impinges through the nozzle on the workpiece. A part of the cell is under hydrostatic pressure, causing the electrolyte to exit via the 500 µm diam. nozzle in the form of a free-standing jet directed onto the anode workpiece. A platinum sheet, with a central hole through which the Argon laser beam is directed, served as the cathode. Air bubbles trapped in the fluid compartment are eliminated by a bleeder valve. The anode workpiece is mounted on a sample holder that is attached to a computer-controlled precision arm capable of micron movement in the X–Y–Z directions. The microprocessor also controls the on-off switch of the power supply as well as the on-off gating of the argon laser.

A CW argon ion laser beam with a constant output power of 22 W was passed through a beam expander and focused with a 75 mm focal length lens to a point near the center of the jet orifice. Sodium chloride (5 M) and sodium nitrate (5 M) solutions were used as the electrolyte. The linear flow velocity of the electrolyte was maintained constant at 1,000 cm/s which corresponded to a Reynolds number of 3,700 at the nozzle orifice. The nozzle-anode spacing was 0.3 cm. Experiments consisted of micromachining holes as a function of applied current density and determining the weight loss of the nickel and iron samples in anode samples. The results indicated that the incorporation of a laser beam into the jet stream did not significantly influence the current efficiency for nickel and iron dissolution in chloride solutions. On the other hand, the holes micromachined by a laser jet were much deeper than those obtained by a jet without the laser beam. Figure 9.13 compares the SEM microphotographs of two micromachined holes in nickel with and without a laser beam. The hole micromachined with a laser jet is much deeper and its outer diameter is relatively small compared to a shallower and a wider hole obtained by a jet without the laser. The stray current effects were characterized by determining the volumetric material removal (v) per unit depth (d) for each machined hole. Decreasing value of v/d indicates smaller stray current and decreased overcutting. The results showed that the v/d values are indeed smaller for laser jet experiments, the effects being more pronounced for steel than for nickel. These studies were extended to generate complicated patterns in sheets of other technically important materials.

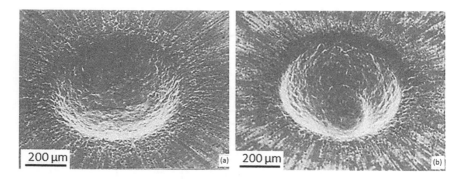

FIGURE 9.13 SEM microphotographs of holes micromachined by EMM on nickel in 5 M NaCl using (a) a jet and (b) a 22 W laser jet [6].

In nitrate electrolyte, the presence of a laser beam tremendously reduced the current efficiency for nickel and steel dissolution indicating that oxygen evolution is the dominant anodic reaction. The observed enhancement of the oxygen evolution reaction by the laser beam is due to photoelectrochemical and thermal effects caused by the absorption of laser energy. The passivating oxide films that are present at the anode even at high current densities in nitrate electrolyte are highly sensitive to absorption of light, which generates excited electrons and excited vacant sites (holes), causing photoelectrochemical effects at the oxide-electrolyte interface. Such effects together with increased localized temperature may indeed create favorable conditions for anodic oxygen evolution.

Experiments conducted by Pajak et al. [27] on a range of materials revealed that LJEMM can improve the volumetric removal rate of 20%, 25%, 33%, and 54% for Hastelloy, titanium alloy, stainless steel and aluminum alloy, respectively. They also found that the laser assists in breaking the oxide layer that occurs on the surface of materials such as titanium alloy or aluminum alloy. Measurements of current and current density have proved that the laser improves the electrolyte conductivity giving higher current compared to JEMM. For materials such as titanium and aluminum alloys that form an oxide layer, current characteristics show an initial and then a boost stage that refers to the moment of oxide layer breakage. On the other hand, laser assistance does not have much influence on the improvement of current efficiency for any of the materials investigated rather they were maintained approximately at the same level. Higher laser energy provided enhanced localization and increased shape accuracy.

9.5.2 Vibration/Ultrasonic-Assisted Wire Electrochemical Micromachining

WEMM is a promising approach for the fabrication of high-quality microcomponents. However, its industrial applications remain limited owing to its relatively low machining efficiency, which is significantly affected by the efficiency with which electrolyte can be refreshed in the narrow machining gap. He et al. [28] investigated the use of axial vibration-assisted multiwire electrodes with high traveling speed to improve upon the machining efficiency of WEMM. A schematic of the multiwire EMM is shown in Figure 9.14 along with the photographs of the machined microstructures. The authors also developed a flow-field model to simulate the flow field in the machining gap with the traveling tool. Experimental results revealed that microstructures with negligible taper, high aspect ratio, and good consistency can be effectively fabricated using this method.

Figure 9.14b (A and B) shows a multiple-slit microstructure with a slit width of 148 μm and an aspect ratio of 20 fabricated successfully using 15-wire electrodes. The structures were fabricated at a pulse voltage of 18 V, a pulse frequency of 50 kHz, a pulse duty cycle of 35%, a feed rate of 5.0 m/s, a vibration frequency of 2 Hz, and vibration amplitude of 10 mm. The length of each slit is 2.0 mm. The fabrication was accomplished by a three-stage machining process in which, after each machining stage was completed, the multiwire tool was shifted 333.3 μm in the X-direction,

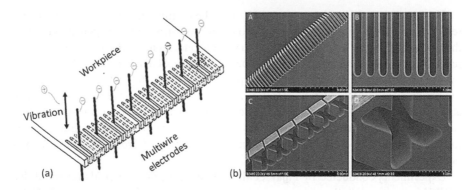

FIGURE 9.14 (a) Schematic diagram of WEMM using axial vibration-assisted multiwire electrodes with high traveling speed and (b) its application in fabricating a series of high precision microstructures [28].

and the machining process was performed again. The machining rate was 75.0 μm/s (4.5 mm/min), and the machining time was about 20 minutes. Figure 9.14b (C and D) shows multiple microcomponents fabricated by 7-wire electrodes. Compared with WEMM, the use of axial flushing of the electrolyte at high speed provides a faster machining rate and minimized tapering.

Wang et al. [29] developed a WEMM system with vibrating microtool. Their experimental results indicated that low frequency and small amplitude vibration significantly improves process stability, overcut, machining accuracy, and repeatability of WEMM cutting. The machining stability worsens as the workpiece thickness increases.

Mitchell-Smith and Clare [30] explored the use of ultrasonic for the removal of the passivating layer during machining processes. The formation of passivating layers in some materials inhibits their machinability by electrochemical processes. These, relatively inert coatings, prevent direct metal dissolution in the electrolyte. This is particularly problematic in JEMM. Experiments presented by Mitchell-Smith and Clare [30] showed that the passivating scale which forms on the workpiece can be partially removed by ultrasonic assistance in JEMM thus enhancing its overall machining performance. The use of ultrasonic during machining is demonstrated to favorably enhance the feature aspect ratio by increasing depth and minimizing kerf. Surface textures of machined faces using ultrasonic assistance also demonstrated a significant reduction in Ra. These results suggest that there are significant advantages for using ultrasonic assistance during the JEMM of Ti alloys.

9.5.3   ABRASIVE-ASSISTED JET ELECTROCHEMICAL MICROMACHINING

Abrasive jet electrochemical micromachining (AJEMM) is a nonconventional process that combines abrasive jet machining (AJM) and JEMM concurrently. A micro-jet of an electrolytic solution containing abrasive particles is made to impinge on the target while applying a DC potential between the jet nozzle and the workpiece. Difficult to machine materials can be machined by AJEMM through a combination of electrochemical dissolution, erosion, and synergistic effects. Liu et al. [31] investigated AJEMM of tungsten carbide (WC) using a neutral pH NaCl electrolyte rather than an alkaline solution which is more commonly used in the electrochemical processing of WC. Their study showed that the erosion due to AJM alone was not able to dissolve the WC and that the electrochemical dissolution by JEMM was slow and produced unacceptably wide channels. The combined AJEMM process, however, was found to involve erosion of the developed oxide layer and subsequent exposure of the bare WC surface, leading to a much higher machining current density, dissolution rate, and machining localization than using JEMM alone. It was also found that the total abrasive kinetic energy, working voltage, and solution concentration strongly affected the machining current density, MRR, and aspect ratio (depth-to-width ratio). The results indicated that AJEMM has a high potential to machine difficult-to-machine metals efficiently and economically.

One of the problems associated with WEMM is the presence of anodic products in the micron-level machining gap which get deposited on the surface of the wire cathode and are very difficult to remove. Usually, an acidic or alkaline solution is chosen as the electrolyte to keep the electrolytic products dissolved in solution during machining, but they pose operational safety issues and are not environmentally friendly. To solve the problem, Jiang et al. [32] used boron carbide (B4C) particles in a neutral pH $NaNO_3$ electrolyte to machine microgrooves by vibration-assisted WEMM. They studied the effects of the B4C particles on the deposition occurring on a wire cathode surface during machining and their role in reducing the accumulation of bubbles. Additionally, the effects of the amplitude, frequency of the wire cathode vibration, and particle concentration on the maximum feed rate and profiles of the microgrooves were examined. The experimental results showed that adding B4C particles not only significantly reduced the electrolytic products deposited on the surface of the wire cathode and prevented bubbles from accumulating in the machining gap but also improved the surface quality of the microgrooves.

9.6 EMM CASE STUDIES

9.6.1 WIRE ELECTROCHEMICAL MICROMACHINING OF ALUMINUM RINGS FOR THE FABRICATION OF CORRUGATED HORNS

Corrugated horns are widely used in space guidance, microwave remote sensing, satellite communication, and target detection. Mechanical milling, electroforming, and additive manufacturing are the typical methods for producing corrugated horns [33]. Corrugated horns in the millimeter waveband are primarily made of copper using an electroforming process. An electroforming mandrel is usually made from an aluminum rod or is assembled from a set of aluminum rings. With the increase of working frequency in short-millimeter-band networks, the feature sizes of corrugated grooves and teeth in a horn, and the corresponding mandrel size are reduced to micrometer dimensions, thus posing challenges for their fabrication [33,34]. Several efforts have been made to produce microparts with corrugated microstructures on the horn's cylindrical surface. Application of focused ion beam and wire electro-discharge machining left surface burrs which could not be used in the electroforming process.

WEMM removes materials at the atomic level with no contact between the tool electrode and the workpiece. Therefore, it has certain inherent advantages, such as causing no burrs, no residual stress, no lateral damage, and no metallurgical defects. WEMM can machine metals regardless of its mechanical properties. All of these features make WEMM a reliable and flexible micromachining method for fabricating aluminum rings that are used to electroform a corrugated horn sample. Fang et al. [33] employed a rotary helical electrode during the WEMM of aluminum foil stack using a 15 g/L sodium nitrate electrolyte. To improve the electrolytic product removal efficiency, glutamic diacetic acid (GLDA) was added to the sodium nitrate electrolyte. GLDA is an environmentally friendly complexing agent and is produced using an environmentally friendly process involving the fermentation of readily available corn sugars. Key operating parameters such as voltage, GLDA concentration, and electrode diameter were optimized to get the desired machining performance. Thus, the optimum conditions included: an electrolyte consisting of 15 g/L sodium nitrate solution with 15 g/L GLDA; the pulsed voltage of 8.5 V in amplitude, 10 kHz in frequency, with a 10% duty ratio; 0.3 mm diameter spindle at a speed of 2,000 rpm; and the electrode feed rate of 1.2 μm/s.

Figure 9.15 shows a schematic diagram of WEMM utilizing a rotary helical electrode to fabricate aluminum rings for the assembled mandrel in electroforming a corrugated horn antenna [33]. First, the rotary electrode was utilized to drill a hole through the aluminum layers to create a starting point for follow-up trajectory cutting. Afterward, the electrode was programmed to proceed along a circular path. These two steps are both EMM processes. Furthermore, anodic workpieces used in the machining were stacked together with several aluminum layers, among which the first layer takes the role of a sacrificial layer to protect other layers from machining

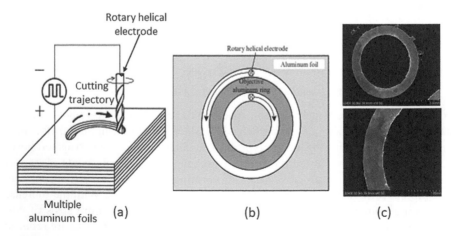

FIGURE 9.15 A schematic diagram of WEMM of aluminum foil stack using a rotary helical electrode to fabricate aluminum rings (a). The feed trajectories of the outer circle and inner circle (b). SEM photographs of the machined aluminum ring (c) [33].

defects (if any). As a result, several aluminum rings could be machined at the same time, thus the machine efficiency was improved.

The fabrication of aluminum rings consisted of inner edge machining and outer edge machining. The slits machined by a rotary helical electrode had good unilateral consistency with better side results occurring to the left of the electrode feed direction. To obtain aluminum rings of high quality, the outer edge and inner edge were machined in line with the better slit sides. Figure 9.15 illustrates the feed trajectories of outer and inner machining edges. The inner edge was machined in a clockwise direction, while the outer was machined in an anticlockwise direction. Employing the experimentally determined optimum WEMM parameters mentioned above, 100 μm thick aluminum rings with a good edge and surface qualities were successfully machined using a helical electrode as shown in Figure 9.15. Finally, these microrings were applied to electroform a corrugated horn sample.

9.6.2 ELECTROCHEMICAL SAW USING PULSATING VOLTAGE

EMM using pulsed voltage/current can be advantageously used for precision cutting of metals and other conducting materials. Indeed, due to the small effective surface area of the cutting tool, one can use relatively small and inexpensive pulse generators and extremely simple electrolyte pumping devices. This should make the process particularly suited for small-scale applications involving the cutting of hard materials such as refractory metals or for the cutting of single crystals in metallography laboratories where one wants to avoid any alteration of the crystal structure during the cutting process. Indeed, among the existing methods of single crystal cutting, electro-discharge machining and other mechanical cutting methods such as diamond saw cutting lead to severe surface alteration of the cut area due to thermal effects and/ or mechanical surface deformation. Also, the resulting cut surfaces are often rough and require chemical or electrochemical posttreatment.

Chemical machining using an acid attack is another method for cutting single crystals. However, compared to an electrochemical process the rate and precision of the chemical attack are much lower and its applicability to different metals is restricted.

Datta and Landolt [35] developed an electrochemical saw which is based on the principle of PECM. The electrochemical cutting assembly shown in Figure 9.16 has the following features: (i) application of a pulsating voltage at a constant advance rate of the anode, (ii) a thin blade cathode moving perpendicular to the advancing anode, and (iii) electrolyte contact only in the machining gap. The operating principle is illustrated schematically in Figure 9.16. The cathode blade is a thin plate which is covered by insulation on all sides except that facing the anode. The electrolyte is admitted to the interelectrode gap through two capillary tubes placed on both sides of the workpiece. Supply of fresh electrolyte to the interelectrode gap and removal of reaction products and heat is accomplished by the movement of the blade. The electrochemical saw has been employed for cutting nickel, steel, and stainless-steel bars using 5 M sodium chloride and 5 M sodium nitrate electrolytes. Optimum operating conditions were obtained from cutting experiments by varying cell voltage, pulse time, duty cycle, and feed rate. Under otherwise similar conditions, cutting in sodium

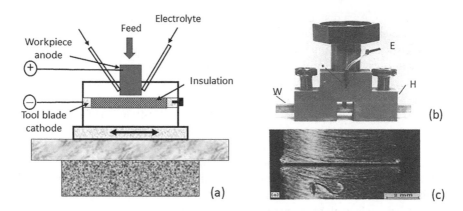

FIGURE 9.16 (a) Schematic diagram of the electrochemical saw assembly. (b) The workpiece holder assembly: W: workpiece, E: electrolyte entry, H: holder. (c) Photograph showing cut line [35].

nitrate electrolyte gave minimum stray cut. Figure 9.16c shows the photograph of a cut on nickel bar in 5 M $NaNO_3$ using a cell voltage of 10 V, 1 ms pulse time, 20% duty cycle, and a feed rate of 150 μm/min. The electrochemical saw has also been successfully employed to cut sintered and doped V_2O_3, an electron conducting ceramic material. Compared to other materials, the cutting rate for V_2O_3 under optimum conditions is relatively small (30 μm/min). A higher cutting rate could be achieved at a higher voltage but resulted in poor precision.

REFERENCES

1. C. van Osenbruggen, C. de Regt, *Philips Tech. Rev.*, 42 (1), 22 (1985).
2. M. Datta, L. T. Romankiw, *J. Electrochem. Soc.*, 136, 285C (1989).
3. M. Datta, Electrochemical Micromachining, *in Electrochemical Technology; Innovations and New Developments*, N. Masuko, T. Osaka, Y. Ito, eds., pp. 137–158, Gordon and Breach, Amsterdam, (1996).
4. B. Bhattacharyya, *Electrochemical Micromachining for Nanofabrication, MEMS and Nanotechnology*, William Andrew (Elsevier), New York, (2015).
5. K. K. Saxena, J. Qian, D. Reynaerts, *Int. J. Mach. Tools Manuf.*, 127, 28 (2018).
6. M. Datta, L. T. Romankiw, D. R. Viglioti, R. J. von Gutfeld, *J. Electrochem. Soc.*, 136, 2251 (1989).
7. C. Clerc, M. Datta, L. T. Romankiw, Electrochemical Society, *in Patterning Science and Technology*, R. Gleason, J. Haffron, L. K. White, eds., vol. 152, pp. PV 90–PV 91, Electrochemical Society, New Jersey, (1990).
8. T. Kendall, P. Bartolo, D. Gillen, C. Diver, *Int. J. Adv. Manuf. Technol.*, 105, 651 (2019).
9. C.-J. Chia, F. Giralt, O. Trass, *Ind. Eng. Chem. Fundam.*, 16, 28 (1977).
10. V. E. Nakoryakov, B. G. Pokuasev, E. N. Troyan, *Int. J. Heat Mass Transfer*, 21, 1175 (1978).
11. D.-T. Chin, C.-H. Tsang, *J. Electrochem. Soc.*, 125, 1461 (1978).
12. D.-T. Chin, K.-L. Hseuh, *Electrochim. Acta*, 31, 561 (1986).

13. M. Datta, Micromachining by Electrochemical Dissolution, *in Micromachining of Engineering Materials*, J. A. McGeough, ed., pp. 239–276, Marcel Dekker, New York, (2002).
14. J. Kozak, K. P. Rajurkar, R. Balkrishna, *J. Manuf. Sci. Eng., Trans. ASME*, 118, 490 (1996).
15. M. Hackert-Oschätzchen, G. Meichsner, M. Zinecker, A. Martin, A. Schubert, *Precis. Eng.*, 36 (4), 612 (2012).
16. X. Lu, Y. Leng, *J. Mater. Proc. Technol.*, 169, 173 (2005).
17. M. Sen, H. S. Shan, *Int. J. Mach. Tools Manuf.*, 45, 137 (2005).
18. Z. Rahman, A. K. Das, S. Chattopadhyaya, *Mater. Manuf. Processes*, 33 (13), 1379 (2017).
19. J. Wang, W. Chen, F. Gao, F. Han, *Int. J. Adv. Manuf. Technol.*, 75, 21 (2014).
20. D. Zhu, H. Y. Xu, *J. Mater. Process. Technol.*, 129, 15 (2002).
21. M. Wang, Y. Zhang, Z. He, W. Peng, J. Matls, *Proc. Technol.*, 229, 475 (2016).
22. Z. W. Fan, L. W. Hourng, *Int. J. Adv. Manuf. Technol.*, 52, 555 (2011).
23. F. Xiaolong, W. Xindi, W. Wei, Q. Ningsong, L. Hansong, *J. Mater. Proc. Technol.*, 247, 40 (2017).
24. K. K. Saxena, J. Qian, D. Reynaerts, *Int. J. Mach. Tools Manuf.*, 127, 28 (2018).
25. F. Xiaolong, L. Peng, Z. Yongbin, Z. Di, *Procedia CIRP*, 42, 423 (2016).
26. Y. Zeng, Q. Yu, X. Fang, K. Xu, H. Li, N. Qu, *Int. J. Adv. Manuf. Technol.*, 78 (5–8), 1251 (2015).
27. P. T. Pajak, A. K. M. Desilva, D. K. Harrison, J. A. McGeough, *Precis. Eng.*, 30, 288 (2006).
28. H. He, Y. Zenga, Y. Yao, N. Qu, *J. Manuf. Proc.*, 25, 452 (2017).
29. S. Wang, D. Zhu, Y. Zeng, Y. Liu, *Int. J. Adv. Manuf. Technol.*, 53, 535 (2011).
30. J. Mitchell-Smith, A. T. Clare, *Procedia CIRP*, 42, 379 (2016).
31. Z. Liu, H. Nouraei, J. K. Spelt, M. Papini, *Precis. Eng.*, 40, 189 (2015).
32. K. Jiang, X. Wu, J. Lei, Z. Wu, W. Wu, W. Li, D. Dia, *Int. J. Adv. Manuf. Technol.*, 97 (9–12), 3565 (2018).
33. X. Fang, X. Wang, J. Zhu, Y. Zeng, N. Qu, *Micromachines*, 11 (1), 122 (2020).
34. D. P. Adams, M. J. Vasile, A. Krishnan, *Precis. Eng.*, 24, 347 (2000).
35. M. Datta, D. Landolt, *J. Appl. Electrochem.*, 13, 795 (1983).

10 Through-Mask Electrochemical Micromachining

10.1 INTRODUCTION

Material removal techniques are among the key processing technologies that are employed in the fabrication of microstructures for their use in many applications including microelectronics, microelectromechanical systems (MEMS), and biomedical components. These methods are popularly known as etching techniques [1]. Dry vacuum processes for thin-film etching are based on plasma-assisted processes and include ion etching, plasma etching, and reactive ion etching. These processes are particularly employed in the semiconductor industry for ultra-large-scale integration because of their ability to remove the material with precision. However, some of the disadvantages that are inherent in dry-etching techniques include high equipment cost, lack of selectivity, and problems arising from redeposition on the sample and deposition on the vacuum chamber. Furthermore, there are concerns about the safety, environmental impact, and disposal of the toxic gases used in plasma-assisted dry etching.

Wet chemical etching involves the removal of unwanted material by the exposure of the workpiece to an etchant. The exposed material is oxidized by the reactivity of the etchant to yield reaction products that are transported from the surface by the medium. Wet chemical etching baths contain chemicals that are generally aggressive and toxic, thus posing safety and disposal problems. In many wet-etching manufacturing processes, waste treatment and disposal costs often surpass the actual etching processing costs [2].

Electrochemical metal removal is an alternative wet etching process where the workpiece is made an anode in an electrolytic cell in which controlled metal removal takes place by application of an external current. Several nonconventional machining processes such as electrochemical machining (ECM) and electropolishing are based on the principle of electrochemical metal removal. Most of the thin films of metals and alloys, including conducting ceramics and highly corrosion-resistant alloys, that are of interest in the microfabrication of advanced products can be anodically dissolved in salt solution of sodium nitrate, sulfate, or chloride. In these electrolytes, the dissolved metal ions form hydroxide precipitates which remain in suspended form in solution and can be easily filtered, thus significantly minimizing problems of safety and waste disposal. These aspects make the electrochemical micromachining (EMM) a simpler and more environmentally friendly manufacturing process.

Through-mask electrochemical micromachining (TMEMM) involves selective metal dissolution from unprotected areas of a photoresist patterned workpiece. Compared to chemical etching, electrochemical dissolution offers higher rates and better control on a micro- and macroscale of shape and surface texture of anodically dissolved materials. TMEMM may involve microstructuring of one-sided or two-sided photoresist patterned workpiece [3]. Through-mask metal removal by a wet process is accompanied by undercutting of the photoresist and is generally isotropic in nature. In isotropic etching, the material is removed both vertically and laterally at the same rate. This is particularly the case in chemical etching, where the etch boundary usually recedes at a 45° angle relative to the surface. In TMEMM, however, the metal removal rate in the lateral direction may be significantly reduced through proper consideration of mass transport and current distribution. Dissolution in the lateral direction under the photoresist is expressed by the etch factor, which is defined as the ratio of the amount of straight-through etch to the amount of undercut [3]. This is schematically shown in Figure 10.1 for both one-sided and two-sided TMEMM. At the initial stage, a photoresist pattern can be visualized as a cavity with its bottom being the conducting metallic flat surface to be dissolved and the cavity walls being the insulating photoresist material. The dissolution process in TMEMM is initiated from the conducting flat surface of the cavity. In one-sided TMEMM, the evolution of the dissolved structure takes place in two phases as indicated in Figure 10.1a. Phase one (P1) involves vertical as well lateral dissolution thereby creating a close to semicircular shape with a certain amount of photoresist undercut until the underlying insulator substrate is exposed.

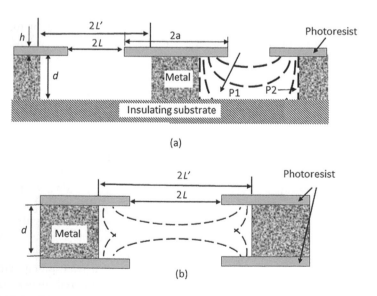

(a)

(b)

FIGURE 10.1 Schematic diagram of shape evolution showing photoresist undercut and etch factor (EF = dissolution depth/undercut) in (a) one-sided TMEMM and (b) two-sided TMEMM. P1: phase 1 involving vertical and lateral dissolution, and P2: phase 2 involving lateral dissolution leading to wall profile.

In phase two (P2), only lateral dissolution takes place thus providing the possibility to control the sidewall angle and the photoresist undercut. Similar processing principles are applicable in a two-sided TMEMM. For applications requiring a high aspect ratio, minimized undercutting of the photoresist and a high value of the etch factor are desirable. Numerical simulation and experimental investigation of the TMEMM process has been conducted by several authors and the topic has been reviewed on several occasions [4–8].

The metal removal rate, microfeature profile, surface finish, and uniformity of metal removal are some of the performance criteria that determine the technical feasibility of a metal removal process. In EMM, these criteria are dependent on the ability of the system to provide desired mass transport rates, current distribution, and surface film properties at the active surface [5]. In through-mask processes, additional issues related to lithography processing are critical to achieving the desired performance. Production of the master artwork, surface preparation, choice of proper photoresist, and imaging are extremely important in the successful implementation of a TMEMM process. Since parts produced by this process are a direct reflection of the master artwork, it is essential that all aspects of preparing the artwork are understood. These include a priori knowledge of the metal removal rate and etch factor. Imaging is another important step, the objective of which is to reproduce the artwork features as closely as possible onto the workpiece. The imaging process capability is measured by its resolution. In TMEMM, careful design of the walls and height of the photoresist provide opportunities to alter current distribution that reduce the photoresist undercutting. The conductivity of the substrate material is also important in influencing the current distribution of a dissolving thin film.

10.2 PHOTORESIST AND LITHOGRAPHY

Photoresists and lithography are the key ingredient and technology for the production of precision micro/nanostructures such as integrated circuits, MEMS, and other intricate structures and devices that are formed by defining patterns in various metallic layers. Pattern transfer in TMEMM consists of two parts: a photoprocess, whereby the desired pattern is photographically transferred from an optical plate to a photosensitive film coating the wafer, and an EMM process of removing materials to create the pattern. The photolithography relies on photosensitive polymers known as photoresist whose structure changes upon exposure to ultraviolet (UV) light [9]. The light-sensitive photoresist materials are of two types, positive and negative, that react very differently when exposed to UV light.

The exposure of a positive photoresist to the UV light changes the chemical structure which becomes more soluble in the photoresist developer. These exposed areas are then washed away with the photoresist developer solvent, leaving the underlying material. The areas of the photoresist that are not exposed to the UV light are left insoluble to the photoresist developer.

With negative photoresists, exposure to UV light causes the chemical structure of the material to polymerize, which makes it extremely difficult to dissolve during development. As a result, the UV exposed negative resist remains on the surface while the photoresist developer solution works to remove the unexposed areas. This leaves a mask that consists of an inverse pattern of the original, which is applied on the wafer.

Both positive and negative photoresists are used in micro and nanomanufacturing. Positive photoresists are preferred in precision structuring due to their higher resolution capabilities. Compared to positive photoresists, negative photoresists have a faster photo speed, wider process latitude, and a significantly lower operating cost. However, due to their lower resolution capabilities, the negative photoresists are mostly used in applications where high resolution is not required.

Photoresists are commercially available in liquid or laminate form which can be positive or negative type. Liquid photoresists are solvent-based and are generally applied by spin coating to a thickness of 1–5 μm per coating. For thicker photoresist applications, multiple coatings are employed. Photoresist laminates are also known as dry film resists and are generally thicker in the range of 10–50 μm.

TMEMM process requires careful consideration and implementation of different lithographic steps that include the production of the master artwork, surface preparation, choice of proper photoresist, and imaging. The basics of photolithography steps are shown in Figure 10.2. A film of light-sensitive organic photoresist is first deposited on the surface of the metal to be patterned. A glass plate with the patterned chrome layer on it, known as the mask, is held above the metal surface. A source of UV light exposes only the areas of the UV light-sensitive photoresist layer that are not covered by chrome. Depending on the type of photoresist used, either the exposed or unexposed regions dissolve away in the developer solution. Therefore, either a positive or negative image of the mask is transferred to the photoresist. The photoresist pattern so formed is used to fabricate microstructures in the underlying metallic films by TMEMM. Yellow light is used in the lithography area to avoid unwanted exposure of the UV-sensitive photoresist.

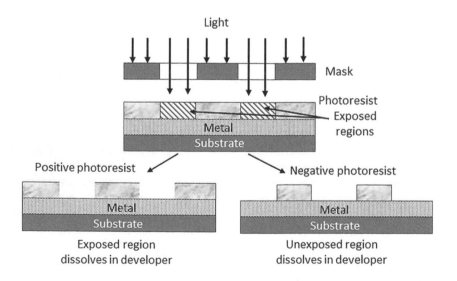

FIGURE 10.2   Schematic diagram showing photolithography process steps for resist patterning.

10.3   MASS TRANSPORT IN A CAVITY

Mass transport processes play a significant role in influencing an electrodissolution process. The maximum rate of an electrodissolution reaction is determined by the prevailing mass transport conditions on the dissolving anode which gives rise to a limiting current. Mass transport processes influence the macroscopic and microscopic current distribution on the workpiece [10]. Also, mass transport-controlled dissolution reactions provide smooth electropolished surfaces. An understanding of mass transport effects is, therefore, a prerequisite for the development of TMEMM processes.

Localized dissolution of a photoresist patterned metal surface leads to the formation of a cavity, the evolving shape of which is dependent on the convective mass transport within the cavity as established by the local hydrodynamic conditions. Such processes influence the supply of reactant to the surface as well as removal of the dissolution products from the surface. In general, the highest transport is observed at the point where fresh solution first contacts the surface. The electrolyte flow in a cavity is characterized by the Reynolds number and the aspect ratio (depth/width). At the initial stage of the TMEMM process in a shallow cavity (aspect ratio <0.3), with a conducting metal bottom and inert sidewalls, the highest mass transfer rate is in the upstream region where the external flow penetrates to the cavity base. For a deep cavity (aspect ratio >0.5) in a laminar flow field, the highest mass transfer rate occurs in the downstream region where inertia carries the freshest electrolyte to its point of first contact with the surface. In both shallow and deep rectangular cavities, the transport rate at the interior corners is low owing to weak recirculation of corner eddies [11,12]. The effect of fluid flow on the convective mass transport in a cavity is expressed by the Peclet number, Pe, which is defined as:

$$Pe = \frac{vL}{D}$$

$$(10.1)$$

where v is the flow velocity, the characteristic length, L, is the width of the cavity, and D is the diffusion coefficient. The influence of the Peclet number on the average mass transfer rate in a cavity has been correlated by Alkire et al. [11,12] based on experimentally measured average mass transfer coefficients during TMEMM of patterned Cu samples. The following empirical correlation was obtained:

$$Sh_{avg} = 0.3 \left(\frac{L}{H} \right)^{0.83} Pe^{0.33}$$

$$(10.2)$$

where Sh_{avg} is the average mass transport rate and H is the cavity depth. Equation 10.2 was found to agree with the average mass transfer rates calculated by solving the equations for Stokes flow using the finite-element method and by a combination of the boundary integral method and Lighthill boundary-layer analysis [11,12].

The high current densities needed to achieve surface brightening during high-rate anodic dissolution in salt solutions may lead to electrolyte heating in the interelectrode space and therefore high forced convection rates are needed to avoid a significant rise in temperature. However, in many microfabrication applications involving delicate parts, high flow rates are not desirable. In such cases, pulsating current allows the application of high instantaneous current for a short duration which permits to exceed the solubility limit during the pulse-on time while keeping the average current density relatively low. In pulse dissolution, this situation can be achieved even at relatively low electrolyte flow rates.

10.4 SHAPE EVOLUTION MODELING

The shape evolution of an anodically dissolving metal depends on the prevailing current distribution on the anode. For conditions where mass transport effects are negligible, the current distribution is obtained from the potential distribution and depends on the geometrical dimensions of the system and electrode kinetics. In the presence of mass transport limitations, the surface concentrations at the anode differ from the bulk which must be included in the current distribution modeling. At the limiting current, the local dissolution rate is entirely controlled by mass transport and the potential distribution can be neglected in the first approximation. In ECM, EMM, and TMEMM where high current densities are employed, the current distribution problems can be described by the primary current distribution, provided there are no significant mass transport limitations.

The current distribution at the dissolving electrode in TMEMM is dictated by the photoresist artwork parameters. Any variations in these parameters affect the current distribution at the metal surface and the spatial distribution of the metal removal rate. For shape evolution modeling, it is essential to consider the current distribution both at the initial electrode surface and within the evolving cavity. The parameters that characterize the shape evolution during TMEMM are shown in Figure 10.1: (i) aspect ratio, h/L, (ii) spacing to opening ratio, a/L, (iii) metal film thickness ratio, d/L, and (iv) mask wall angle, θ. The aspect ratio, spacing to opening ratio, and the mask wall angle influence the current distribution at the electrode. On the other hand, the metal film thickness ratio determines the extent to which TMEMM must proceed before the underlying insulator is exposed. The influence of these parameters on the undercut and the etch factor are of interest. The etch factor is defined as the ratio of the maximum etched depth to the undercut ($d/L' - L$). Once the insulator is exposed, metal dissolution continues in the lateral direction leading to a change in metal wall angle. Fine-line TMEMM of metal films on insulator requires minimized undercut (hence increased etch factor), and straight walls.

The shape evolution of a dissolving cavity during TMEMM has been simulated by several authors [13–17] under the conditions of primary current distribution by solving Laplace's equation using the boundary element method. The model is primarily used to predict the shape of the evolving cavity and the etch factor for a 90° mask angle. Simulation results presented in Figure 10.3 show that with progressing dissolution, a given flat surface evolves into hemispherical shape. In phase 1 of the TMEMM process, the cavities grow in both vertical and lateral direction (Figure 10.3a) until the insulator is exposed [15]. In phase 2, the lateral dissolution leads to the evolution of the metal wall profile as shown in Figure 10.3b [16]. The current density along the exposed metal surface is a function of the photoresist geometry (aspect ratio, h/L). For aspect ratios greater than 0.5 (relatively

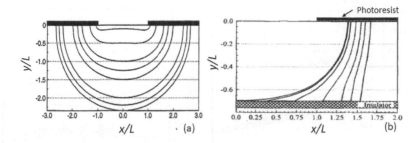

FIGURE 10.3 Mathematically simulated shape evolution of the cavity formed by anodic dissolution (a) during phase 1 of the TMEMM process and (b) the evolution of metal wall profile during phase 2 of the TMEMM process [16].

thick photoresist), the maximum vertical displacement of the metal surface is at the center. The etch factor is independent of the spacing to the opening ratio (a/L) and decreases as the cavity evolves. Figure 10.3a shows the shape evolution for an aspect ratio of 0.095, for spacing to the opening ratio of 2.15 and for a film thickness ratio of 1.175. The initial current density distribution is highly nonuniform and the maximum in the current density distribution occurs at the edges of the feature. This leads to the maximum vertical displacement (or maximum dissolved depth) to be at the edges. However, as the dissolution proceeds, the metal surface under the photoresist draws some of the current and the maximum in the current density distribution at the evolving surface starts to move toward the center of the feature. The numerical results of such studies permit the estimation of the undercutting required to achieve a given wall steepness. The shape of the evolving cavity has been shown starting from the point just before the exposure of the insulator under the metal film. At the point where the normalized maximum etched depth (d/L) is equal to the metal film thickness ratio (d/L), the metal surface evolves in the form of a hemispherical cavity. As the metal removal progresses, the insulator under the metal film is exposed. Figure 10.3b shows the evolution of the metal wall profile. The metal film closer to the insulator is removed faster than that under the photoresist and the metal film profile evolves toward a straight wall. The extent of the metal removal necessary to obtain a certain metal wall angle can be expressed in terms of the ratio of the undercut to the metal film thickness.

Shenoy et al. [16,17] investigated the influence of photoresist angle (θ) and photoresist aspect ratio (h/L) on the etch factor (Figure 10.4). The simulation results indicated that the primary current distribution at the initial electrode surface is very sensitive to the mask wall angle. During TMEMM, the shape of the evolving cavity causes a significant redistribution of the current along the electrode surface. As the shape evolves, the current distribution becomes more uniform. For mask wall angle of $\theta > 90°$, the current distribution at the initial surface has a maximum at the edge of the feature. The maximum in the current distribution moves toward the center of the feature as the metal removal progresses. For mask wall angle of $\theta < 90°$, the current distribution has a maximum at the center of the feature even at the initial stage of dissolution. As the TMEMM proceeds, the maximum in the current distribution remains at the center of the feature. The mask wall angle influences the etch factor

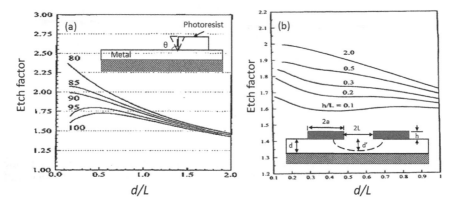

FIGURE 10.4 (a) Etch factor derived from simulated shape evolution of cavities as a function of dissolution depth for various photoresist mask angles; (b) etch factor as a function of the etched depth for different aspect ratios [16,17].

as shown in Figure 10.4a [16]. The etch factor is defined as the ratio of the undercut to the maximum vertical depth. Figure 10.4a shows that the influence of the wall angle is greatest in the initial stages of the metal removal process and diminishes as the metal removal proceeds. With increasing etch depth, the etch factor for different mask wall angles approaches the etch factor for straight vertical mask walls. Acute angled masks have been shown to improve the directionality of through-mask EMM by reducing the undercut for thin metal films. The undercut to obtain a desired metal wall profile is lower for the acute-angled masks than for right-angled or obtuse-angled masks. However, the influence of the mask wall angle on the evolution of the metal wall profile has been shown to diminish with increasing metal film thickness ratio.

Figure 10.4b shows the effect of aspect ratio on the etch factor as the metal dissolution process continues [17]. The etch factor increases with increasing aspect ratio, and the maximum etch factor for a given maximum etched depth is observed for aspect ratios ≥1. Thus, maximum etch factors under conditions of ohmic control are obtained for recessed electrodes which start with uniform current density at the initial metal surface. The subsequent shape evolution leads to the maximum of the current density as well as the vertical displacement of the electrode surface at the center of the feature thereby leading to the higher etch factor.

The shape evolution model assuming the primary current distribution is only applicable to cases where the concentration gradients are not significant. On the other hand, in some cases where flow around the feature is significantly perturbed, the concentration gradient may not be negligible. The calculation of shape evolution under such conditions must include changes in mass transport effects. Higdon [18] simulated the fluid flow around a cavity using the boundary integral method and Occhialini and Higdon [19] solved the convective mass transport problem using the spectral-element method. The results indicated that for shallow cavities, the external flow penetrates the cavity with weakly recirculating flow patterns at the corners. This facilitates a high rate of mass transport at the center of the cavity than at the edges. Such a phenomenon is beneficial as it can improve the anisotropy of the

TMMM process. For deep cavities, a single weakly circulating flow pattern limits the penetration of the external flow into the cavity. Since perturbation from the external flow is minimal for deep cavities, the diffusion layer approximation can be used to determine the undercut and shape of the evolving cavity.

10.5 TMEMM CHALLENGES

Several interesting challenges are encountered during the patterning of thin films by TMEMM which require attention for effective application of the process. The challenges include: the ability to provide processing uniformity, minimized undercutting (anisotropy) for fine features, and elimination of the problem due to loss of electrical contact in one-sided TMEMM. Current distribution and shape evolution modeling studies have proved to be very effective in understanding and overcoming these challenges.

10.5.1 UNIFORMITY CONSIDERATIONS

One of the key advantages of TMEMM is that it enables the fabrication of hundreds and thousands of patterned microstructures consisting of multiple features in a workpiece in one run. However, the production of individual patterns with the same precision along the workpiece is challenging and depends on the uniformity of mass transport and current distribution from corner to corner of the workpiece. Three different scales are to be distinguished when considering current distribution and mass transport on patterned electrodes: the workpiece scale, the pattern scale, and the feature scale [20]. At the workpiece scale, the current distribution depends primarily on cell geometry. The processing tool design must involve hydrodynamic considerations to provide desired mass transport and current distributions conditions. These and other engineering aspects of precision tools for TMEMM processing are discussed under a different section in this chapter.

On the pattern scale, the current distribution depends strongly on the spacing of the features and on their geometry. West et al. [6] analyzed the primary current distribution on the pattern scale on a photoresist covered anode containing features at a different distance from each other. The current density on features spaced farther apart was found to be higher leading to an increased etch rate of these features. The results confirmed previous experimental data of Rosset et al. [21,22]. On the feature scale, the current distribution is governed by the concentration field of the reacting species and/or the potential field within the feature. To estimate the normalized current distribution on an individual feature of a pattern, it is necessary to account for the influences of the nearest neighbor features. For the prediction of the variation of the average current density from one feature to another, one must account for the influence of each of the features of the pattern, but detailed knowledge of the normalized current distribution is not needed.

10.5.2 ANISOTROPY CONSIDERATIONS

As mentioned above, one of the problems of through-mask etching by wet processing is its isotropic nature due to photoresist undercut. Photoresist undercutting and etch factor determine the extent of anisotropy. For improved anisotropy, smaller undercutting and larger etch factor values are desirable. While impingement of electrolyte on the dissolving cavity helps in directional EMM thus providing a certain degree of anisotropy, an understanding of the current distribution and the factors influencing the shape evolution enables optimization of the process parameters including photoresist mask design and the conditions leading to surface films. Anisotropic etching depends in part on the nonuniform convective mass transport caused by local hydrodynamic eddies. Alkire et al. [11,12] simulated the effect of hydrodynamics on local mass transport rates in rectangular cavities exposed to laminar channel flow. Experiments performed with copper in sulfuric acid suggested that anisotropic etching can be achieved at intermediate values of the Peclet number provided salt film forms at the surface and dissolution is mass transport controlled. If the Peclet number is too small, diffusion dominates leading to isotropic etching, and if the Peclet number is too large, the salt film may be swept away by fluid flow. Convective mass transport studies by Higdon and coworkers that improvement in anisotropy can be achieved under conditions where the flow is significantly perturbed within the feature. Formation of weakly recirculating flow patterns at the corners of a cavity at the cavity edges facilitates a high rate of mass transport at the center of the cavity than at the edges, thus improving the anisotropy of the TMMM process.

Photoresist angle (θ) and photoresist aspect ratio (h/L) influence the etch factor and hence the anisotropy during TMEMM [16,17]. As noted earlier, acute-angled masks improve the directionality of TMEMM by reducing the undercut. However, the influence of the mask wall angle on the evolution of the metal wall profile diminishes with increasing metal film thickness ratio. Therefore, the angular dependence of the anisotropy improvement is mainly restricted to thin-film structuring.

A certain degree of anisotropy is achieved by adjusting the photoresist aspect ratio. As shown in Figure 10.4b, the etch factor increases with increasing aspect ratio, and the maximum etch factor for a given maximum etched depth is observed for aspect ratios ≥ 1. Thus, maximum etch factors under conditions when the shape evolution leads to the maximum of the current density as well as the vertical displacement of the electrode surface at the center of the feature thereby leading to the higher etch factor.

10.5.3   ISLAND-FORMATION ISSUES IN ONE-SIDED TMEMM

Loss of electrical contact is one of the key concerns in the pattering of metallic films by TMEMM. Loss of electrical contact leads to a premature stoppage of the TMEMM process thus leaving an island of undissolved metal as shown in Figure 10.5. An understanding of the conditions that lead to island formation and loss of electrical contact can be obtained from the shape evolution modeling results obtained at different stages of TMEMM [7,15–17]. At the initial stage of EMM, the current is concentrated more at the edges than in the middle. As EMM proceeds, the current concentration shifts gradually from the edges to the middle of the feature. The latter condition is desirable since this ensures complete removal of the material from the center. For a given photoresist and metal film thickness, the shift in the current concentration is dependent on the initial feature dimension that is exposed to the current, as shown in Figure 10.5b and c. For small openings (Figure 10.5b), the current concentration moves to the center of the feature before the metal film is completely

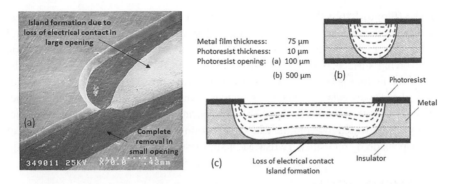

Metal film thickness: 75 μm
Photoresist thickness: 10 μm
Photoresist opening: (a) 100 μm
(b) 500 μm

FIGURE 10.5 (a) Microphotograph of a feature micromachined by TMEMM showing unetched material remaining in large opening (island formation problem while the small opening is completely etched. (b and c) Mathematically modeled shape evolution during EMM of different photoresist openings; large opening shows an island of material that remains unetched [7,17].

etched. For large openings (Figure 10.5c), the metal film may not be thick enough for the current concentration to shift to the middle. Under these conditions, because of the faster rate of dissolution at the edges, an island of material remains unetched, losing electrical contact with the rest of the metallic surface and thus bringing the EMM process to an end. The problem of island formation can, in principle, be resolved by converting the situation of Figure 10.5c into several narrow openings, as shown in Figure 10.5b. This can be achieved by introducing dummy photoresist artwork [23] as shown in Figure 10.6. Incorporation of dummy features reduces the photoresist opening (L), thus increasing the aspect ratio (h/L) and increases the film thickness ratio (d/L). These features work toward moving the current maximum to the center of the feature thereby eliminating the island formation hence the loss of electrical contact during TMEMM. It is essential, however, that the dimensions of the dummy photoresist artwork match the undercutting of the photoresist from both sides so that upon completion of EMM the dummy photoresist is removed, yielding a clean, and straight-walled micromachined trench.

For thin-film patterning, the problem of island formation can also be avoided by adjusting the photoresist artwork parameters. One way of lowering the critical metal film thickness needed to avoid island formation is to increase photoresist

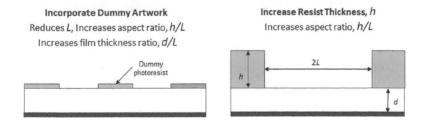

FIGURE 10.6 Solutions to prevent island formation in TMEMM by incorporating dummy artwork or by increasing resist thickness [17].

thickness (Figure 10.6b). An increase in photoresist thickness leads to a higher aspect ratio, making the current distribution more uniform at the initial surface of the metal film. Thus, the metal film thickness needed for the maximum in the current distribution at the evolving surface as well as the maximum vertical displacement to move to the center of the feature is lower. However, it is not always possible to increase the photoresist thickness particularly for features that are more than 100 μm wide. On the other hand, modifying the photoresist design by incorporating dummy artwork has been successfully implemented to avoid the island formation problem during the fabrication of slider suspension by one-sided TMEMM of stainless-steel film on polyimide [23,24].

10.6   TMEMM TOOLS

The development of an effective TMEMM process requires careful design and fabrication of a tool that provides desired current-distribution and mass transport conditions at the dissolving surface. The electrolyte delivery system is one of the main considerations in the design of a precision tool. Different electrolyte delivery systems that are applicable in EMM include channel flow, electrolytic jet, slotted jet, and multinozzle systems [5,25–27]. Electrolyte composition and temperature control, and provisions for filtration are some of the other important design aspects that must be addressed. The tool design must consider all these aspects to build an EMM tool for a specific application.

10.6.1   ONE-SIDED TMEMM TOOL

Two types of TMEMM tools for the fabrication of microstructures of photoresist patterned metallic films/foils/plates are shown in Figure 10.7. A simplified schematic of a one-sided TEMM tool is shown in Figure 10.7a [5,28]. It consists of a multinozzle electrolyte delivery system that is attached to an X–Y driving mechanism. The sample anode is held on a stationary plate facing downward on top of the multinozzle electrolyte delivery systems which is scanned at a constant speed past the photoresist patterned

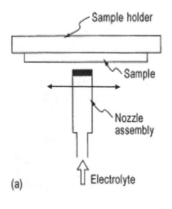
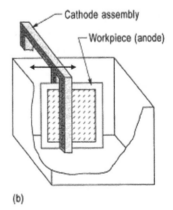

FIGURE 10.7   Schematic diagram of (a) one-sided and (b) two-sided TMEMM tools [28].

sample. The multinozzle plate also acts as the cathode. The interelectrode spacing can be adjusted and kept at a constant small distance in the mm range. The multinozzle flow assembly provides high-speed electrolyte impingement at the dissolving surface, thus permitting effective removal of the dissolved products and the heat generated by joule heating. Other accessories, not shown in the figure, include an electrolyte reservoir and electrolyte pumping and filtration units. The tool can be used for EMM of samples of different sizes. The active area at any given time during micromachining is defined by the electrolyte in contact with the sample. Within the interelectrode gap, the electrolyte emanating from the multinozzle cathode impinges on the workpiece and is directed downwards flowing on the sides of the cathode into the reservoir. At a given point this yields a nonuniform current distribution at the portion of the workpiece that is in contact with the electrolyte. The current density, and hence the metal removal rate, attains a high value in the impingement region while a gradual drop of current as a function of the distance away from the impingement region leads to low metal removal rates in these regions. However, scanning the cathode (or the sample) serves to equalize the distribution of metal removal rate by compensating for the stray current effect, since every part of the sample is exposed to both high current and stray current regions in a cycle. This approach has been successfully implemented to build precision EMM tools that provide wafer-level uniformity.

10.6.2   Two-Sided TMEMM Tool

Figure 10.7b shows a two-sided EMM tool for the fabrication of microstructures by TMEMM of metallic foils or plates that coated are with precisely aligned photoresist patterns on both sides [5,28]. The tool uses the concept of localized dissolution induced by scanning two cathode assemblies over a vertically held workpiece providing movement of the electrolyte on both sides of the workpiece. Highly localized dissolution by using a small cathode width and an extremely small interelectrode spacing provide directionality of metal removal and overall uniformity of current distribution. The electrolyte flows between the moving cathode and the mask anode. Depending on the aspect ratio of features desired, two different flow types can be used in the tool: a shear flow from top to bottom of the cathode, and an impinging flow directed into the anode workpiece. The cathodes on both sides of the anode are scanned back and forth across the vertically held workpiece at a constant speed. Unlike many spray systems that are pressurized and involve solution spilling, the TMEMM tool is a nonpressurized system with electrolyte flowing in the downward direction. The tool can be designed to handle different ranges of sample size, interelectrode spacing, and electrolyte flow.

10.6.3   Precision Electroetching Tool for Semiconductor Processing

Electrochemical processing tools for semiconductor processing include electrolytic plating, electroless plating, chemical etching, electroetching, cleaning, and chemical mechanical polishing. Complexity, configuration, and sophistication of these tools differ depending on their application. For semiconductor wafer processing, such tools must meet some basic requirements that are essential in a fab-processing environment [29]. Electrical contact to the wafer, terminal effect, edge exclusion zone management, backside contamination requirements, and thickness uniformity are some of the key issues that dictate the design of tools for semiconductor wafer processing. Sample orientation, electrolyte heating, steady-state bath control, and provisions for filtration are some of the other important engineering aspects that need to be considered in the design and fabrication of precision electrochemical processing systems. In the present context, a wafer-level electroetching tool used in the semiconductor processing is described. It must be noted that while large-scale electroetching equipment have found limited application so far in the electronics industry, different versions of electroetching tool for the removal of Cr/phased CrCu under bump metallurgy (UBM) layers have been patented and implemented in manufacturing for the fabrication of flip-chip bumps by electrochemical processing [30]. These electroetching tools are applicable in the patterning and blanket layer removal of thin films for microelectronic packaging, MEMS, and other applications.

Figure 10.8 shows a schematic diagram of the electroetching tool, in which four wafers are mounted vertically, two wafers on each side held back-to-back in a wafer holder. An alternative configuration of the tool, in which the wafers are mounted horizontally, has also been demonstrated [30]. An electrolyte delivery system in the form of a multinozzle assembly is attached to a linear motion table and is scanned from one end of the wafer to the other end during electroetching. The wafer is connected as an anode. A stainless-steel plate in the electrolyte delivery system acts as the cathode. The stainless-steel plate is engraved with carefully designed small nozzles through which an electrolyte impinges onto the wafer. In the vertical mode, the multinozzle assembly moves over the wafer surface in an upward direction. The scanning rate is adjusted according to the dissolution rate of the material stack and the thickness of the material to be removed.

The electrochemical fabrication of flip-chip bumps involves both plating and etching [31,32]. After the through-mask electrodeposition of the bumps, the UBM layers between the bumps must be removed to electrically disconnect them. Depending on the UBM metal stack, this may involve chemical or electroetching operation. For UBM stack consisting of Cu/phased CrCu/Cr, electroetching is preferable since

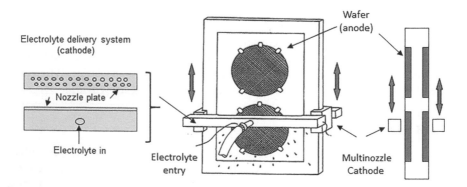

FIGURE 10.8 Schematic diagram of a vertical electroetching tool for semiconductor processing of four wafers. Also shown are the details of the nozzle plate used in the electrolyte delivery system [32].

the phased CrCu layer is not easy to etch by chemical etching. Using a mixture of 0.4 M potassium sulfate and 1.5 M glycerol as the electrolyte, a stack of Cr/phased CrCu/Cu has been successfully etched. The use of glycerol in the electrolyte helped in lowering the limiting current of Cu, thus allowing minimized Cu undercut during the dissolution of the phased CrCu and Cr layers. For the UBM stack consisting of TiW/phased CrCu/Cu, the electroetching process was used to remove only the top phased CrCu/Cu layers. The TiW layer remained completely passivated in the potential region where CrCu and Cu layers anodically dissolved at high rates. This behavior served as an auto stop for the electroetching process and the electroetching stopped at the TiW layer, which was removed in a separate hydrogen peroxide-based chemical etching process. The minimized undercutting of the copper pad during EMM provided the desired bump size and robustness after reflowing. The developed EMM processes and tools thus enabled the electrochemical fabrication of robust flip-chip bumps in high volume manufacturing [31,32].

10.7 TMEMM APPLICATIONS: CASE STUDIES

The feasibility of TMEMM in the fabrication of components with intricate microstructures that are applicable in a variety of industries including microelectronics, biomedical, MEMS, and microfluidics has been demonstrated in several published works [3–5,31–39]. In the following, some selected case studies are presented as examples of the application of TMEMM.

10.7.1 Fabrication of Ink-Jet Nozzle Plates

Ink-jet nozzle plates are used in inkjet printers in which the driving force which causes the ink to be ejected from the nozzle is the formation of a high-pressure bubble, which is formed by rapid heating and vaporization of the ink constituents [33]. This is schematically shown in Figure 10.9 in which the terms entry hole and exit hole refer to the entry and exit of the ink drop during its ejection through the nozzle [34]. Inkjet printers typically have nozzle diameters in the 40–100 μm range [33]. Among different nozzle-fabrication techniques which include mechanical drilling, etching of silicon and photosensitive polyimide, and electroforming, TMEMM

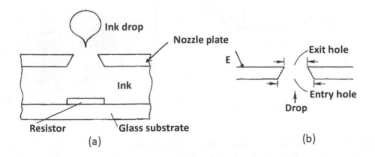

FIGURE 10.9 Schematic diagram of an inkjet assembly showing (a) ejection of an ink drop through a nozzle plate and (b) entry and exit holes of a nozzle plate [34].

has been demonstrated to be a cost-effective, high-speed process for the fabrication of precision nozzles for inkjet printers [34]. The process is applicable to a variety of materials including high-strength, corrosion-resistant materials such as conducting ceramics. TMEMM, therefore, provides the possibility of fabricating high-nozzle-density plates employing mechanically stable foil materials.

Datta [34] conducted a detailed investigation of the electrochemical parameters involved in the nozzle plate fabrication. The TEMM processing involved the following steps. A sample typically consisted of a 25 µm thick sheet of copper or stainless-steel foil which was cleaned and laminated with 25 µm thick photoresist on both sides. The photoresist on one side was exposed and developed to define the initial pattern, consisting of a series of photoresist patterned nozzle plates, each containing thousands of exposed vias to be micromachined. The blanket photoresist on the backside of the foil served as a protective insulating layer. The photoresist pattern consisted of an array of circular openings, 72 µm in diameter. Direct and pulsed-voltage TMEMM experiments were performed using a neutral salt solution of 3 M sodium nitrate or 4 M NaCl. The photoresist was then stripped, and the sample was inspected for entry and exit holes. The one-sided TMEMM tool described above was used.

Figure 10.10 shows a current-cell voltage curve for copper dissolution in 3 M $NaNO_3$ + 100 ppm FC-98 in the one-sided TMEMM tool using a 3 mm interelectrode gap and an electrolyte flow: of 7.5 L/m [34]. The additive FC-98 acted as a surfactant to ensure complete wetting of the surface which was essential for the reproducibility of nozzle fabrication. The current-cell voltage curve shows a well-defined limiting current. For the fabrication of nozzles in copper foils, TMEMM experiments with photoresist patterned samples were performed at different cell voltages corresponding to conditions below the limiting current plateau and at the limiting

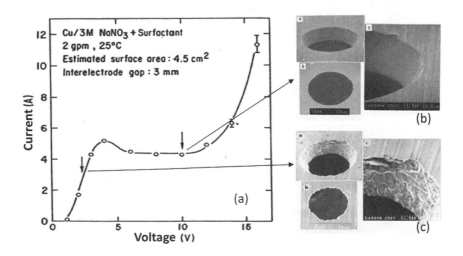

FIGURE 10.10   (a) Current-cell voltage curve for copper dissolution in 3 M $NaNO_3$ + 100 ppm FC-98 in the one-sided TMEMM tool. Interelectrode gap: 3 mm, electrolyte flow: 7.5 L/m. Also shown are the SEM photographs of TMEMM surfaces dissolved (b) at the limiting current and (c) below the limiting current [34].

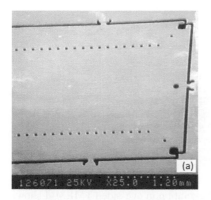
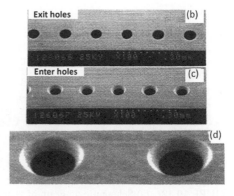

FIGURE 10.11   SEM photographs of nozzles fabricated in a 25 μm thick stainless-steel foil using one-sided TMEMM: (a) a part of the nozzle plate showing an array of nozzles, (b and c) uniformity of exit and entry holes, and (d) details of nozzles showing precision and surface smoothness [34].

current plateau as shown in Figure 10.10. After completion of the EMM experiments, the photoresist was removed, and the shape and morphology of the micromachined nozzles were investigated. Dissolution below the limiting current yielded irregularly shaped nozzles with rough surfaces exhibiting crystallographic features. Circular-shaped, microsmooth nozzles were obtained on dissolution at a voltage corresponding to the limiting current plateau or at a higher voltage. Similar experiments were performed to obtain nozzles in stainless steel foils. Since the anodic dissolution of stainless in nitrate electrolyte may be accompanied by oxygen evolution, which could remain entrapped in the undercut region, a nonpassivating 4 M NaCl electrolyte was employed for TMEMM of stainless-steel foils. Figure 10.11 shows the photographs of nozzles fabricated in a 25 μm thick stainless-steel foil [34]. A part of the nozzle plate is shown in Figure 10.11a which shows an array of nozzles. Figure 10.11b and c shows the uniformity of exit and entry holes, and Figure 10.11d shows the surface smoothness of the nozzles. The extent of micromachining could be controlled by adjusting the scanning speed of the nozzle so that the targeted exit hole dimension and desired nozzle angle could be achieved. Pulsating-voltage EMM was found to be effective in providing dimensional uniformity of nozzles. The use of extremely high peak currents provided directionality and enabled breakdown and elimination of inhibiting layers, thus facilitating activation of all of the openings at the same time. The feasibility of fabricating an array of hundreds to thousands of precision nozzles with microsmooth surfaces in copper and stainless-steel foils was thus demonstrated.

10.7.2   FABRICATION OF CONE CONNECTORS

Cone connectors find application in microelectronics packaging applications including pad-on-pad cable connectors for flex, chip burn-in pads, and high-performance boards. Effective cone-connector structures are characterized by small tips, tall cones, and strong material of fabrication. The cones are fabricated by laser ablation of polymeric films, followed by metallization. This technique produces relatively

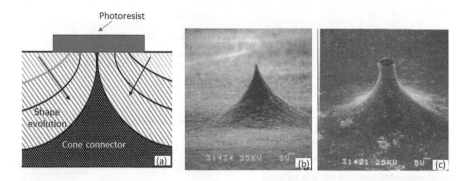

FIGURE 10.12 (a) Schematic diagram of shape evolution in one-sided TMEMM showing the formation of cone connector. SEM photographs of cones with (b) fine tips and (c) flat tips [28,35].

good-quality cones but involves several steps, thus making the process expensive. Furthermore, the cones fabricated by this method lack the desired mechanical strength. A high-speed process of fabricating cones by TMEMM has been developed and patented [28,35]. The process is applicable to a variety of metals and alloys and is independent of the hardness of the material.

A photoresist pattern in the form of evenly spaced dots is generated on a metallic material that is suitably selected for the pad-on-pad connector. During TMEMM, the anode material dissolves in the areas which remain unprotected by the photoresist. As anodic dissolution continues, the removal of material between the photoresist dots leads to the formation of cavities and finally to the formation of cones, as shown in Figure 10.12 [28,35]. Preferential dissolution in the vertical direction is achieved by employing a TMEMM tool in which an impinging electrolyte flow provides directionality. Cones on copper and hardened stainless steel (Fe-13Cr) sheets have been generated by this method. Figure 10.12b and c shows SEM microphotographs of the cones fabricated on a hardened stainless-steel sheet. The desired size and shape of cones could be obtained by proper design of the photoresist dimensions and by properly controlling the amount of charge passed during EMM.

10.7.3 FABRICATION OF METAL MASKS

In the microelectronics industry, metal masks are used for pattern definition in conductor screening and evaporation processes. Large volumes of patterned metal sheets are also used in the production of aperture masks for color CRTs. A variety of metal and alloy sheets are used, including iron, stainless steel, copper, brass, and nickel-iron alloys. For the most demanding applications in screening and evaporation processes in microelectronics, molybdenum is the material of choice. Fabrication of metal masks involves through-patterning by etching of a foil that is coated with perfectly aligned patterned photoresist on two sides. Molybdenum masks are generally chemically etched in a spray etcher using heated alkaline potassium ferricyanide solution which loses its activity due to its reduction to ferrocyanide. The etchant is regenerated electrochemically or by using a chemical oxidizer such as ozone. The spent etchant is

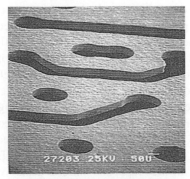
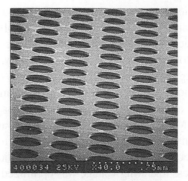

Evaporation molybdenum mask   Stainless steel screening mask

FIGURE 10.13 SEM photographs showing precision features with smooth surfaces in evaporation molybdenum mask and stainless-steel screening mask fabricated by two-sided TM EMM [28,35].

disposed of as a hazardous waste. In addition, the rinse water from the etching operation is also segregated and treated as a hazardous waste stream. Waste disposal, therefore, is a major effort that makes the chemical etching process extremely expensive.

An alternative, "greener" process for high-speed fabrication of molybdenum masks is the use of TMEMM employing a salt solution as the electrolyte. Datta and Harris [2] developed and employed a two-sided precision tool for the fabrication of metal masks. Evaporation masks consisted of 25×25 cm molybdenum sheets of thickness 50 or 100 μm with photoresist feature varying between 50 and 200 μm. The moly sheets with as many as 120,000 features were etched to a precision of 10% of the total feature size using a 3 M $NaNO_3$ electrolyte (pH 3) at ambient temperature. SEM photographs in Figure 10.13 show the straight-walled features with smooth surfaces fabricated in evaporation molybdenum masks [2,5]. The microfabrication data of TMEMM were compared with those obtained by the conventional chemical etching process using a ferricyanide solution. Performance criteria included machining rate, surface finish, aspect ratio, and simplicity of operation. The metal removal rate in TMEMM was found to be several times higher than that in chemical etching. As indicated above, TMEMM provided microsmooth surfaces and pattern uniformity. A higher aspect ratio was achieved by using an electrolyte impinging cathode. The two-sided EMM tool was also employed to fabricate stainless steel screening metal plates that are used in the manufacturing of PC boards and PC cards [2].

10.7.4 TMEMM OF TITANIUM FOR BIOLOGICAL APPLICATIONS

Titanium and titanium alloys have attracted considerable interest in the biomedical industry due to their good mechanical properties, excellent corrosion resistance, and biocompatibility. The chemical stability of titanium results from the presence of a few nanometer thick, stable surface oxide film. The biological performance of implantable titanium devices in medicine and dentistry depends critically on their surface topography in the micrometer and nanometer range [36]. For many such applications, the surface topography must be carefully controlled to achieve optimum cell adhesion and differentiation. To obtain precise information about the effects of surface topography on cell behavior, one should be able to study independently the microtopography, the nanotopography, and the superposition of both, all without changing the chemical characteristics of the surfaces. Zinger et al. [37] developed methods for the fabrication of titanium model surfaces with well-defined micrometer- and nanometer-scale topographies to be used for biological *in vitro* testing.

TMEMM of photoresist patterned titanium was performed in a methanol-based 3 M sulfuric acid electropolishing (EP) electrolyte. In order to minimize electrolyte heating, a specially designed sample holder which is cooled from inside was used. TMEMM process operating voltage was optimized that allowed rapid and uniform depassivation of the oxide film while minimizing Joule heating effects. Hexagonal arrays of closely spaced smooth hemispherical cavities of 10–100 mm diam were fabricated using controlled dissolution and precise charge control. Figure 10.14 shows the SEM picture of a dense array of 30 μm diameter cavity microstructure fabricated by TMEMM of titanium surfaces [37]. The figure illustrates that using the TEMM process, uniform and reproducible cavities with sharp borders and smooth surfaces are achieved.

The surfaces of the micropatterned cavities fabricated by TMEMM were further processed to superimpose submicrometer structures on the surface. The submicrometer structures were produced by two different methods: porous anodization and chemical etching. Porous anodization was performed in an electrolyte consisting of either 1 M H_2SO_4 or 2 M H_3PO_4. The anodization procedure consisted of sweeping the potential from 0 V to a final value of between 30 and 175 V at a scan rate of 20 V/s. This fast sweeping rate prevented the blistering of the oxide film. For the chemical etching option, the microstructured workpiece was immersed for a few minutes in a mixture of concentrated HCl and H_2SO_4 heated above 100°C. Figure 10.14b and c shows SEM pictures of an anodized titanium surface showing 30 μm cavities produced by EMM with superimposed nanopores resulting from anodic oxidation [37]. Anodization, in this case, was carried out in 1 M H_2SO_4 by sweeping the potential at 20 V/s to a final value of 125 V. These studies on electrochemical fabrication of titanium surfaces with scale-resolved topography on the micrometer and nanometer scales provide the framework for *in vitro* experiments aimed at the elucidation of how surface topography of implant materials affects cell response.

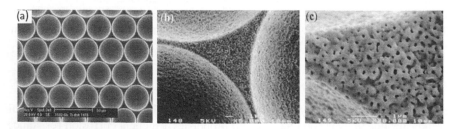

FIGURE 10.14   (a) SEM pictures of microstructured titanium fabricated by TMEMM showing 30 μm diameter hemispherical cavities with smooth surfaces and sharp borders. (b and c) SEM pictures of an anodized titanium surface showing 30 mm cavities produced by TMEMM and nanopores resulting from anodic oxidation [37].

10.7.5   TMEMM OF TITANIUM USING LASER PATTERNED OXIDE FILM MASKS

The TMEMM process using photoresist patterning is generally limited to planar surfaces. Certain applications, for example, biomedical implants, involve nonplanar surfaces to which a photoresist cannot be easily applied. Furthermore, the precision loss by light diffraction in contact mask UV illumination due to poor contact between the mask and the workpiece or the limited depth of field in the mask projection imaging strongly limit the patterning process. Chauvy et al. [38,39] demonstrated that in TMEMM of titanium, a patterned anodic oxide film can assume the function of the photoresist. Patterning of anodized titanium can be achieved by laser irradiation which permits the fabrication Ti microstructures by EMM without the need for an expensive photoresist application and patterning process.

Figure 10.15 shows the different steps involved in the TMEMM process using a laser patterned oxide film mask and its use in the fabrication of microstructure on a cylindrical titanium surface [38,39]. Oxide films were formed on titanium surfaces in a 0.5 M H_2SO_4 acid solution at 25°C using potentiodynamic control at a potential scanning rate of 20 V/s. The optimized maximum voltages were 100 V for mechanically polished substrates and 40 V for electropolished substrates corresponding to film thickness of about 250 and 100 nm, respectively. For laser patterning, a short-pulse (20 ns) laser setup with working fluence set at 1,000 mJ/cm^2 was used. The repetition rate of the excimer laser was 16 Hz and the advance rate of the sample was 40 µm/s.

TMEMM of the oxide-patterned titanium anodes was performed in an electropolishing electrolyte containing 3 M sulfuric acid in methanol. Dissolution experiments were conducted at a constant potential of 20 V. The electrolyte temperature was maintained at −10°C by an external cooling system. The desired volume of dissolved metal was determined by the amount of electric charge passed using Faraday's law and a dissolution valence of 4. Under the applied conditions, anodic dissolution

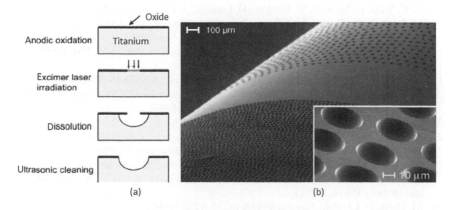

(a)   (b)

FIGURE 10.15   (a) Process steps involved in THEMM of titanium using oxide film lithography. (b) SEM photograph of microstructures fabricated on a cylindrical titanium surface using oxide film lithography, inset shows the detailed definition of features with a smooth surface [39].

was mass transport controlled which yielded a smooth polished surface. Using these process conditions, Chauvey et al. [39] fabricated two types of 3D model structures: surface structuring of a cylinder, and micromachining of a complex two-level structure. Figure 10.15b shows the SEM micrograph of the micromachined cylinder exhibiting the uniformity of pattering along the curved surface. The insert in Figure 10.15b provides a detailed definition of the cavities at higher magnification showing their reproducibility and their sharpness and smooth surfaces.

In another work, Chauvey et al. [38] employed a laser patterned oxide film mask to fabricate a test microstructure consisting of 26 mm long, 30 μm wide, and 15 μm deep microchannels for microfluidic applications. The width and depth dimensions did not show any significant variations over the etched length. An advantage of the use of laser patterned oxide film masks in such applications is the possibility of processing in the air without the need for an ultraclean and yellow light environment.

REFERENCES

1. J. L. Vossen, W. Kern, *Thin Film Processes*, Academic Press, Inc., New York, (1991).
2. M. Datta, D. Harris, *Electrochim. Acta*, 42, 3007 (1997).
3. M. Datta, *Interface*, 4 (2), 32 (1995).
4. M. Datta, *Micromachining and Microfabrication Process Technology III*, vol. 3223, p. 178, SPIE, (1997).
5. M. Datta, *IBM J. Res. Dev.*, 42 (5), 655 (1998).
6. V. M. Volgin, T. B. Kabanova, A. D. Davydov, *J. Appl. Electrochem.*, 45, 679 (2015).
7. M. Datta, *ECS Trans.*, 66 (22), 1 (2015).
8. T. Baldhoff, V. Nock, A. T. Marshall, *J. Electrochem. Soc.*, 165, E841 (2018).
9. C. A. Mack, *Fundamental Principles of Optical Lithography: The Science of Microfabrication*, John Wiley & Sons, London, (2007).
10. D. Landolt, P.-F. Chauvy, O. Zinger, *Electrochim. Acta*, 48, 3185 (2003).
11. R. C. Alkire, H. Deligianni, *J. Electrochem. Soc.*, 135, 1093 (1988).
12. R. C. Alkire, H. Deligianni, J.-B. Ju, *J. Electrochem. Soc.*, 137, 818 (1990).
13. A. C. West, M. Matlosz, D. Landolt, *J. Electrochem. Soc.*, 138, 728 (1991).
14. A. C. West, C. Madore, M. Matlosz, D. Landolt, *J. Electrochem. Soc.*, 139, 499 (1992).
15. M. Datta, R. V. Shenoy, L. T. Romankiw, *J. Eng. Ind., Trans. ASME*, 118, 29 (1996).
16. R. V. Shenoy, M. Datta, *J. Electrochem. Soc.*, 143, 544 (1996).
17. R. V. Shenoy, M. Datta, L. T. Romankiw, *J. Electrochem. Soc.*, 143, 2305 (1996).
18. J. J. L. Higdon, *J. Fluid Mech.*, 159, 195 (1985).
19. J. M. Occhialini, J. J. L. Higdon, *J. Electrochem. Soc.*, 139, 2845 (1992).
20. J. O. Dukovic, *in Advances in Electrochemical Science and Engineering*, H. Gerischer, C. W. Tobias eds., vol. 3, p. 117, VCH, Weinheim, (1994).
21. E. Rosset, D. Landolt, *Precis. Eng.*, 11 (2), 79 (1989).
22. E. Rosset, M. Datta, D. Landolt, *J. Appl. Electrochem.*, 20, 69 (1990).
23. M. Datta, L. T. Romankiw, U.S. Patent 5,284,554, February 8, 1994.
24. M. Datta, Electrochemical Micromachining, *in Electrochemical Technology: Innovations and New Developments*, N. Masuko, T. Osaka, Y. Ito eds., KodanshaiGordon and Breach, Tokyo, (1996).
25. M. Datta, D. Landolt, *Electrochim. Acta*, 25, 1255 (1980).
26. D. T. Chin, C.-H. Tsang, *J. Electrochem. Soc.*, 125, 1461 (1978).
27. R. C. Alkire, T.-J. Chen, *J. Electrochem. Soc.*, 129, 2424 (1982).

28. M. Datta, Micromachining by Electrochemical Dissolution, *in Micromachining of Engineering Materials*, J. McGeough ed., p. 239, Marcel Dekker, Inc., New York, (2002).

29. T. Ritzdorf, D. Fulton, Electrochemical Deposition Equipment, *in Microelectronic Packaging*, M. Datta, T. Osaka, W. J. Schultze eds., p. 471, CRC Press, Boca Raton, FL, (2005).

30. M. Datta, R. V. Shenoy, US Patent No. 5,486,282, January 23, 1996; US Patent No. 5,543,032, August 6, 1996.

31. M. Datta, R. V. Shenoy, C. Jahnes, P. Andricacos, J. Horkans, J. Dukovic, L. T. Romankiw, J. Roeder, H. Deligianni, H. Nye, B. Agarwala, H. M. Tong, P. Totta, *J. Electrochem. Soc.*, 142, 3779 (1995).

32. M. Datta, *Electrochim. Acta*, 48, 2975 (2003).

33. W. J. Lloyd and H. H. Taub, *in* Output Hard Copy Devices, R. C. Durbeck, S. Sherr, eds., Academic Press, Inc., New York, (1988).

34. M. Datta, *J. Electrochem. Soc.*, 142, 3801 (1995).

35. M. Datta, D. E. King, A. D. Knight, C. J. Sambucetti, U.S. Patent 5,105,537, April 21, (1992).

36. A. Curtis, C. Wilkinson, *in* Cell Behavior: Control and Mechanism of Motility, Portland Press, Ltd., London, p. 15 (1999).

37. O. Zinger, P.-F. Chauvy, D. Landolt, *J. Electrochem. Soc.*, 150, B495 (2003).

38. P.-F. Chauvy, P. Hoffmann, D. Landolt, *Electrochem. Solid-State Lett.*, 4(5), C31 (2001).

39. P.-F. Chauvy, P. Hoffmann, D. Landolt, *Appl. Surf. Sci.*, 165, 208–209, (2003).

11 Electropolishing in Practice

11.1 INTRODUCTION

Metal fabrication processes such as stamping, grinding, machining, and heat treating all are employed to convert a metal bar, casting, or sheet into a finished part. As the metal is bent, ground, heated, and altered, the metal surfaces alter significantly. These alterations take the form of burrs, surface deformation and irregularities, scale, and embedded particulate, thus making the fabricated part venerable to degradation by corrosion. These imperfections may inhibit the functionality of the component. In many cases, these surface imperfections can be a focal point for infection or metal contamination. A combination of different metal finishing techniques is used to obtain a metal surface that is devoid of imperfections and contaminants. The essential criteria of the finishing operation are to provide improved surface smoothness, brightness, reflectivity, and cleanliness. Among the different finishing methods, mechanical and electrochemical polishing techniques are the two predominant techniques used in the metal finishing industry.

Mechanical polishing is widely used in industries where it is mostly performed as a manual process on individual parts. It uses brushes, belts, and wheels with various abrasives to remove layers of material from metal parts. Being a labor-intensive and operator-controlled process, mechanical polishing very often produces irreproducible results. Complicated shaped parts and small features are not easily polished by mechanical polishing. Mechanically polished metal surfaces contain scratches, metal debris, and embedded abrasives. For high-purity applications such as medical, food services, semiconductor, and aerospace industries, these contaminants can be dangerous. Furthermore, surface distortions and surface contaminants that remain after mechanical polishing make the parts susceptible to corrosion. It is obvious, therefore, that mechanical polishing is not applicable in these sensitive applications requiring ultraclean surfaces.

Electropolishing involves anodic metal dissolution from an electrically conducting workpiece which is made an anode in an electrolytic cell. It is generally employed as a finishing operation to remove surface roughness from a workpiece, and therefore requires the removal of only small amounts of material. Electropolishing is usually carried out in concentrated acids with little or no electrolyte agitation, at current densities between 0.01 and 0.5 A/cm^2. The interelectrode spacing is generally not critical in electropolishing. Electropolishing is a combination of macrosmoothing (leveling) and microsmoothing (brightening) processes. The process is independent of the hardness of the material so that parts with soft materials as well as hard materials are polished with equal ease. Complex shaped parts and parts with small features are easily polished by EP. Compared to mechanical polishing, EP offers several advantages. Being an anodic process, it leaves a passivating oxide film on the surface thus making the part resistant to corrosion. Electropolishing removes a layer of surface material to create a smooth surface free of burrs, debris, or other contaminants thus providing an ultraclean surface. Through the choice of optimized process conditions, EP can give reproducible results even on challenging parts. Electropolished parts have a bright, shiny appearance that is uniform and lasts indefinitely.

11.2 ELECTROPOLISHING PROCESS DESCRIPTION

The first patent on electropolishing was assigned to Jacquet [1] in the 1930s. Since then, EP has been practiced commercially primarily for enhancing cosmetic appeal to consumer goods. With the realization of the virtues of the process, applications of EP widened, and it is now used in many industries including semiconductor, food, medical, pharmaceutical, automotive, and aerospace [2–5].

The workpiece to be electropolished is connected as an anode in an electrolytic cell in which an inert metal serves as the cathode. The workpiece and the cathode are immersed in an acidic solution in the electrolytic cell through which a constant current or voltage is supplied by a power supply. Electrolytes used for electropolishing are most often concentrated aqueous or nonaqueous acid solutions having a high viscosity [6,7]. As the current flows through the circuit, metal dissolution takes place from the anode workpiece. Depending on the operating parameter and the metal/electrolyte system metal dissolution reaction may be accompanied by oxygen evolution at the anode. The cathodic reaction is generally hydrogen evolution. These evolving gases form a dense gaseous layer in the cell. The optimum operation is accomplished through proper control of the electrolyte composition, temperature, current density, and processing.

The process steps involved in an industrial electropolishing operation is shown in Figure 11.1 [2–5]. Metal preparation steps aim at producing a reproducible clean surface before being subjected to the electropolishing process. The metal preparation steps include alkaline or organic solvent cleaning to remove oil, grease, fingerprints, dirt, etc. that may have originated during the fabrication and transport of the parts. The rinsing process follows, which removes residual cleaning chemicals on the surface. The pickling step removes oxide and alkaline films from the surface which is followed by another rinsing step. The freshly cleaned part is transported to the electropolishing tank which contains a high viscous electrolyte consisting of concentrated acids. After electropolishing, the parts are neutralized in sodium hydroxide

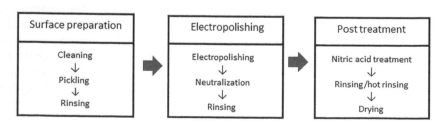

FIGURE 11.1 Process steps involved in a high-volume industrial electropolishing line.

FIGURE 11.2 Electropolishing line with electric hoist line and automatic lid with exhaust hose. (Reproduced with permission from Roman Buergler, CTO, GALVABAU A.G., Ref. [8].)

solution and rinsed in water. The posttreatment of parts consists of a nitric acid treatment to remove the electropolished surface layers that may remain even after water rinsing. The parts then go through cold water rinsing and hot water rinsing followed by drying in heated chambers.

An electropolishing operation line is similar to an electroplating line which consists of a series of wet processing tanks as shown in Figure 11.2 [8]. The parts are mounted on a jig, the design of which depends on the size and shape of the part to be electropolished. In some cases, involving small parts, titanium basket anodes are used to hold several parts. The jigs or baskets are held in an electrical hoist which moves the parts from the tank to tank. While electropolishing is generally performed in a tank-type process as described, some challenging parts may require other methods. For electropolishing of large vessels or pipes, the parts can themselves become the processing tank for electropolishing of the internal surfaces. In such cases, innovative concepts of cathode design and its placement may be needed. The electropolishing tanks carry an automatic lid with exhaust hose as shown in Figure 11.2.

11.2.1 Operating Conditions

Operating conditions including the electrolyte composition, temperature, current density, and time vary depending on the part material, shape, size, and the required surface specification.

Electrolyte: The electrolyte generally consists of a mixture of equal proportion of concentrated sulfuric acid and phosphoric acid or some other formulations such as mixtures of perchloric acid with acetic anhydride or acetic acid, or methanolic solutions of sulfuric acid [6,7].

Temperature: Aqueous electrolytes are generally operated at high temperatures ranging between 40°C and 80°C. In applications where nonaqueous electrolytes are used the EP systems operate at low temperatures generally in the range of 0°C–10°C.

Current density: The current density varies between 5 and 25 A/dm². Electropolishing of large parts can require dc currents of several thousand amperes, necessitating the use of large rectifiers, heavy cables, and/or buss. In such cases, cooling systems are required for the electrical connectors. In some specialized applications, pulsed current or pulsed-reverse current may be required.

Processing time: Electropolishing is a relatively quick process. Typical times range from 1 to 20 minutes, although longer times are sometimes necessary. Metal removal for finishing a plane surface is generally in the range of 2–3 μm. However, for finishing sharp edges or for deburring the amount of material removal depth can be up to 20–30 μm.

11.2.2 MONITORING AND CONTROL

Several critical process parameters in an electropolishing process must be controlled to produce a quality surface. Process design involves the identification of critical parameters and assigning a specification window to each parameter. The successful operation of an electropolishing process depends on careful monitoring and control of each parameter. Any deviation from the specified range of one of the parameters may seriously influence the process performance.

Monitoring of bath components is one of the key critical parameters which is essential to maintain the electrolyte composition within the specified range. Dissolved metal products and the water content of the electrolyte, if not controlled, can affect the electropolishing performance. With successive usage of the process, the concentration of dissolved metal products increases in the electrolyte which can be controlled by adding fresh electrolytes. Also, depending on the electrolyte temperature and the type of electrolyte used, water may be lost due to evaporation or if hygroscopic it may take on water. Addition or cooking-off water may be required to keep the water content within the specified range. Frequent monitoring of electrolyte temperature, density, and pH must a part of the process log sheet and these parameters must be adjusted as needed.

Another critical parameter is the quantity of charge (in ampere-second per square centimeter), which determines the amount of metal removal. Both current density and the time can be controlled to obtain a reproducible and reliably controlled surface finish. In most of the EP operations, material removal is generally in the 10–20 μm range. However, with precise monitoring, the electropolishing process can reliably remove as small as 2 μm. In many operations, the longer EP time, involving more material removal, leads to nonuniform surface finish. In such cases, prolonged oxygen evolution and its bubbles sticking to the surface or release of inclusions may create conducive conditions for pitting and streaks.

Current distribution on the workpiece is an important criterium in electrochemical processing. Fortunately, electropolishing is performed at the limited current density where the process is mass transport controlled, and the tertiary current distribution prevailing on the anode ensures uniform current distribution over the workpiece independent of the distance of the cathode. However, for EP of a complicated shaped workpiece, adjustment of the cathode placement and/or proper electrolyte agitation may be required to reach the inaccessible parts of the workpiece.

Maintaining firm electrical contact by clamping the parts is essential for high process repeatability. Spring-loaded clam-shell clamping is commonly used in electropolishing. Fully automated EP systems must ensure firm contact throughout the electrical portion of the cycle and in some instances must assure that parts do not fall from racks into the solutions. Maintaining contact cleanliness and configuration throughout the repeated operation is yet another important consideration.

Besides electrolyte management, controlling solution drag-out is essential to obtaining part quality and bath integrity. Wiper systems, brushes, and air blow-offs are used to minimize solution losses. Careful control of drag-out results in more precise process control and reduced rinse water and chemical consumption.

11.2.3 SURFACE EVALUATION

While visual observations allow one to get an aesthetic sense of an electropolished part such as shine and reflectivity, more involved examination of surfaces is required to quantify the surface roughness, surface features, and chemical nature of the surface. Electropolished surfaces are measured by either contact or noncontact methods. Contact methods use a profilometer in which a stylus is dragged across the surface. Noncontact methods include optical microscopy, electron microscopy, atomic force microscopy, interferometry, confocal microscopy, and electrical capacitance measurements.

Surface roughness is the most commonly used parameter to characterize an electropolished surface. Surface roughness is quantified by the deviations in the direction of the normal vector of a real surface from its ideal form. The surface is rough when the deviations are large; if they are small, the surface is smooth. Roughness plays an important role in determining how a real object will interact with its environment. Rough surfaces usually wear more quickly and have higher friction coefficients than smooth surfaces. Roughness is often a good predictor of the performance of a mechanical component, since irregularities on the surface may form nucleation sites for cracks or corrosion. On the other hand, roughness may promote adhesion. A roughness value can either be measured on a profile (line) or a surface (area). Ra and RMS are representations of surface roughness, but each is calculated differently. Ra is calculated as the Roughness Average of a surface's measured microscopic peaks and valleys. RMS is calculated as the root mean square of a surface's measured microscopic peaks and valleys. A single large peak or flaw within the microscopic surface texture will affect (raise) the RMS value more than the Ra value. The profile roughness parameter (Ra) is a more commonly used standard in the industry.

11.2.4 QUALITY CONTROL

The reliability and repeatability of an electropolished surface depend on the precise maintenance and control of the critical electropolishing parameters described above. An important criterium that impacts the EP surface quality is the history of the material including its composition, fabrication history, machining history, thermal treatment history, and surface contaminants. As described above, some of these issues must be considered during the choice of the EP electrolyte and during surface preparation before the EP operation. A complete electropolishing process must include quality control measures to ensure that the finished products consistently meet or exceed the specified standards. Quality control activities include the following [2–5]:

- Creating an inspection system that establishes in-coming, in-process, and final inspection criteria.
- Evaluating process risk via the use of process failure mode effects analysis (PFMEA).
- Conducting daily tests for process maintenance and making corrections as and when needed.
- Conducting visual (optical microscopy) and surface roughness tests to ensure the products to be within specifications.
- Establishing engineering capabilities to provide tailored services in response to specific project requirements (e.g., designing custom fixtures and cathodes for complicated shaped objects).
- Establishing a surface analytical facility for high-end products that require microscopic surface information. The analytical facility should include instruments such as scanning electron microscopy (SEM), Auger electron spectroscopy, x-ray photoelectron spectroscopy (XPS), and atomic force microscopy (AFM).

11.2.5 Environmental Issues

Electropolishing stations continue to use the acid electrolytes until they become too contaminated. The large amounts of metals and other compounds dissolved into the acid solutions lead to degradation of the electropolishing performance. At that point, tanks are drained, and the spent acid is disposed of for off-site treatment and disposal. The spent acid is extremely hazardous due to its pH and its metals content. Some shops choose to use in-house waste treatment [2] consisting of chemical reduction, waste stream blending, coagulation, flocculation, pH adjustment, settling microfiltration, ion exchange, and dewatering. This results in the total separation of the water from the waste material. The waste material is dried, formed into compact cakes that are disposed or sold for further use.

Using acid purification chemistry, the zero-discharge operation has been reported by clean-economical electropolishing [3]. They use sodium hydroxide to maintain pH in its normally acidic rinses between 6.5 and 8.5. The chemistry and filtration remove organics such as greases, oils, and chelating agents, and keep metal concentrations low in the rinses. Also, salts are picked up in the filter paste at a rate that keeps their concentration low enough for effective rinsing. Filter cartridges are scraped when they become caked with sludge. The cartridges are reused four to six times before disposal. Implementing the closed-loop processing of rinse waters thus allows waste reduction, improves operations, and reduces operating costs.

11.3   APPLICATIONS

Generally, any metal may be electropolished but the most-commonly electropolished metals are stainless steels, copper and copper alloys, nickel, aluminum, and titanium [2–5]. It is commonly used in the postproduction of large metal pieces such as those used in drums of washing machines, bodies of ocean vessels and aircraft, and automobiles. The process plays an essential role in the food, medical, and pharmaceutical industries due to its ability to create a microsmooth, corrosion-resistant, and oxygen-rich surface that inhibits the growth of bacteria. Ultrahigh vacuum components are typically electropolished in order to have a smoother surface for improved vacuum pressures, out-gassing rates, and pumping speed. Electropolishing is commonly used to prepare thin metal samples for transmission electron microscopy. Some key application areas are outlined below:

Semiconductor industry: Electropolishing has become a common process in the production of semiconductor equipment due to its ease of operation and its usefulness in polishing irregularly shaped objects. The semiconductor industry has some of the most demanding surface finish requirements. Due to the corrosive nature of the liquids and gases used in their manufacturing processes, the industry has developed rigid specifications that provide standards for the wetted surface characterization requirements and acceptable finish quality. The wetted surfaces require a 0.25 μm Ra finish or better after electropolishing and for 316L stainless steel the chromium oxide enrichment of minimum chromium oxide to iron oxide ratio of 2:1. The results are verified through optical examination, ESCA, Auger, and SEM analyses. Electropolishing can comply with these specifications.

Electronics: Electropolished components with a smooth finish, and free of burrs and other surface imperfections significantly reduces the chance of performance failures like arcing or current flow interruption in electronic components. Electropolishing of copper-beryllium contact parts removes all surface contaminants embedded in the metal, allowing the gold plating to adhere.

Medical devices: Due to its ability to create a microsmooth, corrosion-resistant, and oxygen-rich surface that inhibits the growth of bacteria and reduces the risk of inflammatory and allergic reactions in patients, electropolishing is often used for implantable medical devices. Critical, reusable medical devices like implants require high-level disinfection to eliminate all microorganisms in or on the device, which is made easier with electropolishing. Surgical tools and implantable devices are microfinished by electropolishing.

Food and beverage industry: The biggest concern in the food and beverage industry is the buildup of bacterial biofilms and how to prevent it from occurring. When the bacteria accumulate, it becomes increasingly difficult to remove through normal chemical washing. The smooth, electropolished surface finish improves the ability to keep the processing equipment free of contaminants such as *Salmonella*, and it offers a way to prevent potentially devastating contamination with reduced use of chemicals and their environmental impact.

Automobile industry: Automakers are particularly concerned with burrs, small pieces of metal, and other contamination that can enter the fuel and brake systems and cause premature failure. Electropolishing auto parts have proven to be an effective method of deburring and removing other surface contaminants that may break loose after assembly. In addition, electropolishing of the auto parts significantly improves their corrosion resistance, which is especially critical for welded assemblies exposed to corrosive elements like road salt. Electropolishing is widely used for finishing and deburring of interior and exterior auto parts.

Aerospace industry: The aerospace industry uses electropolishing for fatigue life improvement of components. The metal parts used in helicopters and airplanes must be able to withstand stress and corrosion with long-term use and must meet highly demanding cycle-life specifications. Removing the surface defects and deposits by electropolishing enables the part to pass stringent bench testing. Materials that are electropolished for aerospace industries include ultra-high-strength steel, stainless steel, titanium, and aluminum.

11.4   CASE STUDIES

11.4.1   ELECTROPOLISHING OF NITINOL STENTS

Nitinol (NiTi) is a shape memory alloy that is made of nickel and titanium in almost equal concentrations. The alloy has been increasingly used and accepted within the medical device industry [9]. A shape memory alloy can restore its original shape after deformation. Nitinol is popular in medical devices due to its biocompatibility, superelasticity, and fatigue, and kink resistance [10]. Among many different applications, the major use of Nitinol is in stents. Stents are structures that support or hold open circulatory vessels [11]. The use of a stent is shown in Figure 11.3 which illustrates three different blood flow situations: a normal artery showing free blood flow (a); in peripheral artery disease, the accumulation of plaque obstructs the blood flow (b); and during angioplasty, a stent is inserted to keep the artery open and restore the blood flow (c).

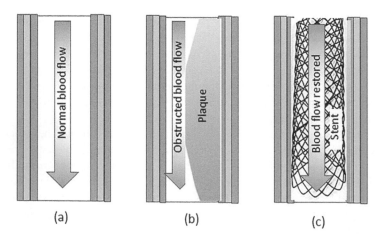

FIGURE 11.3   Use of a stent to treat a peripheral artery (schematic). (a) Normal blood flow in an artery, (b) plaque obstructs the blood flow in the artery, and (c) stent is inserted to restore the blood flow.

Stents are produced by laser cutting tube or braiding wire. The stent fabrication process consists of laser cutting a Nitinol tube into a strut pattern [9]. The laser-cut pattern is subjected to deburring and slag removal by sand or bead blasting. The resulting frame is then expanded or shape-set in a multistep operation using progressively larger mandrels until the stent reaches the required size. The inner surface is honed to smooth out the inner diameter and remove defects. The stent is then electropolished and passivated.

Electropolishing effectively removes surface defects such as burrs, sharp edges, and heat-affected zones; reduces the surface roughness with minimal material removal; and yields a well-defined thin TiO_2 layer without any free Ni-ions on the surface [12]. As a result, there is an increase in the corrosion resistance of the electropolished material. In a study involving the comparison of the probability of pitting corrosion for NiTi alloys with different surface treatments, electropolishing was found to have a significantly lower chance for pitting corrosion [13].

Although electropolishing is the state-of-the-art surface finishing process for NiTi biomedical devices, the process details of the technology such as the composition of the electrolyte, the applied voltage, temperature, and processing times are usually kept strictly confidential by the industry. Some of the electropolishing systems mentioned in the open literature include methanolic sulfuric acid [14,15] and perchloric acid-acetic acid electrolytes [16]. Inman et al. [17] patented a pulse/pulse reverse system for electropolishing and through-mask electroetching of Nitinol materials in 30 wt.% aqueous sulfuric acid electrolyte. For through-mask electroetching, short anodic pulses of relatively high voltage amplitude with an interspersed cathodic pulse of moderately high amplitude were used to depassivate the passive surface oxides. They obtained mirror-like finish with Ra ~ 0.1 μm and good pattern fidelity. They claim that the use of pulse/pulse reverse can provide a robust, cost-effective electropolishing process for Nitinol materials.

Kassab et al. [16] compared the corrosion behavior of as-received versus electropolished NiTi stents. They performed electropolishing of the NiTi stents in a 79 vol.% acetic acid and a 21 vol.% perchloric acid electrolyte at 20°C, at 10 V. The time range of the electropolishing process varied from 15 to 120 seconds. Before electropolishing, the stent surface was covered with a blue oxide, resulting from the shape-setting heat treatment. The electropolishing process removed the thermal oxide, leading to a bright and shiny surface.

The corrosion-resistant properties were evaluated by anodic polarization measurements performed with as-received and electropolished nitinol stent wires in a 0.9 NaCl solution [16]. As seen in Figure 11.4, the electropolished wires exhibited significantly wider passive range compared to the as-received condition. Localized corrosion occurred for as-received samples with a mean breakdown potential of 460 mV. For electropolished samples, on the other hand, the current density remained in the passive region up to 1 V. SEM photographs of the surfaces of the stents following the corrosion experiments are shown in Figure 11.5. The as-received samples revealed the presence of numerous pits on the surface (Figure 11.5a and b), while no evidence of the localized electrochemical attack was detected on the electropolished stent surfaces (Figure 11.5c and d). The increased corrosion resistance of electropolished Nitinol surface enhances the biocompatibility of stents, thus enabling seamless integration into the body with increased health and longevity of the patient.

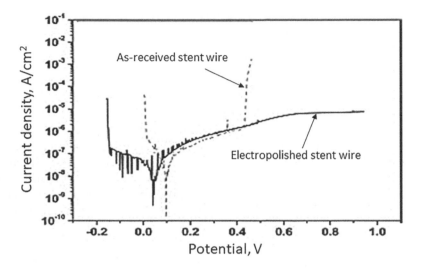

FIGURE 11.4 Potentiodynamic polarization curves for as-received and electropolished nitinol stent wires in 0.9% NaCl solution [16].

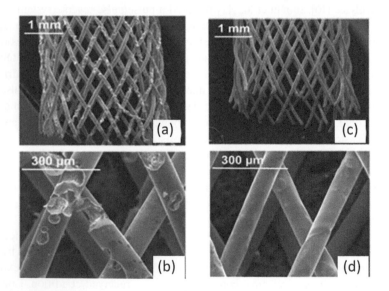

FIGURE 11.5 SEM photographs of as-received braided stents (a and b) and electropolished stents (c and d) after potentiodynamic polarization in 0.9% NaCl solution at two different magnifications [16].

11.4.2 FABRICATION OF STM PROBES

Electropolishing is widely used to fabricate probes for scanning tunneling microscopy (STM) and AFM. The characteristics of the probe, such as its apex radius, the contour, and aspect ratio, play an important role in the resolution and sensitivity of

STM [18,19]. Very smooth tips with a wide range of radii are necessary for many applications. The aspect ratio of a tip is also an important parameter. The tips for STM require low aspect ratios to minimize vibrations, while AFM tips with high aspect ratios are preferred to scan deep grooves/trenches.

Tungsten is the most commonly used material for preparing STM/AFM probes. Among the common methods used to produce high-quality tips with atomic resolution, electropolishing is employed as a fast and convenient method to obtain cost-effective and reliable tips. Several researchers have contributed to the improvement of the electropolishing equipment and the optimization of parameters, including wire diameter, electrolyte concentration and composition, voltage and waveforms applied to the electrodes, immersion depth of the tungsten wire, and a cut-off time of the electrical circuit [18,19].

A typical electropolishing setup is schematically shown in Figure 11.6. A tungsten wire anode is clamped at the center of a ring of an inert metal (generally platinum or gold) which acts as the cathode in an electrolytic cell. The Z direction movement of the tungsten wire is controlled by a precision motorized stage. The vertical positioning of the tungsten wire in the Z direction is extremely important for the final probe contour. The X–Y stage enables the desired positioning of the immersed tungsten wire within the cathode ring. The electrolyte may be NaOH or KOH.

A meniscus forms around the wire once it is placed into the electrolyte, as shown schematically in Figure 11.6b. When the potential is applied, tungsten dissolves at the surface of the wire anode inside the meniscus and below the nominal air/electrolyte interface. The meniscus that forms around the anode is critical for controlling the geometry of tips produced by the electropolishing process. There are two methodologies to achieve the desired anodic tip: static and dynamic methods [18]. These methods are based on the meniscus profiles. In the static electropolishing, the meniscus acts on an original position to get an exponentially shaped probe. On the other hand,

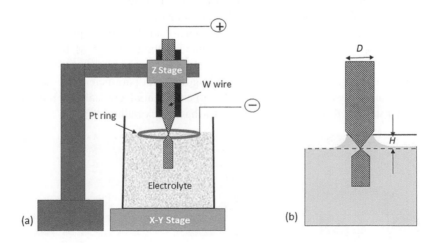

FIGURE 11.6 Schematic of (a) a tool for fabrication of a tungsten STM probe and (b) the meniscus at the air/electrolyte interface, where H is the height of the meniscus and D is the diameter of the tungsten wire.

in the dynamic process, the meniscus imposes the effect to the tungsten wire in the entire lifting-up process. During the lifting-up process, the meniscus drops down continuously due to the thinning of the immersed wire and this mechanism creates a long and cone-shaped probe. By controlling the lifting speed and distances of various shapes and radial dimensions of probes can be customized.

Xu et al. [19] conducted electropolishing experiments using 0.2 mm diameter tungsten wire anodes that were placed at the center of a platinum ring cathode of 1 cm diameter. The electrolyte used was a 1.78 M potassium hydroxide solution and the applied voltage was 4 V. The process was monitored through an optical microscope. The radii of all tips were characterized using images obtained from a scanning electron microscope. After placing the wire inside the meniscus and at different levels above the nominal air/electrolyte interface, tips with different radii were obtained until the rupture of the meniscus occurred, which automatically stopped the process. Figure 11.7 shows a series of tips that were obtained by terminating the etching at different times following the drop-off [19]. Figure 11.7a shows an irregular and rough tip obtained by shutting off the electropolishing process approximately 0.5 seconds after the drop-off. When the process was terminated 60 seconds after the drop-off (Figure 11.7b) the resulting surface was still rough. Figure 11.7c shows that the tip with a smooth surface was obtained by shutting off the process 240 seconds after the drop-off. Apparently, this is the time span required for establishing the electropolishing conditions at the anode that yields smooth surfaces.

Ge et al. [18] performed electropolishing experiments using a 250-μm diameter tungsten wire in a 1.5 M KOH. They observed that the meniscus behavior is the core factor that determines the probe profile. Static meniscus results in an exponential shape, while the continuous dynamic movement of meniscus leads to the conical shape. Using step dynamic movement of meniscus multidiameter shape probes can be obtained. The aspect ratio of the probe tip is determined by the effective contact time between the electrolyte and tungsten as well as the neck-in position of the immersed wire. The sharp tip is formed by breaking off at the neck-in position when it cannot sustain the weight of the immersed portion of the wire. Figure 11.8 shows an SEM image of a fabricated exponential shape probe using 1.5 M KOH electrolyte, 2 V, 700 seconds electropolishing time, and 850 μm immersion depth [18]. The aspect ratio of the probe is 40:1. By controlling different parameters, Ge et al. were able to produce tungsten probes with profiles of exponential, conical, and multidiameter with a stabilized stylus contour, and ultrasharp apex radius [18].

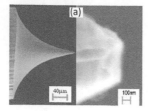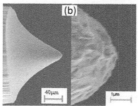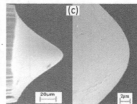

FIGURE 11.7 SEM images of tips fabricated in 1.78 M KOH at 4 V by terminating the electropolishing process after (a) 0.5, (b) 60, and (c) 240 seconds once the lower portion has dropped off. (Reproduced from Ref. [19] with the permission of AIP Publishing.)

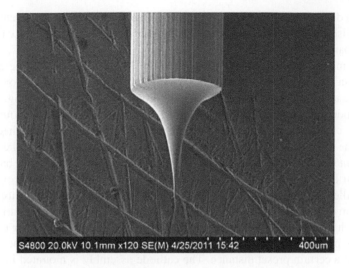

S4800 20.0kV 10.1mm x120 SE(M) 4/25/2011 15:42   400um

FIGURE 11.8 SEM image of an exponential shape probe fabricated by electropolishing in 1.5 M KOH at 2 V. Time: 700 seconds, immersion depth 850 µm, aspect ratio 40:1 [18].

11.4.3 ELECTROPOLISHING OF PRINT BANDS

The print bands for high-speed printers are made of hardened stainless steels. To ensure proper print registration, the precise location of all the characters on a print band relative to any given hammer is accomplished by the timing marks on the band [20]. The key steps involved in the manufacturing of print bands consist of photochemical etching to reveal the characters and the timing marks on stainless steel plates. The etched sheets are cut to specified width of print bands followed by laser welding to make the loop and baking to release stresses in the band. Finally, the print bands undergo surface finishing, which is a critical step that defines the quality and application of print bands. The characters on the print bands must have special characteristics to meet the desired trade-off between ribbon life and print quality. Bands with characters that have rounded edges minimize tearing and snagging of the ribbon and thus yield longer ribbon life. On the other hand, sharper corners on characters lead to better print quality. Though good print quality is obviously desired, ribbon life is an equally important criterion that requires surface finishing to yield a certain degree of character rounding.

Mechanical buffing, used for final finishing of print bands, is highly irreproducible and labor-intensive. Although buffing removes the original surface roughness to a great extent, it introduces numerous scratches that are unevenly distributed over the surface. Buffed print bands, therefore, have a dull appearance. Moreover, mechanically induced stresses are present in the surface following buffing. As an alternative method, Datta et al. [21–23] developed an electrochemical tool and a process for the surface finishing of print bands. In order to achieve character rounding and electropolishing of print bands, the surface finishing tool has provisions for (i) uniform microfinishing of the entire print band surface, and (ii) selective metal removal from the corners of characters and timing marks. For selective metal removal, the tool employs a directional electroetching technique that can be used concurrently during electropolishing or the electroetching and the electropolishing processes can be done in two separate steps.

A schematic diagram of the tool is shown in Figure 11.9 [21,22]. The electropolishing unit (Figure 11.9a) consists of a driving mechanism for band movement, an electrolyte tank, a cathode assembly, and facilities for rinsing and drying. The print band is mounted on four Teflon pulleys, P1–P4, which in turn are mounted on a motor-driven aluminum plate D that can be moved up and down. The wheel and the shaft of the top pulley P2 are made of stainless steel where the electrical connection to the print band is made as an anode. To minimize heating of the print band which may occur at electrical connection, a pair of compressed air jets are provided over the top pulley P2. The cathode assembly consists of graphite blocks that are arranged in the form of a semicylinder. The graphite blocks are connected to a stainless-steel plate which is connected to the negative pole of the power supply. During electropolishing, the print band moves around the bottom pulley P3, which faces the stationary cathode at a certain preset distance. The cathode assembly is mounted in polyvinyl chloride tank filled with the electrolyte. The level of the electrolyte in the tank is carefully maintained so as to match the desired surface area of the print band to be dipped in the electrolyte and to obtain a certain desired residence time during

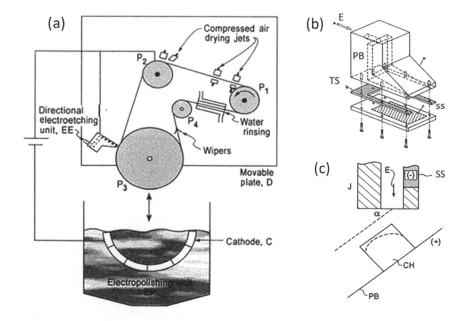

FIGURE 11.9   (a) Schematic diagram of a tool for electropolishing and character rounding of print bands. (b) Schematic diagram of the directional electroetching unit; (PB) Plexiglas block with flow openings, (TS) Teflon spacer, (SS) stainless steel plate cathode, (E) electrolyte inlet. (c) Directional etching by the positioning of the electrolytic jet, J. (SS) Cathode SS plate, (PB) print band, (CH) character [22].

polishing. The electrolyte is circulated with a small centrifugal pump which helps to maintain a well-mixed electrolyte and to minimize temperature increase. A rinsing system is provided to remove electrolytes from the print band when it leaves the electrolyte tank. The system includes a pair of Viton wipers and a water rinse which applies a stream of water. Several compressed air-drying jets are provided to dry the print band after rinsing.

An electroetching unit is incorporated in the tool that preferentially etches the characters in the print band. An electrolytic jet emanating from a rectangular slot is directed at the moving print band. A schematic of the slotted-jet assembly is shown in Figure 11.9b and c. A stainless-steel plate on one side of the slot serves as the cathode. A 200-μm thick Teflon spacer, which is cut in the middle, forms the rectangular slot. The positioning of the jet and its size is very critical in obtaining the desired degree of rounding of the leading and the trailing edges. In order to get directional etching, the jet is placed close to the character at an angle such that one edge of the character is preferentially etched as shown in Figure 11.9c. Figure 11.10 shows the picture of a manufacturing prototype tool built for electropolishing of hardened stainless-steel print bands for high-speed printers [22].

The electropolishing tool was evaluated using an electrolyte consisting of a mixture of two parts by volume of phosphoric acid, one part by volume of sulfuric acid, and one part by volume of glycerol [23]. Print bands were made of hardened Fe-13Cr

FIGURE 11.10 Photograph of a manufacturing prototype electropolishing tool showing details of the print band driving mechanism. The drive pulley P1 is attached to a stepping motor (not shown) which is mounted behind the aluminum plate. Electrical connection (E) to the print band is made at the stainless-steel wheel and shaft of the pulley P2. P3 and P4 are other pulleys for print band support. Also shown are (A) directional etching assembly, (B) rinsing unit, (C) water drain, D is the moveable mounting plate and (J) compressed air knife-jets for drying [22].

stainless steel. They were 0.16 cm thick and 1.8 cm wide. The length of the print bands in the form of a loop was typically on the order of 122, 132, or 162 cm. During electropolishing, the length of the print band dipped in electrolyte was kept constant at 16 cm. The optimum print band movement speed was found to be 2.5 cm/s, and the applied current density was found to be 1 A/cm². With one pass defined as the length of the print band, the usual number of passes applied to electropolish print bands varied between one and three.

Figure 11.11 compares the scanning electron microphotographs of surfaces of an as-received, a buffed, and an electropolished print band. The as-received surface of Figure 11.11a is highly textured and is extremely rough. While buffing removes the original surface roughness to a certain extent, it introduces numerous scratches that are uniformly distributed all over the surface (Figure 11.11b). The electropolished surface (Figure 11.11c), on the other hand, is uniformly flat even at a microscopic scale except for a few micropits.

For character rounding, a directional electroetching technique was employed which can be used concurrently during electropolishing or the electroetching and the electropolishing processes can be accomplished in separate steps. In concurrent etching and polishing, the same electrolyte is used for both character rounding and microfinishing. However, several passes are required to obtain the desired degree of character rounding. A sequential two-step process consisting of directional electroetching in 5 M NaCl at 55 A/cm² followed by electropolishing in a mixture of phosphoric acid, sulfuric acid, and glycerol was employed to obtain the desired degree of character rounding and mirror finishing of the print band.

The character roundness value was determined by comparing the profilometer traces or optical photograph of the cross section of a character with the arc of a circle of different diameters ranging up to 1 mm using a proper magnification factor. Figure 11.12 shows SEM photographs of vertical stroke edges of a character H before and after electropolishing. The leading edge of a character, which is more critical in determining the effectiveness of a print band, was found to be within the desired range between 0.1 and 0.16 mm. The results demonstrated the feasibility of the electropolishing tool and process to provide shiny print bands with the desired character rounding.

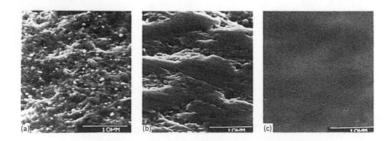

FIGURE 11.11 SEM photographs of (a) unbuffed, (b) buffed, and (c) electropolished print bands [22].

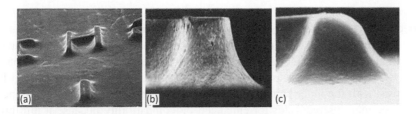

FIGURE 11.12 SEM photographs of character H (a), and the section of a vertical stroke showing leading and trailing edges before (b), and after (c) directional etching of a print band [22].

11.5 ASSISTED ELECTROPOLISHING

11.5.1 MAGNETO-ELECTROPOLISHING

Magneto-electropolishing (MEP) is electropolishing in the presence of a magnetic field, which is used to further improve the surface properties of alloys. For paramagnetic materials, like austenitic steels, titanium, and Nitinol, the MEP process enriches the surface layer with the chromium in austenitic steel or titanium compounds in Nitinol. Furthermore, the surface layers of biomaterials after the MEP process contain much lower amounts of carcinogenic compounds like chromium VI oxidation stage (Cr^{6+}) in austenitic stainless steel and nickel compounds in austenitic stainless steel and Nitinol.

The MEP setup is the same as the convention electropolishing. The MEP is additionally equipped with a neodymium magnet [24,25]. The forms of magnets that can be used for the MEP include a flat cylindrical one or a ring type. There are no significant cost implications and the safety level is comparable to that of the conventional electropolishing. However, the process applies only to nonferromagnetic materials such as austenitic or duplex stainless-steel alloys, titanium, Nitinol, and tantalum. Ferromagnetic steels in a strong magnetic field rapidly magnetize up, so their interaction with the source of the magnetic field is extremely large and unsafe.

Hryniewicz et al. [24] compared the conventional EP and MEP using titanium and Nitinol in a mixture of sulfuric, hydrofluoric, and nitric acids. They concluded that the application MEP provides super-critical refinement of surface properties to the new high level required for medical implant devices. The improved corrosion resistance of magneto-electropolished titanium surface is caused by a more homogenized amorphous mixture of titanium oxides and hydroxides compared with very crystalline titanium oxide mainly in rutile form on conventionally electropolished surface. Compared to conventional EP, the MEP process yielded smoother surfaces with lower surface roughness, better corrosion resistant properties 27% better resistant to fracture.

Gill et al. [26] analyzed the surface characteristics of MEP Nitinol alloys (NiTi, NiTiCr, and NiTiTa) by SEM and XPS; the wettability by contact angle measurements, the mechanical properties by nanoindentation and potentiodynamic polarization tests to determine the corrosion susceptibility. Improved corrosion resistance and cellular viability were observed with MEP surface treated alloys. They concluded that MEP reduced the surface asperities on Nitinol alloys and provided stable oxides on the surface, which significantly improved the corrosion resistance and biocompatibility of the alloys. NiTiCr was found to be more resistant to corrosion and displayed a negative hysteresis, whereas, that of NiTi and NiTiTa displayed a positive hysteresis, with the latter being more prone to pitting. High corrosion resistance is one of the main prerequisites for enhanced biocompatibility because it minimizes Ni ion release. The MEP treated alloys formed nickel oxide on their surfaces with no elemental Ni indicating the nontoxic behavior after surface treatment, thereby improving biocompatibility. The cytotoxicity and cellular growth tests with endothelial cells confirmed that MEP provides a low surface free energy layer that is appropriate for endothelial cell growth. As a result, MEP reduces the risk of Ni sensitivity and enhances the biocompatibility of Nitinol alloys.

11.5.2 PLASMA ELECTROPOLISHING

Plasma electropolishing is a combination of electrochemical and physical removal of material from a workpiece. The same as in electropolishing, the part that is to be polished is electrically contacted to be used as an anode and immersed in an electrolytic bath. However, there are some basic differences. The electrolyte used is an aqueous salt solution, usually a low concentration (<10%) ammonium sulfate with typical conductivity values ranging between 80 and 40 mS/cm [27,28]. The process involves the formation of a plasma skin around the anode surface immersed in the electrolyte. The plasma skin is formed under atmospheric pressure and at a high applied DC voltage ranging between 200 and 400 V. A high voltage is needed to ignite the plasma under water. The current density is in the range of 0.2 A/cm^2, comparable to electropolishing. The development of the vapor skin that results in a plasma zone in combination with the electrochemical processes at the anode produce an exceptional surface quality of the workpiece.

Glossy surfaces with surface roughness down to 0.01 μm are achievable by plasma electropolishing. Typical material removal in plasma electropolishing is between 2 and 8 μm/min. By adjusting the voltage, the removal rate can be increased up to ten times but at the cost of the gloss effect. The use of nontoxic, low-concentration salt solution makes it an environmentally friendly process, avoids problems due to hazardous working conditions, and minimizes waste disposal concerns.

Plasma electropolishing is particularly suitable for surface finishing of bio manufactured parts where ultrasmooth and ultraclean surfaces are essential to avoid bacteria growth. The process can be used as a reliable and reproducible finishing process for surgical instruments and other medical devices made from stainless steels. Plasma electropolishing can also be used to achieve defect-free and smooth surfaces of cobalt chrome alloys that are commonly used for implants such as joints for hip or knee surgery. With some modifications in the electrolyte composition, the process can be adapted for microfinishing of titanium and titanium alloys.

11.5.3 PULSE AND PULSE REVERSE ELECTROPOLISHING

The use of pulsating current to modify the mass transport conditions at the anode by varying pulse parameters has been discussed in Chapter 4. Since electropolishing is a mass transport-controlled process, the use of pulsed current can lead to process simplification in some applications. Indeed, pulsed polishing has been advantageously used in several applications particularly for deburring [29,30]. Taylor et al. [29] employed pulsed current to develop a simpler and faster finishing process for deburring of automotive planetary gears made of cast iron (SAE 1010 steel). The use of pulsed current enabled deburring of gears in a simple 12% NaCl aqueous solution thus replacing a highly resistive ethylene glycol-based DC electropolishing process. For microfinishing of SKD61 steel molds, Hocheng and Pa [30] compared the performance of DC versus pulsed electropolishing. They observed that pulsed current instead of continuous current improves microfinishing since during the pulse-off period the system relaxes, and the residues are removed. A longer off-period is more effective, although at the cost of the increasing polishing time.

The use of pulse reverse current has been very effective in developing simpler and safer electropolishing processes involving metals that easily passivate in aqueous environments. For passivating metals such as nickel, stainless steels, titanium, niobium, and their alloys, conventional electropolishing uses aqueous or nonaqueous concentrated electrolyte solutions. For example, in the case of niobium, hydrofluoric acid is added to the sulfuric acid-based electrolyte to depassivate the surface. The use of such an electrolyte mixture poses electrolyte handling, safety, and environmental issues. Inman et al. [31] described a pulse reverse electrochemical surface finishing process that eliminates the need for aggressive and hazardous acids. The surface finishing process is based on pulse/pulse reverse electrolysis utilizing simple, easy to control aqueous electrolytes. The pulsating current/voltage scheme consists of a combination of anodic pulses, cathodic pulses, and pulse-off times. The anodic pulse (peak current/voltage and pulse-time) is chosen to be comfortably in the transport limited dissolution process at the end of the pulse. The cathodic pulse acts to depassivate the surface thus eliminates the need for hydrofluoric acid or other oxide removing chemical additions. The off-time allows the system to relax and facilitates the replenishment of reacting species and removal of byproducts and heat. The material removal rate during pulse/pulse-reverse electropolishing is generally higher than or equal to that obtained under the direct current (DC) electropolishing. This is because the instantaneous anodic pulse current is much higher than the steady-state current obtained under DC conditions and compensates for off-times and cathodic periods such that the average material removal rate (average anodic current density) is equivalent to or greater than DC electropolishing. Using pulse reverse scheme, electropolishing in dilute sulfuring acid (10% H_2SO_4) of niobium yielded a surface roughness, Ra, in the range of 0.0040 μm and for the nitinol (nickel-titanium alloys) samples, the Ra value was in the range of 0.12 μm. Due to the absence of hydrofluoric acid and other chemical additives, the pulse/pulse reverse electropolishing is claimed to be robust, low cost, and safe [31].

REFERENCES

1. H. Figour, P. A. Jacquet, French Patent No. 707526, 1930.
2. https://www.ableelectropolishing.com/industry-solutions.
3. https://www.pfonline.com/articles/clean-economical-electropolishing.
4. https://www.delstar.com/characteristics-of-the-electropolishing-process.
5. https://www.besttechnologyinc.com/electropolishing-equipment/.
6. W. J. McTegart, *The Electrolytic and Chemical Polishing of Metals*, Pergamon Press, London, (1956).
7. P. V. Shigolev, *Electrolytic and Chemical Polishing of Metals*, 2nd edition, Freund Pub, Tel-Aviv, Israel, (1974).
8. GALVABAU AG, www.galvabau.swiss, Switzerland.
9. D. Kapoor, *Johnson Matthey Technol. Rev.*, 61 (1), 66 (2017).
10. T. Duerig, A. Pelton, D. Stöckel, *Mater. Sci. Eng. A*, 149, 273–275 (1999).
11. *Peripheral Artery Disease*, Harvard Health Publications, Harvard Medical School, Boston, Massachusetts, (2012).
12. B. Thierry, M. Tabrizian, C. Trepanier, O. Savadogo, L. H. Yahia, *J. Biomed. Mater. Res.*, 51, 685 (2000).
13. O. Cissé, O. Savadogo, M. Wu, L. H. Yahia, *J. Biomed. Mater. Res.*, 6, 339 (2002).
14. K. Fushimi, M. Stratmann, A. W. Hassel, *Electrochim. Acta*, 52, 1290 (2006).
15. L. Neelakantan, A. W. Hassel, *Electrochim. Acta*, 53, 915 (2007).
16. E. Kassab, A. Marquardt, L. Neelakantan, M. Frotscher, F. Schreiber, T. Gries, S. Jockenhoevel, J. Gomes, G. Eggeler, *Materialwiss. Werkstofftech.*, 45 (10), 920 (2014).
17. M. Inman, E. J. Taylor, A. Alonso-Morales, H. Garich, T. Hall, Electrochemical System and method for Machining Strongly Passivating Metals, US Patent Appl. No. 61/353,934, June 11, 2010.
18. Y. Ge, W. Zhang, Y. Chen, C. Jin, B. Ju, *J. Mater. Process. Technol.*, 213, 11 (2013).
19. D. Xu, K. M. Liechti, K. Ravi-Chandar, *Rev. Sci. Instrum.*, 78, 073707 (2007).
20. J. L. Zable, Impact Printing: Introduction and Historical Perspective *in Output Hardcopy Devices*, R. C. Durbeck, S. Sherr eds., p. 117, Academic Press, New York, (1988).
21. M. Datta, J. C. Andreshak, L. T. Romankiw, L. F. Vega, U.S. Pat. 5,066,370, 1991.
22. M. Datta, J. C. Andreshak, L T. Romankiw, L. F. Vega, *J. Electrochem. Soc.*, 145, 3047 (1998).
23. M. Datta, L. T. Romankiw, *J. Electrochem. Soc.*, 145, 3052 (1998).
24. T. Hryniewicz, R. Rokicki, K. Rokosz, Magneto-electropolished Titanium Biomaterial *in Biomaterials Science and Engineering*, R. Pignatello, ed., p. 227, InTech, London, (2011).
25. T. Hryniewicz, K. Rokosz, *Front. Mater./Corros. Res.*, 1, 1 (2014).
26. P. Gill, V. Musaramthota, N. Munroe, A. Datye, R. Dua, W. Haider, A. McGoron, R. Rokicki, *Mater. Sci. Eng. C*, 50, 37 (2015).
27. K. Nestler, F. Böttger-Hiller, W. Adamitzki, G. Glowa, H. Zeidlera, A. Schubert, *18th CIRP Conference on Electro Physical and Chemical Machining (ISEM XVIII), Procedia CIRP*, vol. 42, p. 503, (2016).
28. H. Zeidlera, F. Boettger-Hiller, J. Edelmann, A. Schubert, *The Second CIRP Conference on Biomanufacturing, Procedia CIRP*, vol. 49, p. 83, (2016).
29. E. J. Taylor, M. E. Inman, *Electrochem. Soc. Interface*, 23 (3), 57 (2014).
30. H. Hocheng, P. S. Pa, *Int. J. Adv. Manuf. Technol.*, 21, 338 (2003).
31. M. Inman, E. J. Taylor, T. Hall, S. Snyder, S. Lucatero, *Proceedings of SRF 2015*, Pre-Press, Whistler, BC, Canada, p. MOPB101, (2015).

12 Electrochemical Planarization of Copper Interconnects

12.1 INTRODUCTION

A paradigm shift in interconnect technology took place when the vacuum-deposited Al was changed to electrodeposited copper [1]. Relative to comparable Al interconnect, Cu interconnect has the advantages of significantly low resistance, higher current carrying capability, and increased scalability [2]. For similar dimensions, the electromigration lifetime of Cu interconnects is more than 100 times longer than for Al lines. Cu metallization, therefore, supports much higher current density specifications and makes it extendible to finer dimensions and pitches. Furthermore, near bulk resistivity for Cu metallization can be obtained in submicron interconnects. Varying needs of interconnect, from low capacitance to low RC to low resistance, requires a hierarchical interconnect scheme consisting of lower interconnects at minimum possible pitch and thickness to minimize capacitance and maximize interconnect density [3]. The upper-level interconnects are scaled uniformly, both vertically and horizontally, to maintain a constant capacitance while reducing resistance. However, since it is difficult to etch copper by dry processing, the dual-Damascene process was introduced for fabrication copper interconnects. The dual-Damascene process is referred to as a metallization patterning process by which two insulators (dielectric) levels are patterned, filled, and planarized to create a metal layer consisting of vias and lines [4]. A sandwich of two levels of the insulator and etch stop layers are patterned as holes for vias and troughs for lines. They are then filled with copper in a single electrodeposition step. Finally, the excess material is removed, and the wafer is planarized by chemical mechanical polishing (CMP). An essential feature of CMP is its ability to planarize multiple materials in one step with global planarity in the nanometer size range. Although the CMP process is a powerful planarization method, it has several disadvantages. Processing parameters in CMP depend on feature size and pattern density. The chemical concentrations in the slurry often vary during CMP, and the endpoint detection is difficult [5]. During wafer processing, several defects such as dishing, erosion, and delamination may be commonly introduced by CMP as shown in Figure 12.1. Dishing of metals and erosion of dielectrics [6,7] are very common problems encountered in the CMP process (Figure 12.1a). Dishing occurs during overburden Cu removal and its extent is primarily a function of linewidth but is generally insensitive to pattern density [8]. Erosion is the unwanted removal of dielectric, which typically occurs in regions of high pattern density [9–12]. Post-CMP concerns include the variety of defects that are introduced during

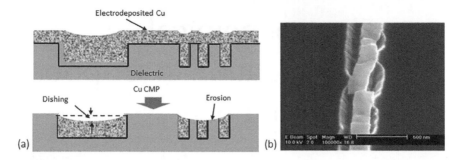

FIGURE 12.1 CMP defects: (a) Schematic of dielectric erosion and copper dishing during Cu CMP and (b) SEM photograph showing CMP damage of Cu lines and delamination of the low-k dielectric [13].

the process, including scratches, slurry particles, metal particles, and other particles of unknown origin. Some of these particles may be embedded on the wafer surface. While brush cleaning is a widely used method, wafer/particle cleaning remains an extremely complex process. Another important issue arises due to the differences in the hardness between copper interconnects and the fragile porous interlayer dielectric (ILD) materials. Due to the tremendous hardness mismatch between adjacent layers, the CMP process may damage the Cu lines and create delamination of low-k/ULK dielectric. Figure 12.1b shows post CMP damage of Cu lines and the delamination of low-k dielectric [13]. The yield of a semiconductor device is strongly affected by the above-mentioned defects generated in the CMP process. New CMP approaches involving low abrasive and low pressure have been developed to address some of the challenges. While very low-pressure CMP has been somewhat successful, it is an extremely low throughput process. Due to these reasons, the semiconductor industry explored the development of alternative planarization methods involving no physical contact or minimal contact pressure to the wafer, processes such as electropolishing, and electrochemical mechanical polishing (ECMP).

12.2 PLANARIZATION BY ELECTROPOLISHING

Electropolishing involves anodic dissolution to remove and microfinish metal layers without introducing physical stress on the surface. In a typical wafer electropolishing process, the wafer with metallization layers is made the anode in an electrolytic cell where it faces a cathode. The commonly used electrolyte is concentrated phosphoric acid, although other electrolytes have also been reported. The metal dissolution rate is determined by the current density, temperature, hydrodynamic conditions, and the nature of the electrolyte. The advantages of Cu electropolishing relative to CMP include its simplicity, easier endpoint detection, and the absence of particle contamination. Furthermore, EP being a noncontact process, the issues due to delamination and dielectric damage are generally absent.

Schematics of electropolishing tools for wafer processing are shown in Figure 12.2. In its simple form, the electropolishing tool consists of an anode facing a cathode in

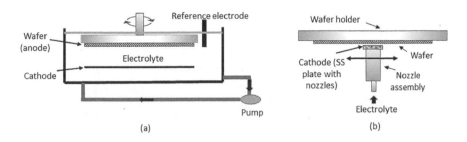

FIGURE 12.2 Schematic of electropolishing tools: (a) rotating wafer-type and (b) moving multinozzle cathode-type.

the form of a parallel plate structure where the electrodes are dipped in a tank filled with the electrolyte as shown in Figure 12.2a. The electrodes may be held vertically or horizontally in the electrolyte with an electrolyte circulating system using a pump. For the experimental investigation, the electrolytic cell may include a reference electrode near the anode. For wafer processing, the electropolishing cell design may include a stationary or a rotating wafer anode with a stationary or a rotating cathode. Datta et al. [14,15] developed a moving multinozzle cathode-type tool (Figure 12.2b) in which a stainless-steel cathode plate, placed at a narrow interelectrode spacing from the anode wafer, is scanned along the wafer surface at an adjustable rate. The cathode plate is engraved with a series of carefully designed small nozzles. The plate is attached to an electrolyte delivery assembly which locally impinges electrolyte onto the wafer as it moves past the wafer. The scanning rate is adjusted according to the dissolution rate of the material stack and the thickness of the material to be removed.

In a typical phosphoric acid-based system, it is well established that the electropolishing action takes place in the mass transport-controlled regime where a salt film or a passivation layer forms on the surface. Figure 12.3 shows a typical polarization curve for anodic dissolution of copper in a concentrated phosphoric acid electrolyte. Note that such types of curves are common for anodic dissolution of copper in other concentrated acid electrolytes as well as in salt solutions. At low potentials, a steep increase in current density corresponds to active dissolution, with metallurgical aspects such as crystallographic orientation, inclusions, etc. playing an important role in the surface morphology resulting from dissolution. In the current plateau region, a surface film forms which randomizes the dissolution process making it independent of the surface defects leading to microfinishing. The current plateau or the limiting current is mass transport controlled and its value is dependent on the diffusion layer thickness which can be calculated based on the hydrodynamic conditions at the anode as described in Chapter 4. At a potential higher than the limiting current, the transpassive dissolution of copper may involve simultaneous anodic oxygen evolution in some electrolytes.

The early use of electrochemical dissolution as a means of planarization was reported in the 1990s [14,17,18]. Since then several exploratory investigations have been reported on Cu electropolishing as an alternative to CMP [19–22].

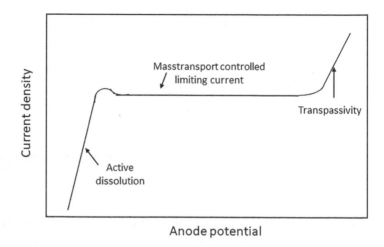

FIGURE 12.3 A schematic of the anodic polarization curve for copper in concentrated phosphoric acid.

Chang et al. [19] investigated the electropolishing process in concentrated phosphoric acid in the mass transport control regime. After electropolishing, the average roughness of polished surfaces was measured to be 1.1 nm. The electrochemical impedance spectroscopy (EIS) suggested the existence of a passivation film on the polished surface which contributed to the microfinishing effect of Cu electropolishing. The current-time curve during electropolishing could be used for end-point detection. The end-point was reached when the current began to fall, due to a change in the film's resistance, at which time the barrier layer was exposed. By applying the end-point technique and the optimal electropolishing process, planarized trenches with dishing below 50 nm were obtained even with a 50 μm wide trench. In another work, Chang et al. [20] investigated the influence of acid additives (such as citric acid) on the planarization efficiency during electropolishing in phosphoric acid. They also used polyethylene glycol as an additional additive to suppress oxygen evolution during electropolishing. With this so-called super-polishing electrolyte, they claimed to get a smooth surface with complete step height elimination in a 70 μm wide trench. However, the planarization results of Suni and Du indicated that the improvement provided by this electrolyte is relatively small [16]. Padhi et al. [21] contended that a finite amount of copper must be anodically dissolved to create the effective electropolishing conditions for the planarization of thin films where the disparity in the topography is significant as compared to the thickness of the film. Electropolishing is shown to effectively remove the bulk of electrodeposited copper layers used in ULSI metallization schemes without application of mechanical force and to planarize local topography. They investigated the effects of changes in current density and rotational speed of wafer on the extent of planarization. Under optimal galvanostatic and hydrodynamic conditions, the disparity in the topography over wide trenches adjacent to dense features decreased by 60%. Huo et al. [22] presented a novel ECP system consisting of a rotating anode and a circular sectorial cathode with very close interelectrode distance which they claim to produce excellent planarization. However, the demonstration of this approach to planarize a full wafer is missing.

12.2.1 ELECTROPOLISHING MECHANISM

Electropolishing involves anodic leveling and anodic brightening of a surface [23]. Anodic leveling is due to a nonuniform current distribution on protruding and receding parts of a rough surface. From a mathematical point of view, leveling under primary current distribution conditions and leveling under mass transport control in the presence of a stagnant diffusion layer are equivalent. These conditions correspond to the maximum achievable rate in anodic leveling [24–26]. The theoretical models of anodic leveling based on current distribution considerations do not account for the possible effects of crystallographic orientation on dissolution kinetics. Such effects on a microscopic scale are responsible for the fact that anodic dissolution under activation control usually leads to an optically dull surface appearance due to crystallographic etching. Surface brightening is achieved only under conditions where the dissolution mechanism is independent of crystallographic structure, and metallurgical defects. This is achieved under conditions where the rate of anodic dissolution is mass transport controlled [21–35]. Different transport mechanisms can limit the anodic dissolution reaction under electropolishing conditions. For electrodissolution involving salt solutions and also several electropolishing systems, it is widely accepted that the transport of dissolved metal ions from the anode is rate-limiting which leads to the formation of a salt film at the anode. In the salt-film mechanism, dissolved metallic ions accumulate near the electrode surface until at the limiting current the concentration of the dissolving metal ions at the surface corresponds to the saturation concentration of the salt formed with the electrolyte anions. However, electropolishing in concentrated acid solutions, particularly in the copper/phosphoric system, there is some confusion about the rate-limiting species since these electrolytes contain a low amount of water. While some investigations agree with the salt-type mechanism [29–33], several other studies have proposed an acceptor-type mechanism [21,34,35], according to which the dissolution process is limited by transport of an acceptor species such as water or a complexing ion to the electrode surface. These species react with the dissolving metal ions to form complexed or hydrated species. At the limiting current, the surface concentration of the acceptor is zero. While the precise mechanism that creates a limiting current density in the copper/phosphoric electropolishing system is debatable, a detailed mechanism does not impact the anodic leveling simulation studies, as long as it is assumed that the dissolution rate is limited by mass transport in the electrolyte. Furthermore, a stagnant diffusion-layer approximation is often adequate for the simulation.

12.2.2 PLANARIZATION ISSUES

The use of electropolishing to planarize copper is aimed at removing the copper overburden that arises in dual-Damascene processing. Shrinking feature size continues to pose problems due to the formation of dishing and mounds thus making the planarization process even more challenging. Two types of planarization issues are to be considered: wafer-scale and feature-scale. The wafer-level uniformity requires uniform accessibility of the anode surface across the wafer. This is achieved in a rotating electrode system (RDE) for which the diffusion layer (δ) is expressed by the well-known Levich equation described in Chapter 4; the value of δ depends on the rotation speed, diffusion coefficient and the kinematic viscosity of the electrolyte. The uniqueness of the RDE system is that the diffusion layer thickness does not depend on the radius hence the mass transport rate is uniform all over the surface. Thus, by ensuring a rugged rotating system without eccentricity, scaling up to large-size wafers is applicable. Another method is to use a scanning cathode at a narrow electrode spacing which locally removes copper at an unequal rate but averages out to provide uniformity as it scans over the entire wafer. Using this approach, manufacturing tools have been successfully employed to uniformly remove seed layers in the manufacturing of flip-chip (C4) bumps [13,15]. The approach was also employed for the planarization of multilayer interconnects in thin-film modules [14]. A similar approach, that of using a segmented cathode, has been proposed to obtain wafer-level uniformity but the feasibility of this concept to a full wafer has not been established [22].

Feature-scale planarization is the key issue in electropolishing. Nonuniformity of planarization on a feature scale arises due to the differences in the diffusion layer profile on such features. Isolated narrow features can be planarized since the diffusion layer is typically much thicker than the feature width. On the other hand, features that are wider than the diffusion layer thickness are extremely difficult to planarize since the diffusion layer follows the feature profile thus leading to conformal copper removal. Several authors have investigated to alleviate the problem [19–22]. As described above, the use of additives [20] that may preferentially sit at the bottom of features and minimize its dissolution was found to give marginal improvement [16]. Based on mathematical simulation and experimental studies, several authors conclude that electropolishing in its conventional form cannot provide planarization of wafers with wide features and therefore, its application in the ULSI application remained doubtful.

12.3 PLANARIZATION BY ELECTROCHEMICAL AND MECHANICAL ACTIONS

As noted above, while electropolishing is a cost-effective method of Cu overburden removal, it does not have significant planarization capability, especially for large features with low aspect ratio. A combination of electrochemical and mechanical means of Cu removal has been studied as a novel planarization technique. This has involved mainly two types of planarization techniques as shown in Figure 12.4 [36,37]. In one approach, the planarization is achieved during electroplating using electrochemical mechanical deposition (ECMD) which is followed by electropolishing to remove the overburden, while in the other approach electroplating is followed by ECMP. Both approaches emphasize electropolishing as the key metal removal method, thus positioning electropolishing at the center stage in the planarization of interconnect structures.

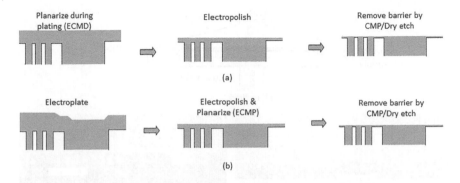

FIGURE 12.4 Two types of planarization techniques: one involving (a) ECMD followed by electropolishing and the other (b) involving electroplating followed by ECMP [36].

12.3.1 ELECTROCHEMICAL MECHANICAL DEPOSITION

ECMD process described by Basol et al. [37–40] and Jeong et al. [41] involves simultaneous electrochemical metal deposition and removal. Mechanical action during electrodeposition is accomplished by the sweeping action of a pad on the wafer surface. The plating bath does not contain any abrasive slurry. Therefore, the mechanical action is purely due to the sweeping action of the pad. Sweeping of the pad removes Cu from the top surface, thus allowing Cu growth rate to be higher in the features. Chemical modifications of the plating bath in combination with ECMD may make the planarization action more effective without the need to apply pressure to the pad.

Basol et al. [39] investigated the planarization efficiency of the ECMD technique as a function of various process parameters. Planarization was found to be a strong function of the nature of the organic additives in the copper plating bath. While no planarization was observed in an additive-free bath and a bath containing only accelerators, a small degree of planarization was found in a bath containing only suppressors. The best copper layer planarization results and efficiencies over 70% were observed in plating electrolytes containing both accelerators and suppressors. This is due to a differential in the relative surface coverage of additives on the top surface vs. the cavities. Sweeping by the planarization pad reduces additive coverage at the top surface but leaves the cavities with preferential coverage at the bottom. During the plating period, more of the current flows into the cavities where the accelerator-to-suppressor surface coverage ratio is higher compared to the top surface. The addition of levelers into the plating bath was found to be detrimental for planarization efficiency. Planarization efficiency was found to increase with increased accelerator concentration within the process window used in this study. Higher planarization was observed in the high-acid electrolyte compared to the low-acid electrolyte.

Figure 12.5 shows an example of planar deposition by ECMD followed by electropolishing to remove copper down to 100 nm thickness [40]. In principle, ECMD has unique capabilities as compared with the standard electrodeposition process. Metal layers deposited by ECMD are planar and the thickness of the overburden is smaller than the films deposited by standard approaches. ECMD achieves

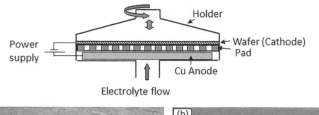

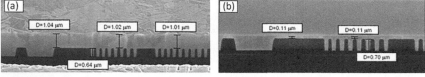

FIGURE 12.5 Top: Schematic diagram of the ECMD tool. Bottom: Cross section of parts planar deposited by ECMD followed by electropolishing. (a) After planar electrodeposition and (b) after copper electropolishing down to 100 nm thickness [40].

this result by enhancing material deposition rate into the cavities while retarding or minimizing deposition on the substrate top surface. The thin and planar copper deposits such as those shown in Figure 12.5 are very attractive for etching, electropolishing, and CMP. For etching and electropolishing, a planar layer offers the possibility of removing the overburden in a planar manner without causing excessive dishing into the large features. For CMP, ECMD offers significant cost advantage due to thinner copper to be removed.

12.3.2 ELECTROCHEMICAL MECHANICAL PLANARIZATION

In electrochemical mechanical planarization (ECMP), metal removal and planarization are accomplished by a combination of the virtues of electropolishing and CMP. Compared to conventional CMP, ECMP allows ten times lower down-force for planarization. A pad is used in conjunction with an electrolyte for electropolishing. Figure 12.6 shows the schematic of a typical ECMP equipment. It consists of a cathode plate between the platen and the pad, while the wafer is attached to a rotating mount which is connected as an anode. There is provision for the supply of the electrolyte above the pad. The holes in the pad are filled with the electrolyte and form electrical contact between the wafer anode and the cathode. The planarization efficiency of ECMP depends on the electrolyte composition. The electrolyte used in ECMP is generally H_3PO_4 based, containing additives that can form a complex on the Cu surface. Electrolytes based on H_2SO_4 or HNO_3 have also been reported. The passive film formed must provide a certain degree of resistance to copper dissolution in the recessed areas but at the same time, they must be soft enough to be removed by the mechanical abrasion of the polishing pad at low down pressures (<0.5 psi).

Copper film on the Si wafer is polarized anodically while the polishing pad mechanically wears the metal surface. At the surface where the pad is in contact with the wafer, more Cu removal takes place due to both electrochemical and mechanical action. On the other hand, only electrochemical action removes material from

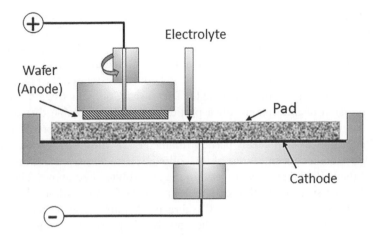

FIGURE 12.6   Schematic diagram of an ECMP tool.

locations of the wafer where contact with the pad is absent. The contact of the polishing pad with the copper surface is maintained at a low-down pressure so that the metal film does not delaminate from the low-k/ULK film. As a result, defects such as dishing, erosion, and scratch that are common in the conventional CMP process are minimized. The applied charge controls the material removal rate (MRR). Electrochemical dissolution leads to the formation of metal ions and passivation film on the metal surface. The low-lying areas of the film are protected by the passivation film while the protruding features are polished by the polishing pad. However, the problem of dishing, particularly in wide features, continues to remain the key concern in the ECMP thus requiring optimization of the process parameters. In the following, some of the published literature related to the influence of electrolyte composition and mechanical factors on the ECMP process is briefly described.

12.3.2.1   Mechanical Factors

The mechanical factors such as the pad material, pad design, and the abrasive in the electrolyte play a significant role in influencing the ECMP performance. A polymeric pad must have mechanical integrity and chemical resistance to survive the rigors of polishing. Mechanically, a polishing pad should have acceptable levels of hardness, and modulus, and good abrasion resistance to endure the Cu ECMP process. Chemically, a polishing pad should be able to survive the electrolyte chemistries, which include either highly alkaline or highly acidic electrolytes. Jeong et al. [42] measured compressibility, elastic recovery, permanent deformation, viscoelastic property, and other time related physical properties of two types of pads: Polyurethane pad (IC 1400 k-groove), polymer impregnated felts pad (Suba 600). The mechanical properties of the polishing pad were measured under three different conditions; dry pad, pad soaked in the deionized water, and pad soaked in the electrolyte containing a mixture of H_3PO_4 6 wt.%, H_2O_2 0.5 wt.%, BTA 0.5 wt.%, glycine 0.5 wt.%, and citric ammonium 5 wt.% for 16 hours. Based on their results, they recommended the use of the polyurethane pad, which has stable viscoelastic behavior and high chemical attack resistance in the electrolyte. Because the hardness of the polyurethane pad was higher than that of the polymer impregnated felts pad, better global uniformity was achieved, and the metal removal rate was also high during the ECMP process. Jeong et al. [42] also investigated the effect of abrasive on the uniformity of the wafer scale. They used Colloidal silica consisting of stable dispersion of amorphous silica particles. To achieve stable dispersion, not affected by gravity, the silica particle size of the order of 20 nm were used (Ludox TM colloidal silica with mean particle diameter: 22 mm, silica concentration: 50 wt.%). Different concentrations of the colloidal silica abrasive ranging from 0% to 50% of the colloidal silica were used in the electrolyte cited above. Their results showed that the surface roughness, which is around 50 nm without the abrasive, goes through a minimum (16 nm) at 10% abrasive. Within the wafer nonuniformity, which is around 10%–12% without the abrasive, improves to 2% at an abrasive concentration of 10% or higher.

ECMP pads require punching holes to ensure electrolyte contact between the cathode and the substrate surface during polishing. The differences in the arrangement of holes on the polishing pad may cause uneven distribution of electrochemical action on the substrate, which may result in different MRRs. Liu et al. [43] investigated the nonuniformity of material removal in ECMP by using track point density distribution. The coefficient of variation of track point density and density distribution of two polishing pads at different speed ratios were simulated to represent the uniformity of electrochemical action across the substrate. The two types of pad designs used by Liu et al. [43] consisted of concentric-type holes and phyllotactic-type holes. The simulation results were verified by experiments. The simulation and experimental results showed that in pads with concentric-type arrangement, the track points produced by the holes have a ring-shaped ripple distribution on the substrate, and the track point density at the peak ring is large. For pads with the phyllotactic arrangement, the track point density distribution is generally uniform. With increasing radius, the track point density decreased slightly, indicating that the electrochemical effect will decrease gradually as the radius increases. Pads with phyllotactic holes showed improvement in substrate flatness. Both the simulation results and the experimental results showed that the material removal nonuniformity of the phyllotactic arrangement polishing pad is better than the concentric arrangement pad, indicating that the uniformity of the ECMP material removal can be represented by the density distribution of the track point on the substrate.

Kondo et al. [44] developed a carbon polishing pad for the ECMP process and conducted experiments on an orbital 300 mm CMP machine. Figure 12.7 shows a schematic diagram of the carbon pad. The pad consisted of a surface carbon layer acting as an anode, an intermediate insulating layer, and an underlying cathode sheet. More than 100 electro-cells were fabricated within this tri-layered structure, which was about 5-mm-thick. The intermediate insulating layer acted as a cushion layer to improve within-wafer nonuniformity. Soft carbon material was chosen so that the copper surface would not be damaged. The power supply was connected with the cathode at the edge of the pad. The carbon pad was stuck to the CMP platen with an adhesive sheet and could be easily replaced. This enabled easy changing a CMP

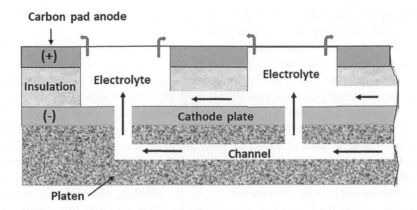

FIGURE 12.7 Schematic diagram of a carbon polishing pad for the ECMP process [44].

system into an ECMP system by replacing the conventional polyurethane CMP pad with a carbon one. Kondo et al. claimed that the use of carbon pad in the ECMP process resolved several issues such as scratching, copper residues, and cathode regeneration. They also claimed to have successfully fabricated 45-nm-node porous low-k/ Cu interconnects using an ECMP process followed by removal of the TaN barrier layer by CMP.

12.3.2.2 Electrolyte Composition

During ECMP, an anodic potential supplies the driving force for the dissolution Cu and transport of copper ions. Planarization in ECMP possibly occurs due to accelerated electrochemical Cu removal from surface elevations, where the passive film is abraded, coupled with little or no removal at surface recesses, where the passive film remains intact. Hence, the nature of the passive film formed during ECMP largely determines the planarization efficiency and the Cu surface quality. The material removal differential in ECMP can be made more effective by selecting conditions whereby surface passivation plays a major role in the electrochemical reaction. The nature of the passive film formed during ECMP is determined by the inhibiting or passivating agents in the electrolyte which in turn determine the copper removal rate and planarization efficiency. Benzotriazole (BTA), which is the most commonly used inhibiting species used in CMP, is ineffective in acid electrolytes and at high anode potentials encountered in ECMP. Several investigations involving the influence of different inhibitors and chelating agents have been reported in the literature [35,45–56]. Among several inhibitors studied, 5 phenyl-1-H-tetrazole (PTA) was found to be the most effective for copper CMP at lower pH [51]. The importance of the use of a chelating agent in the electrolyte is the fact that the passive film thickness in such a system can be modulated by the applied anode potential. Hydroxyethylidenediphosphoric acid (HEDP) and oxalic acid have been found to be effective chelating agents for copper ECMP [49,50]. Another interesting manifestation of the dependence of metal removal on applied current is the possibility of varying applied current density at different radial zone on the wafer, thereby permitting desired differential removal rates at different locations [56].

All of these factors have permitted the planarization of copper during ECMP with significantly minimized to no particulates.

In general, copper is etched in acidic solutions because Cu is readily ionized to Cu^{2+} ions. In many cases the etch rate may be too high to control under very low pH conditions. In order to precisely control the polishing rate and reduce the dishing effect in the ECMP process, Oh et al. [54] used alkali conditions by forming copper oxide or copper ions which are readily ionized to $HCuO_2^-$ or CuO_2^{2-} in an alkali electrolyte. The use of KOH electrolyte produces Cu hydroxide on the Cu surface which on the application of an anodic potential to the Cu wafer ionizes the Cu ions from Cu hydroxide to Cu^{2+} or CuO_2^{2-}, and mechanical force is also applied to the Cu surface employing a soft polymer polishing pad. The mechanical effect induced by the polishing pad increases the uniformity and the polishing rate of the Cu layer. The chemical state of the Cu surface was studied by X-ray photoelectron spectroscopy analysis and found that the oxidation rate was increased by adding H_2O_2, which increases the polishing rate by increasing the surface oxidation of Cu to $Cu(OH)_2$. Furthermore, to prevent the dishing effect, Oh et al. [54] added a BTA inhibitor to the KOH electrolyte. It is generally known that BTA is easily adsorbed on the Cu surface and prevents the excessive etching of Cu during the ECMP process. The ECMP process was performed for the 3 μm trenched wafer with 20 wt.% KOH electrolyte, containing 0.005 M BTA at a current density of 9 mA/cm². Figure 12.8 shows that the Cu film was successfully polished and planarized, with the Cu remaining inside the trenches. These experimental results support that the BTA acts as an inhibitor and suppresses the dishing phenomenon. However, the process is very slow, and it takes 14 minutes to complete the planarization process. The addition of H_2O_2 to the electrolyte containing BTA, increased the etching rate, by augmenting the oxidation rate of the Cu surface to $Cu(OH)_2$. The etch rate increased as the concentration of H_2O_2 was increased from 1 to 7 wt.%. However, the etch rate was drastically reduced in the electrolyte containing 10 wt.% of H_2O_2. This indicated that H_2O_2 promotes the

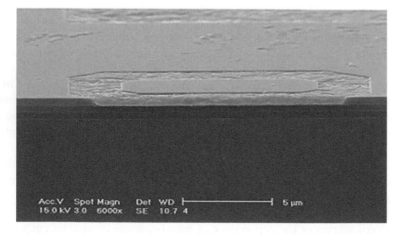

FIGURE 12.8 Cross-sectional SEM images after the ECMP process in 20 wt.% KOH solution containing 0.005 M BTA [54].

generation of $Cu(OH)_2$, but an excess of H_2O_2 induces the generation of CuO rather than $Cu(OH)_2$ and the formation of CuO reduces the ionization of Cu to Cu^{2+}.

Shattuck et al. [57,58] explored the use of phosphate salt as an electrolyte component to conduct the ECMP process at low pH. An electrolyte containing 1.0 M potassium phosphate salt concentration with a pH value of 2 and a BTA concentration of 0.001 M was tested for its planarization capability on patterned Cu structures using a custom-built ECMP tool. Planarization was achieved on patterned Cu structures with an abrasive-free electrolyte. Results indicated that using a low pH phosphate-based electrolyte and BTA, as an inhibitor, ECMP can provide the necessary metal removal rate to polish Cu while providing adequate passivation in the recessed regions to allow for planarization to occur.

Tripathi et al. [53] described electrochemical studies and polishing results during ECMP of blanket and patterned Cu wafers using electrolytes containing HEDP, PTA, and oxalic acid. Voltammetry measurements were made with and without abrasion. The nature of the passive film formed at different potentials was investigated by EIS and electrochemical quartz crystal microbalance (EQCM). Their results indicated that the Cu removal rate and the planarization efficiency during Cu ECMP can be approximated using electrochemical measurements of the Cu removal rate, with and without surface abrasion. These results predicted a 500 mV potential window within which the Cu removal rate is greater than 600 nm/min and the planarization efficiency is greater than 0.90. However, high planarization efficiencies are only obtained when silica abrasives are included within the ECMP electrolyte. In-situ EIS results indicate that the interfacial impedance is increased by the presence of silica, suggesting that silica is incorporated into the PTA-based passive film and is thus needed for effective planarization. EQCM experiments indicate that PTA may provide better Cu surface passivation at a high anodic potential than BTA, which is widely used during Cu CMP.

Choi et al. [55] investigated the influence of copper ion concentration on the kinetics of formation of a protective layer on copper in an acidic solution containing BTA and glycine. They observed that even modest additions of copper ions to the electrolyte impacted the kinetics of the formation of the protective layer on the copper surface. This is due to Cu^{2+} ions complexing with BTA and eventually nucleating and precipitating $Cu(II)BTA_2$. The higher the concentration of Cu^{2+} in the electrolyte, the more easily $Cu(II)BTA_2$ nucleates. Nucleation and growth of $Cu(II)BTA_2$ causes the desorption of BTA species from the adsorbed protective layer. This nucleation of $Cu(II)BTA_2$ followed by the desorption of BTA could be particularly problematic at recessed regions on the wafer surface. At recessed regions there would be less agitation, thus BTA has enough time to be adsorbed onto the copper surface and form protective layers or multilayers. If the electrolyte contains a sufficiently high concentration of Cu^{2+} ions and BTA, however, the adlayer formed on the surface may be destroyed. With the elimination of passivity, the recessed regions of the wafer would not be protected, which could lead to roughened topography, or at least longer times needed to achieve planarization. Even for low concentrations of copper ions in the bulk, high concentrations of copper ions may develop if mass transport of dissolved Cu^{2+} into the bulk is hindered. This suggests that the influence of copper ions on the kinetics of the formation of the protective layer must be considered when formulating new electrolytes or slurries for ECMP or CMP of copper. Also, the concentration of copper ions in the electrolyte must be controlled below a certain level to ensure the integrity of the protective material during the process. This might be achieved with lower residence times for slurries, and passage of the withdrawn slurry through a bed of a cation exchange resin that could replace Cu^{2+} by H^+ ions.

12.4   REMOVAL OF THE BARRIER LAYER

A serious limitation of the above approaches to planarization is their inability to remove the diffusion barrier layer. EP, ECMD, and ECMP processes leave behind a thin layer of Cu and the underlying Ta/TaN barrier layers which need to be removed either by soft CMP or by dry processing.

Diffusion barriers such as layers of tantalum (Ta) and tantalum nitride (TaN) are used to prevent contamination caused by Cu diffusing through ILDs into silicon. However, thin barrier layers like TaN are too resistive to plate Cu effectively, especially in the high-aspect-ratio features. Hence, a continuous thin Cu-seed layer is deposited over Ta/TaN by physical vapor deposition to assure a good Cu electrofill. During the Cu CMP process, the underlying Ta/TaN barrier film remains intact which requires an additional CMP process for its removal to expose the low-k dielectric.

Tantalum metal is well known to form a robust, passivating oxide layer at its surface. The native tantalum oxide, Ta_2O_5, typically has a thickness of 20 Å. The Ta_2O_5 film is chemically inert due to its very limited aqueous solubility and solution reactivity. The robust oxide film hinders the facile removal of the tantalum barrier film in the subsequent CMP step.

The development and implementation of a full-sequence ECMP process, which includes the removal of the barrier layer, are desirable. Tantalum, being a refractory metal, is not attacked by many chemical systems and, hence, poses challenges to the development of ECMP formulations. Du and Suni [59] used cyclic voltammetry to demonstrate the electrochemical dissolution of both Ta and TaN in a 2.5 M HF solution. For Ta dissolution, current maxima typical of passive film formation are seen, and the peak current occurs at about −420 mV vs. a saturated calomel electrode (SCE), while that for TaN dissolution occurs at about +50 mV vs. SCE. Through the appropriate choice of potential, the removal rates of Ta and TaN could be varied. These results demonstrated the possibility of all-electrochemical, noncontact methods for removal of Ta, and TaN barrier materials.

The use of a less aggressive solution is desirable for the removal of barrier layers. Muthukumaran et al. [60] employed dihydroxy benzene sulfonic acid (DBSA) for the electrochemical removal of tantalum under applied pressure conditions. Tantalum, as well as copper samples, was polished at low pressures (0.5 psi) under galvanostatic conditions in DBSA solutions maintained at different pH values. At a current density of 0.5 mA/cm^2 and pH 10, the tantalum removal rate of 200 Å/min with a 1:1 selectivity to copper has been obtained in a 0.3 M DBSA solution containing 1.2 M H_2O_2. The presence of a small amount (0.1%) of colloidal silica particles is required to obtain good removal rates.

Ruthenium (Ru) is another Cu diffusion barrier material for the construction of IC chips. Ru provides a conductive platform for the direct Cu plating that eventually eliminates the need for an additional Cu-seeding layer in the interconnect fabrication [61]. Aoki et al. [62] developed an electrochemical etching method for removing Ru film and TaN film on the wafer. They used a dilute HCl (0.5%) solution at 4 V and obtained etching rates of 50 and 25 nm/min for Ru film and TaN film respectively. They also claimed that Ru films can be etched using not only acid solutions such as HCl, H_2SO_4, and HNO_3 but also an alkaline solution such as NH_4OH. Anodic polarization of Ru films enables its easy dissolution through the formation of ruthenium oxide or ruthenium hydroxide.

12.5 SUMMARY AND FINAL REMARKS

Some of the basic principles and tooling aspects involved in electropolishing and ECMP processes have been discussed in this chapter. Due to the inherent advantages over conventional CMP, these processes have drawn the attention of the ULSI researchers both from industry and academia. While Cu electropolishing has the advantage of being a simple process, its biggest drawback is its feature-scale nonuniformity which makes planarization of wide features extremely challenging. Compared to electropolishing, the ECMP process shows more promise for the ULSI application. Planarization in ECMP is caused by both electrochemical and mechanical actions. Electrochemical action produces a passivating film on the copper surface which is critical to the planarization performance. Mechanical abrasion removes the passive film at Cu protrusions, allowing for electrochemical Cu dissolution, while at Cu recesses, a protective passive film exists, preventing electrochemical dissolution. The nature of the passive film that forms during ECMP can be controlled through the variation of the applied potential, and the slurry composition. The nature and properties of the passive film influence the downward pressure required for passive film removal, the Cu removal rate, and the extent of dishing.

The advantages of ECMP for Cu planarization include the reduction or complete elimination of toxic oxidizing agents from the polishing slurry which makes it an environmentally friendly process. Process control is easier since the Cu removal rate can be varied by changing the voltage or current during the planarization process and since the changes in the slurry composition during processing are reduced. Also, since the ECMP electrolyte replaces expensive slurries and the consumable life is extended, the ECMP process is expected to be relatively inexpensive. However, the ECMP process has several drawbacks that have not been resolved yet. Although the feature size nonuniformity has been reported to be improved, it still exists. The main disadvantage of ECMP is that the disadvantages of CMP are not completely removed. The mechanical action of the slurry and polishing pad is retained, albeit with lower pressure, and some chemicals may also be needed in the slurry which may include some abrasives.

Planarization efficiency remains the most critical issue of EP and ECMP processes. While some scattered results have been reported about improved planarization efficiency, none of the processes discussed above have been developed to an extent that they can provide high yielding planarization for advanced Cu interconnects. Reliable, scalable solutions for the aforesaid drawbacks are needed to be found for electropolishing/ECMP processes to be integrated with the ULSI processing.

REFERENCES

1. D. C. Edelestein, *Tech. Digest. IEEE Intl. Electron Devices Conference*, IBM Res. Magazine, No. 4, 16, p. 773, (1997).
2. R. Rosenberg, D. C. Edelstein, C.-K. Hu, K. P. Rodbell, *Annu. Rev. Mater. Sci.*, 30, 229 (2000).
3. D. C. Edelstein, G. A. Sai-Halasz, M. J. Mii, *IBM J. Res. Dev.*, 39, 383 (1995).
4. P. C. Andricacos, C. Uzoh, J. O. Dukovic, J. Horkans, H. Deligianni, *IBM J. Res. Dev.*, 42, 567 (1998).
5. P. B. Zantye, A. Kumar, A. K. Sikder, *Mater. Sci. Eng. Res.*, 45, 89 (2004).
6. G. Zhang, G. Burdick, F. Dai, T. Bibby, S. Beaudoin, *Thin Solid Films*, 332, 379 (1998).
7. S. Wolf, R. N. Tauber, *Silicon Processing for the VLSI Era*, 2nd edition, vol. 1, Process Technology, Lattice Press, Sunset Beach, CA, (2000).
8. R. J. Gutmann, T. P. Chow, S. Lakshminarayanan, D. T. Price, J. M. Steigerwald, L. You, S. P. Murarka, *Thin Solid Films*, 270, 472 (1995).
9. M. Fayolle, F. Romagna, *Microelectron. Eng.*, 37–38, 135 (1997).
10. S. V. Babu, Y. Li, A. Jindal, *JOM*, 53, 50 (2001).
11. J. M. Steigerwald, R. Zirpoli, S. P. Murarka, D. Price, R. J. Gutmann, *J. Electrochem. Soc.*, 141, 2842 (1994).
12. J. Y. Lai, N. Saka, J. H. Chun, *J. Electrochem. Soc.*, 149, G41 (2002).
13. M. Datta, *Electrochim. Acta*, 48, 2975 (2003).
14. M. Datta, T. R. O'Toole, Electrochemical Metal Removal Technique for Planarization of Surfaces, US Patent # 5,567,300, October 22, 1996.
15. M. Datta, R. Shenoy, Electroetching Process for Seed Removal in Electrochemical Fabrication of Wafers, US Patent # 5,486,282, January 23, 1996.
16. I. I. Suni, B. Du, *IEEE Trans. Semicond. Manuf.*, 18 (3), 341 (2005).
17. R. J. Contolini, A. F. Bernhardt, S. T. Mayer, *J. Electrochem. Soc.*, 141, 2503 (1994).
18. R. J. Contolini, S. T. Mayer, R. T. Graff, L. Tarte, A. F. Bernhardt, *Solid State Technol.*, 10, 155 (1997).
19. S. Chang, J. Shieh, C. Huang, B. Dai, Y. Li, M. S. Feng, *J. Vac. Sci. Technol. B*, 20, 2149 (2002).
20. S. Chang, J. Shieh, B. Dai, M. Feng, Y. Li, C. H. Shih, M. H. Tsai, S. L. Shue, R. S. Liang, Y. Wang, *Electrochem. Solid-State Lett.*, 6 (5), G72 (2003).
21. D. Padhi, J. Yahalom, S. Gandikota, G. Dixit, *J. Electrochem. Soc.*, 150, G10 (2003).
22. J. Huo, R. Solanki, J. McAndrew, *Electrochem. Solid-State Lett.*, 8, C33 (2005).
23. D. Landolt, *Electrochim. Acta*, 32, 1 (1987).
24. R. Sautebin, H. Froidevaux, D. Landolt, *J. Electrochem. Soc.*, 127, 1096 (1980).
25. C. Clerc, M. Datta, D. Landolt, *Electrochim. Acta*, 29, 1477 (1984).
26. C. Clerc, D. Landolt, *Electrochim. Acta*, 32, 1435 (1987).
27. M. Datta, D. Landolt, *J. Electrochem. Soc.*, 122, 1466 (1995).
28. M. Datta, D. Landolt, *Electrochim. Acta*, 25, 1255 (1980); 25, 1263 (1980).
29. W. C. Elmore, *J. Appl. Phys.*, 10, 724 (1939).
30. E. C. Williams, M. A. Barrett, *J. Electrochem. Soc.*, 103, 363 (1956).
31. K. Kojima, C. W. Tobias, *J. Electrochem. Soc.*, 120, 1202 (1973).
32. S. Chang, J. Shieh, C. Huang, B. Dai, Y. Li, M. Feng, *J. Vac. Sci. Technol. B*, 20, 2149 (2002).

33. J. Mendez, R. Akolkar, T. Andryushchenko, U. Landau, *J. Electrochem. Soc.*, 155, D27 (2008).
34. R. Vidal, A. C. West, *J. Electrochem. Soc.*, 142, 2682 (1995).
35. J. Huo, R. Solanki, J. McAndrew, *J. Appl. Electrochem.*, 34, 305 (2004).
36. M. Datta, *Micro Nanosyst.*, 1, 83 (2009).
37. B. M. Basol, C. E. Uzoh, H. Talieh, D. Young, P. Lindquist, T. Wang, M. Cornejo, *Microelectron. Eng.*, 64, 43 (2002).
38. B. M. Basol, *J. Electrochem. Soc.*, 151, C765 (2004).
39. B. M. Basol, S. Erdemli, C. E. Uzoh, T. Wang, *J. Electrochem. Soc.*, 153, C176 (2006).
40. B. M. Basol, C. E. Uzoh, H. Talieh, T. Wang, G. Guo, S. Erdemli, M. Cornejo, J. Bogart, E. C. Basol, *Chem. Eng. Commun.*, 193, 903 (2006).
41. S. Jeong, H. Seo, B. Park, J. Park, S. Park, S. Kim, K. Kim, H. Jeong, *Key Engineering Materials*, vols. 326–328, pp. 389–392, Trans Tech Publications, Switzerland, (2006).
42. S. Jeong, B. Park, J. Bae, H. Lee, Y. Lee, H. Kim, S. Kim, H. Jeong, *Sensor. Actuat. A-Phys.*, 163, 433 (2010).
43. Z. Liu, Z. Jin, D. Wu, J. Guo, *J. Solid State Sci. Technol.*, 8 (5), 3047 (2019).
44. S. Kondo, S. Tominaga, A. Namiki, K. Yamada, D. Abe, K. Fukaya, M. Shimada, N. Kobayashi, *Proc. IEEE, IITC*, p. 203, (2005).
45. F. Zucchi, G. Trabanelli, M. Fonsati, *Corros. Sci.*, 38, 2019 (1996).
46. D. P. Schweinsberg, S. E. Bottle, V. Otieno-Olego, T. Notoya, *J. Appl. Electrochem.*, 27, 161 (1997).
47. N. Huynh, S. E. Bottle, T. Notoya, D. P. Schweinsberg, *Corros. Sci.*, 42, 259 (2000).
48. D. Q. Zhang, L. X. Gao, G. D. Zhou, *Corros. Sci.*, 46, 3031 (2004).
49. V. K. Gorantala, A. Babel, S. Pandija, S. V. Babu, *Electrochem. Solid-State Lett.*, 8 (5), G131 (2005).
50. S. Aksu, *J. Electrochem. Soc.*, 152, G938 (2005).
51. D. Q. Zhang, L. X. Gao, G. D. Zhou, *Appl. Surf. Sci.*, 252, 4975 (2006).
52. S. K. Govindaswamy, A. Tripathi, I. I. Suni, Y. Li, *J. Electrochem. Soc.*, 155, H459 (2008).
53. A. Tripathi, C. Burkhard, I. I. Suni, Y. Li, F. Doniat, A. Barajas, J. McAndrew, *J. Electrochem. Soc.*, 155, H918 (2008).
54. Y. Oh, G. Park, C. Chung, *J. Electrochem. Soc.*, 153, G617 (2006).
55. S. Choi, D. A. Dornfeld, F. M. Doyleb, *J. Electrochem. Soc.*, 160, H653 (2013).
56. L. Economikos, X. Wang, A. Sakamoto, P. Ong, M. Naujod, R. Knarr, L. Chen, Y. Moon, S. Neo, J. Salfelder, A. Duboust, A. Manens, W. Lu, S. Shrauti, F. Liu, S. Tsai, W. Swart, *Proc. IEEE, IITC*, p. 233, (2004).
57. K. G. Shattuck, J. Lin, P. Cojocaru, A. C. West, *Electrochim. Acta*, 53, 8211 (2008).
58. K. G. Shattuck, A. C. West, *J. Appl. Electrochem*, 39, 1719 (2009).
59. B. Du, I. I. Suni, *Electrochem. Solid-State Lett.*, 8 (10), G283 (2005).
60. A. Muthukumaran, N. Venkataraman, S. Raghavan, *J. Electrochem. Soc.*, 155, H184 (2008).
61. O. Chyan, T. N. Arunagiri, T. Ponnuswamy, *J. Electrochem. Soc.*, 150, C347 (2003).
62. H. Aoki, D. Watanabe, N. Ooi, J. Jong-Hyeon, C. Kimura, T. Sugino, *214th ECS Meeting, Abstract #2063*, (2008).

Index